IEE HISTORY OF TECHNOLOGY SERIES 33

Series Editors: Dr. B. Bowers
Dr. C. Hempstead

Spacecraft Technology: The Early Years

Other volumes in this series:

Spacecraft Technology: The Early Years

by Mark Williamson

BSc, CPhys MInstP, CEng MIEE, FBIS

Space Technology Consultant

The Institution of Electrical Engineers

Published by: The Institution of Electrical Engineers, London, United Kingdom

©2006: The Institution of Electrical Engineers

The Institution of Electrical Engineers,
Michael Faraday House,
Six Hills Way, Stevenage,
Herts., SG1 2AY, United Kingdom

www.iee.org

British Library Cataloguing in Publication Data

Williamson, Mark
  Spacecraft technology: the early years.-
  (IEE history of technology series; v.33)
  1. Aerospace engineering - History 2. Astronautics - History
  I. Title II. Institution of Electrical Engineers
  629.1'09

ISBN-10 0 86341 553 9
ISBN-13 978-086341-553-1

Typeset in India by Newgen Imaging Systems (P) Ltd., Chennai, India
Printed in the UK by MPG Books Ltd., Bodmin, Cornwall

To my son,
Thomas, who is still in *his* early years.

Cover Photographs:
Gemini capsule following splashdown [NASA]
Ranger lunar spacecraft [NASA]
Apollo command and service module undergoes final integration [NASA]
Apollo 12 Saturn V leaves VAB for launch pad [NASA]
Telstar communications satellite [used with permission of Lucent Technologies Inc/Bell Labs]

Contents

Foreword

As mankind reaches half a century as a space-faring species, we understandably take the wonders of working in and exploring the space environment very much for granted, since it is now embedded in our everyday lives to an extent that we rarely stop to appreciate. Yet we are still only on the verge of space: apart from the brief sojourn on the Moon, man has not ventured further from the surface of our planet in the last 30 years than the distance between London and Paris. It is true that we have sent robots to explore distant worlds in our stead but, although remarkable achievements, these provide just tantalising glimpses as though looking through a keyhole into a room full of wonders.

We must not become complacent with our introspective exploitation and armchair exploration of space. If mankind is to develop further as an inquisitive species then we must continue our efforts to travel beyond the cosy atmosphere of Earth. This will not be easy, nor will it be cheap, nor will it be without substantial risk – but sooner or later it is our destiny.

Mark Williamson's book brings back into focus the endeavours, trials, tribulations and successes of the past 50 years that formed the building blocks of our human space endeavour. It reminds us of the very different driving forces that caused the Space Age to occur and places them in the context of the prevailing politics and, in some instances, personal as well as national ambitions. We learn a great deal from reviewing the past – not just in order to avoid the repetition of mistakes but, perhaps more importantly, to draw upon the energy and vision of the pioneers who overcame the seemingly insurmountable barriers of the day and drove single-mindedly into the unknown.

The future of space belongs to a new generation who were not yet born when man first set foot on the Moon. The heart-stopping excitement of that event is difficult to communicate to young men and women who now take space for granted. *Spacecraft Technology: The Early Years* is a comprehensive multispectral history of this period and offers a retrospective perspective to those of the first space generation who, although involved first hand, were often blinkered by circumstances. It likewise holds much of value to the next space generation who will follow in our footsteps and take us, hopefully, further afield into the reaches of space to increase our understanding of who we are – and why.

Prof Sir Martin Sweeting, OBE FRS
Director of the Surrey Space Centre and
Chief Executive of Surrey Satellite
Technology Ltd (SSTL)

Preface

I became interested in space technology in the late 1960s, when NASA was preparing to send men to the Moon. As a schoolboy in an English market town, I had no idea that I would one day earn a living from space technology, let alone write a book about its early development. I wish I had. I might have paid more attention to what was happening!

This book is the result of a fascination in the early history of space technology which produced a series of papers for the IEE's annual meetings on the history of electrical engineering. Although the papers stuck out 'like a Dali in a Greek Temple' among the others on thermionic tubes, Second World War radar developments and turn-of-the-century railway traction, they were always well received and I was welcomed back to present more of this 'new-fangled technology'. It was only when I came to turn the papers into a book that I truly began to appreciate the skills of the *bone fide* historian (within whose ranks I shall never stand). Needless to say, there was a lot more work to do.

I should state quite clearly that this is not a book for professional space historians, who tend to prefer a more formal style. It is aimed at space scientists and engineers interested in placing their own expertise in historical perspective and the general reader who enjoys delving into the background of modern technology.

I decided at the outset that I had no interest in attempting to write yet another literal history of events concerned with space exploration, presented in chronological order and covering all the missions of every space-faring nation. For that, you should consult an encyclopedia. And space aficionados should not necessarily expect to find their favourite rocket, satellite or historical figure here – some of them will be, some will not. Nor was I interested in researching a learned political history which places the events in a world context, explaining who did what when and why. Both are perfectly valid approaches, but they fail to place the emphasis on the technology itself, which is my main interest in writing this book.

So, what *is* the book about? It is about the early development of space technology, between the late 1950s and the early 1970s, which encompasses the launch of the first satellites and the landing of men and their machines on the Moon. Why this period? Because, in my opinion, it is one of the most important periods of technological development mankind has ever known: it is the period in which we learned how to leave this planet and explore another, entirely separate and distinct planetary body. The achievement itself is a key point in history, in exploration and in global culture that will be remembered as long as the human race survives. It is also significant in that it has yet to be repeated: since the last Apollo astronaut left the Moon in 1972, no one has travelled more than a few hundred kilometres away from mother Earth.

There is more to the story than Apollo, however. Following a chapter on the pre-history of space exploration, the book discusses the challenge of getting hardware into space, which was met by the development of the space launch vehicle. Then, in separate chapters, it considers a number of space applications – science, Earth observation and communications – which are rooted in the early years of the Space Age. Having covered the basic technologies required for space activity, the reach of exploration is extended towards the Moon and the unmanned probes that prepared the way for the manned missions. Following a chapter on the development of space technology for manned spaceflight in Earth orbit, the book moves on to the hardware designed for the Apollo programme itself: the command and service module, lunar module and lunar roving vehicle. For their roles in transporting men to the Moon, landing them there and transporting them across the lunar surface, these unique spacecraft are granted separate chapters.

The development of space technology was a step-by-step, but interactive process. It occurred across the applications, not simply within them. I see this as the warp and weft of time's tapestry: the parallel tracks of the different space applications are tied together by the development of the technologies (power, propulsion, telemetry, etc.) required for all spacecraft.

This book is unusual in its concentration on space technology, as opposed to events and individuals. It is also unusual in being written by an engineer with experience in space technology and an interest in history, as opposed to an experienced historian or journalist with an interest in space technology. The result is a book which remains true to the importance of technology in the context of the historical events, rather than featuring technology as a necessary evil.

I hope that, having read this book, you will be sufficiently fascinated in the early development of space technology to seek out other books, which dedicate themselves to individual details I can only touch on here. For now, though, enjoy the multi-coloured tapestry of space technology that allowed beings from this planet to stand on the surface of another.

Mark Williamson
Kirkby Thore, Cumbria
January 2006

Acknowledgements

Deciding from the outset that I should conduct my own form of 'peer review' for this book, I obtained the assistance of a number of friends and colleagues, who kindly agreed to review sections of the original manuscript and make appropriate comments. I am indebted to them for the time they gave so freely and would like to acknowledge their contributions, which have improved the book. They are not, of course, accountable for the text as it appears. I have assumed editorial control and any errors are entirely my own responsibility.

I would particularly like to thank David Wade, who willingly read the entire manuscript and made useful comments and corrections. I am also grateful to Michael Ciancone, Kerrie Dougherty, Philippe Jung and Paul van Woerkom, whose contributions were invaluable. Many thanks are also due to Marsha Freeman, James Garry, Roger Launius, Paul Maughan, Doug Millard, Ron Miller and Joe Pelton for their helpful comments.

For assisting my attempts to provide useful illustrations for the book, I am indebted to James Garry, Ron Miller, Fred Ordway, Jody Russell and Irene Willhite. I would also like to acknowledge the wealth of NASA imagery available on the web, particularly the GRIN (Great Images in NASA) database at http://grin.hq.nasa.gov.

Finally, I would like to thank Martin Sweeting for sparing the time to write a foreword to this volume, and of course my wife, Rita, for her usual dedication in support of my writing projects.

MW

Chapter 1

A prologue to the Space Age

Earth is the cradle of mankind, but one cannot remain in the cradle forever

Konstantin Tsiolkovsky

The earliest periods of mankind's technological development – most notably the Stone Age, the Bronze Age and the Iron Age – were characterised by resources used in an organised and prolonged manner. The most recent, continuing example is the Space Age, a period of technology development in which the resource is outer space itself.

Although earlier 'Ages of Man' are difficult to date precisely, because of the lack of documentation, this is not the case for the Space Age. The Space Age can be defined as beginning with the launch of Sputnik 1 – on 4 October 1957 – since, from this point on, organised and prolonged activity in space became possible.

Some historians argue that the first time a rocket flew high enough to cross the notional boundary between the Earth's atmosphere and space marks the beginning of the Space Age, but that was but a prelude to the symphony that followed. Sputnik 1 transmitted its historic 'beep-beep' signal for 21 days and remained in orbit around the Earth for about three months. Although rocket-borne payloads had been flown into space before Sputnik, they were launched on suborbital trajectories (with insufficient energy to reach orbital velocity), allowing them to experience only a few minutes of the space environment before gravity returned them to Earth.

It was the launch of Sputnik 1 which proved that man-made devices could operate in space long enough to perform useful scientific investigations, preparing the way for the many spacecraft that were to follow. As such, it ushered in the Space Age, a new technological 'Age of Man' deserving of its initial capitals.

In some people's minds, the Space Age is a term applied to a short era of space exploration that culminated in the landing of a man on the Moon, but this is probably a media construction based on a misguided impression that nothing much has happened in space since that time. Nevertheless, those first few years were crucial in proving that space was accessible, not only to our technological creations but also to ourselves, and in showing what might be possible in the future.

In terms of the 'Ages of Man', we are still in the Space Age and will remain so as long as spacecraft continue to be launched into space.

Space technology in perspective

Space technology has a unique place in the history of technology, since it concerns the application of technology beyond the confines of the Earth. This application offers many significant challenges to the engineering profession, not least in terms of the ruggedness and reliability of its creations. Spacecraft must not only survive the hostile thermal and radiation environment of space, but must also do so without the physical intervention of their creators since spacecraft beyond the lowest altitude orbits cannot be retrieved or repaired. Indeed, even communicating with a spacecraft can pose a challenge, dependent on its distance from Earth, since it may involve signal travel-times measured in minutes, or even hours for some planetary probes.

Space technology is also unusual because it is both a user of other technologies – such as electrical, electronic, mechanical, power, computing and telecommunications – and a developer of these technologies. Such is the challenge of developing systems that operate reliably in the space environment, and hardware sufficiently light and compact to be launched in the first place, that developments in space technology often require long-term effort and significant expenditure. However, as a spin off, they often lead to improvements in earthbound systems.

The following chapters describe the development of space technology in the late 1950s and 1960s, from the launch of Sputnik 1 in 1957 to the landing of men on the Moon in 1969. They do so by considering the technology required to place satellites in orbit and that of the spacecraft themselves. The former concerns the development of rocketry; the latter encompasses the development of payloads and subsystems for space science, Earth observation and communications satellites, and the further steps required to qualify technology for manned space applications.

First of all, however, it is useful to place these developments in their historical context. Although the launch of Sputnik 1 was the first practical step in *actual* as opposed to *theoretical* space exploration, space technology did not suddenly begin on 4 October 1957. There were many developments in physics, mathematics, chemistry and other sciences, spread over many centuries, which culminated in space exploration. Moreover, beyond the realms of science, creative minds were working towards the opening of what would later be termed 'the final frontier'.

Indeed, space exploration and development is very much a cultural activity, often entered into for political, social and economic reasons rather than the development of technology for its own sake. It is also an endeavour that has fuelled the imagination of many an individual, possibly since our ancestors stared up at the Moon and stars and wondered what they were, where they were ... and, eventually, how one might reach them.

As with other cultural activities, such as art and music, the development of technology is an evolutionary, iterative process influenced by what has gone before. Of course, the leading proponents in a field are blessed with original thoughts, but no one thinks and works in a cultural vacuum. We are all influenced in some way by the knowledge and opinions of the day and the cultural environment in which we operate. Thus the early developers of space technology were influenced by the science, technology, art and literature of their day.

Of course, the greatest advances are made by those with the greatest powers of imagination, those with the ability to see beyond the common understanding of the day and extend the boundaries of knowledge. This is a practice which often leads to ridicule from those who lack imagination – an issue which has certainly affected the developers of space technology. For example, no lesser individual than England's Astronomer-Royal, Sir Richard van der Riet Woolley, was quoted in 1956 as saying 'space travel is utter bilge' [1].

As science writer Arthur C. Clarke's 'third law' states, any sufficiently advanced technology is indistinguishable from magic. Indeed, it is also certain that the majority of those living in the first half of the twentieth century would fail to comprehend what is now occurring in orbit around the Earth on a daily basis. This was the extent of the cultural challenge for those who were laying the foundations for space exploration at the time.

The age of imagination

It seems likely that space travel has been part of human imagination since mankind realised that space was 'somewhere to go'. As a planet, the Earth is unusual in that it has a moon which appears relatively large in its sky, allowing considerable detail to be discerned on its surface even with the naked eye. This alone was bound to have evoked speculation on the possibilities of travelling there, even before the invention of the telescope.

Indeed, the earliest known story of spaceflight – if we interpret this in its broadest sense – is a legend described on Babylonian clay tablets dating from 2350 to 2180 BC: it concerns the epic poem of Etana, who flew on the back of an eagle far into the sky towards Venus, the home of the Babylonian goddess Ishtar, the light bringer [2].

However, the first work of what one might call space fiction is believed to be a story written by the Greek satirist Lucian of Samosata in around 165 AD. The *True History*, often referred to by its Latin translation *Vera Historia*, describes a sailing vessel carrying 50 Greek athletes which is suddenly swept into the heavens by a violent whirlwind; it travels for eight days before reaching the Moon [3].

Naturally enough, there are a number of similar tales from other cultures through-out the ages, most of which used some force of nature or familiar creature as a means of propulsion. Around 1010 AD, for example, the Persian epic *Shah-Nama* describes the mythical King Kai-Kaus, who is lifted heavenwards by a team of four eagles bound to his throne [4].

Most historians of space technology trace the development of the current means of access to space – the rocket – to China, sometime between the eleventh and thirteenth centuries. Although there are reports of the use of 'gunpowder' in creating fireworks from as early as 850 AD [5], the first published gunpowder formulae, written by

Tseng Kung-Liang, have been dated to the mid-eleventh century [6]. Another source says that a formula for gunpowder was published in the *Wu Ching Tsung Yao* of about 1040 AD [7], while yet another believes that the first powder rocket was 'probably the accidental discovery of thirteenth century Chinese Taoist alchemists who were experimenting with various flame-enhancing materials' in their search for 'an elixir of life' [8].

Whatever the rocket's precise heritage, there seems little doubt that, in common with many technological innovations, it was first applied within a military context. For instance, it is known that by 1232 rocketry was sufficiently advanced to enable the Chinese to beat back the Mongol cavalry at the siege of Kai-fung-fu. Reports refer to the use of 'fire arrows' (probably incendiary rockets) and gunpowder bombs that were dropped from the city walls on the heads of the marauding Mongols (Figure 1.1) [7].

It would be many centuries, however, before the idea of using a rocket to fly into space would be conceived. A possible exception was a first attempt at manned rocket flight, allegedly made in about 1500 by Wan-Hoo, a Chinese official whose name can be roughly translated as 'crazy fox'. He is said to have fitted 47 powder rockets to two kites attached to a chair in which he sat. The story goes that, at a pre-arranged signal, 47 coolies lit the 47 rockets...and Wan-Hoo vanished in a brilliant flash and cloud of smoke, thus reaching the heavens a little earlier than intended. However, most authorities consider the story apocryphal, since the first known published reference was dated 1909 and was possibly a fabrication of Europe's Chinoiserie period, which was characterised by a fascination with the Orient [9].

Many early methods of attaining the 'high frontier' were based on variations on the flying chariot, while a favourite destination was the Moon. For instance, in Ludovico Ariosto's *Orlando Furioso* of 1532, a chariot powered by winged horses lifts the hero towards a Moon that has 'rivers, valleys, cities, towns and even castles' [10].

The somewhat better-known Johannes Kepler, German astronomer and author of Kepler's Law's of planetary motion, knew that the Earth's atmosphere did not extend as far as the Moon and could therefore not support any method of flight that he could imagine. Thus in *Somnium* ('Dream'), a story published in 1634, four years after his death, he invokes the supernatural alternative of the demon: unable to withstand the light of the sun, the demons travelled only at night and, sometimes, carried humans with them [11].

The theme of lunar travel continues with English bishop Francis Godwin's *The Man in the Moone: or a Discourse of a Voyage Thither, by Domingo Gonsales* (1638), in which Gonsales trains swan-like birds to carry him on a platform to the Moon. By coincidence, in the same year, another bishop and a member of the Royal Society of London, John Wilkins, published *The Discovery of a World in the Moone* (*Or, a Discourse Tending to Prove that 'tis probable there may be another habitable World in that Planet*). Although they suffered from similarly verbose titles, the difference with Wilkins' work was that it was considered science rather than fiction, which says more about the state of science in the seventeenth century than it does about fiction. Indeed, a later 'corrected and amended' edition, published in 1684, has Wilkins agreeing with Godwin's premise that birds might be trained to carry men to the Moon, where they might, among other things, discover the religion of its inhabitants [12].

火
籠
箭
式

Figure 1.1   The ancient Chinese used rockets, or 'fire arrows', in warfare [Ron Miller]

Cyrano de Bergerac is considered to have been the first to propose that rockets should be used as a form of 'spacecraft propulsion'. In his novel, *Histoire Comique des Etats et Empires de la Lune et du Soleil* (The Comic History of the States and Empires of the Moon and the Sun), which was published in various editions in the 1650s and 1660s, he imagined a machine being raised aloft by firework rockets (Figure 1.2) [13]. However, this was more of a 'throw-away idea' than a design, as shown by other methods included in this 'comic novel'. For example, he also

Figure 1.2   Cyrano de Bergerac imagined using firework rockets to power a flying machine [US Space and Rocket Center]

suggested collecting dew in a number of vials that could be tied to the body, the logic being that since dew is apparently drawn up by the morning sun, a person could be carried up with it.

Throughout the seventeenth and eighteenth centuries, there followed a number of tales of flying chariots, airships and even whole islands, using birds, balloons, springs and even lodestones as means of propulsion. It is evident that the link between rockets and space travel had yet to be accepted. Indeed, rockets in reality had changed little since the siege of Kai-fung-fu in 1232.

This began to change in the early nineteenth century, however, when Englishman William Congreve is credited with initiating the modern process of rocketry research and development. Congreve built the first rocket weapon systems – to challenge Napoleon's invasion plans for England – and it was Congreve rockets used at the bombardment of Fort McHenry, Baltimore in 1814 that gave rise to the line 'And the rockets' red glare…' in the American national anthem [14].

But the use of rockets in spaceflight was still confined to fiction, most famously that of French author Jules Verne. Indeed, Verne is credited with writing the first fictional account of spaceflight based on scientific fact – *De la Terre à la Lune* (From the Earth to the Moon) – which appeared in 1865. Since the bullet was by far the fastest thing in Victorian life, it comes as no surprise that his fictional spacecraft was a bullet-shaped projectile propelled by a cannon, named Columbiad, which was fixed from a 900-ft shaft sunk into the ground (Figure 1.3). Verne's first astronaut crew would have been pulverised by the acceleration, but at least he had recognised the need for speed to escape the Earth's gravity, a concept described by Newton two centuries earlier in his famous *Philosophiae Naturalis Principia Mathematica* (Mathematical Principles of Natural Philosophy).

In fact, it has been suggested that Verne knew a great deal more than his fictional launch system would suggest. The story includes the successful test flight of a scale model containing animal subjects – a possible literary device to encourage the reader to believe in the later manned launch – and a description of weightlessness in space. Verne is also said to have enlisted his cousin, mathematician/astronomer Henri Garcet, to help with calculations for the lunar flight, apparently the first time in history that a trip to the Moon had been granted a mathematical basis [15]. For example, he correctly determined the Earth's escape velocity to be about 7 miles per second.

In a sequel, *Autour de la Lune* (Round the Moon), published in 1870, Verne unknowingly predicts the Apollo lunar missions of a century later: as his projectile approaches the Moon, its crew fire rockets to adjust its trajectory (thus anticipating the spacecraft manoeuvring thruster) and take it within 3000 miles of the surface [16]. An illustration in the book is believed to be the first to depict a rocket operating in space (Figure 1.4). In further predictions of Apollo, Verne's launch site is in Florida and his projectile, having looped around the Moon and returned to Earth, splashes down in the Pacific Ocean to be recovered by the US Navy. Little wonder that Verne's tales of space exploration have attracted much analysis and comment in recent decades.

Strangely, as seems to happen with scientific discoveries and engineering creations in the real world, in the same year that *De la Terre à la Lune* was published, 1865, another author, Achile Eyraud, published a tale of interplanetary spaceflight called *Voyage à Venus* (Voyage to Venus). In it, he describes a spaceship propelled by a 'reaction motor' in the form of a water rocket, but negates the propulsive effect by proposing to collect the water. Whereas Verne produced world famous masterpieces which, to some, qualify him as 'the father of modern science fiction', Eyraud's work has gone largely unnoticed [17].

Another science fiction novel which has avoided widespread publicity is *The Brick Moon* – published in 1869 in *The Atlantic Monthly* ('A magazine of Literature, Science, Art and Politics') – by American author Edward Everett Hale. It is,

Figure 1.3   The Columbiad cannon launches Jules Verne's projectile [US Space and Rocket Center]

nonetheless, an interesting waypoint on the journey towards the Space Age, being the first story about an artificial Earth satellite, the first published description of a navigation satellite and literature's first exposition of a manned space station [18].

The somewhat unlikely, brick-built satellite was 200 ft in diameter, and was placed in a 5000-mile high polar orbit by two huge, contra-rotating flywheels. The launch system involved depositing the satellite between the wheels, which supposedly threw it into space as a result of their rotation. Brick was chosen as the construction material for its refractory properties, which allowed the satellite to withstand the heating effect of atmospheric friction (Figure 1.5) [19]. Although the story was purely fictional, it marked a departure from the tales of lunar exploration and prefigured the development of more utilitarian aspects of space technology. Its application as a navigational aid

Figure 1.4 Verne's spaceship fires a rocket to change its trajectory [Ron Miller]

for sailors could be seen as a precursor to the modern day Global Positioning System (GPS), for example.

Another of spacecraft technology's textbook solutions, though one which has yet to find broad application, was introduced by German law student and inventor Hermann Ganswindt in a lecture describing an interplanetary spaceship in 1881. Realising that artificial gravity would be desirable for the crew, his design incorporated two spacecraft joined by a tether and spun about their common centre of mass [18].

The age of theory

Despite all these ideas of sending men and their machines into space, and despite the fact that rockets had been used to propel fireworks for centuries, writers rarely seemed to think of the rocket as a propulsion device.

Figure 1.5   On the Brick Moon [Ron Miller]

The reason the rocket was largely overlooked is thought to be a result of a widespread ignorance of the reaction principle, the fundamental principle of rocket flight. Despite Newton's formulation of the Third Law of Motion, published in 1687 in *Principia Mathematica*, the misguided rocket propulsion theories of the 1700s were accepted even by some so-called experts well into the twentieth century. In essence, they thought that a rocket moved because its flame pushed against the surrounding air, and since there was no air in space, the rocket would not work [20].

As with Newton, who himself produced a step-change in the understanding of physics to which others would subscribe in due course, it would take a person of special insight to break the mould in rocketry and link it once and for all with spaceflight. The credit for this achievement is usually given to three men in three different countries, separated not only by geography, but also by political systems and their own individual characters. As a result, although their lifetimes and careers overlapped, their work was performed largely independently of and without influence from the others.

The three individuals were a Russian by the name of Konstantin Eduardovich Tsiolkovsky (1857–1935); an American called Robert Hutchings Goddard (1882–1945); and a German known as Hermann Oberth (1894–1989).

Inspired as a boy by the novels of Jules Verne, Tsiolkovsky (Figure 1.6) spent a good deal of his spare time thinking and writing about space travel. In 1883, he shed the first true scientific light on the field of space rocketry when he wrote about reaction motion in a document entitled 'Free Space'. However, although he considered releasing a compressed gas from a barrel which would then be propelled, he does not appear to have equated this with the rocket [21].

According to one of his papers, he became interested in the theory of rockets in 1896, and it was from that point that he began to make significant progress. As early as 1897, he derived a formula relating a rocket's final velocity to the exhaust velocity and the ratio of its full and empty weights, later incorporating the effects of gravity and atmospheric drag [22]. Originally known as the Tsiolkovsky Formula, every astronautics student today knows its descendent, 'The Rocket Equation', arguably *the* fundamental equation for rocket flight.

Figure 1.6   Konstantin Tsiolkovsky [US Space and Rocket Center]

In 1898, by then employed as a schoolteacher, Tsiolkovsky went on to establish, among other things, that the performance of liquid rocket propellants would be higher than the solid propellants which had been used since the birth of rocketry, a fact that remains valid to this day. This led to the publication of his seminal work, 'Investigating Space with Reaction Devices',[1] in the May 1903 issue of the magazine *Survey in*

---

[1] Different translations of this title appear throughout the literature (e.g. *The Exploration of Cosmic Space by Means of Reactive Devices*) and it is sometimes referred to as *A Rocket into Cosmic Space*, the title of part one of the paper; similarly, the magazine title itself is also quoted as *Scientific Review* [24]. The paper was reprinted in 1911 in *Herald of Aeronautics* magazine, this time complete with diagrams, which had been omitted from the first edition [25].

Science [22,23]. A mathematical and descriptive study of rocketry and spaceflight, the paper's chief historical importance was its recognition that the rocket provided the only means of access to space, using the reaction principle on which Tsiolkovsky had worked several years earlier. Interestingly, by way of historical context, the paper was published the same year that Wilbur and Orville Wright made their first powered-aircraft flight.

Significantly, Tsiolkovsky also proposed the now accepted and developed technologies of liquid propellants as coolants (regenerative cooling), hypergolic (self-igniting) propellants, gyroscopic stabilisation using moveable vanes in the rocket's exhaust, and electrically-powered ion engines. In addition to rocket technology, he also provided technical discussions on artificial satellites and interplanetary spacecraft, and wrote about spacesuits, airlocks and umbilical tethers (though using different terms) [26]. Little wonder that, in Russia, he is known as the father of cosmonautics.

Unfortunately, the magazine in which Tsiolkovsky's paper was published also contained a politically revolutionary piece that led to the issue's confiscation by the State and his subsequent writings were self-published in small print runs commensurate with his meagre salary. As a result, his work remained largely unknown outside Russia until the 1920s and 1930s, when his homeland sought technical precedence and international recognition. Inside Russia, however, men such as Valentin Glushko and Sergei Korolev had been sufficiently influenced by his writings to become the nation's leading rocket designers (see Chapter 2). Eventually, in the years prior to his death in 1935, Tsiolkovsky served as an advisor to Russia's Gas Dynamics Laboratory, which developed the first Soviet liquid-propellant rocket engine in the early 1930s.

While Tsiolkovsky was working on the facts regarding rocketry and spaceflight, the western world continued to develop its fascination for fiction, most notably through the pen of one Herbert George Wells (although Tsiolkovsky, himself, published a science fiction story, 'On the Moon', in 1893). H.G. Wells' *The War of the Worlds* was first published in serial form in *Pearson's Magazine* between April and December 1897, and in book form the following year [27]. This time, the spacecraft technology came not from Earth, but from Mars. Although a science fiction classic, the novel's descriptions of the Martian spacecraft were more or less confined to 'cylinders' apparently propelled by green flares (since he was more interested in people than technology). In fact, as shown by Wells' *First Men in the Moon* of 1901, for which he invented the gravity-shielding material 'Cavorite', he was in no way a leading proponent of rocket propulsion.

Meanwhile, elsewhere in the world of entertainment, the technology of 'moving pictures' was beginning to be used for what we now call science fiction movies. Indeed, a number of SF 'shorts', of only a few minutes in length, were made prior to 1900. However, the first feature-length (21-min) SF film was *Le Voyage dans la Lune* (A Voyage to the Moon), made by Georges Méliès in 1902; unsurprisingly, it was inspired by Verne's *From the Earth to the Moon* and H.G. Wells' *First Men in the Moon*. It featured a shell-shaped capsule depicted, tongue-in-cheek, as hitting the man in the moon in the eye! Thus began a long line of space-related films which, for some, culminated in the 1968 classic *2001: A Space Odyssey*, released the year before Neil Armstrong became the first man to set foot on the Moon.

The age of experimentation

Although Tsiolkovsky had been the first to propose a theory of liquid rocket propulsion, it was the American researcher Robert H. Goddard who developed the first liquid-propellant rocket engine. While Tsiolkovsky-the-schoolteacher was a gifted theoretician, Goddard – a professor of physics at Clark University, Worcester, Massachusetts – was very much the dedicated experimenter (Figure 1.7).

Like so many others, Goddard was influenced by the fiction of H.G. Wells: in 1898, at the age of 16, he read a newspaper serialisation of *The War of the Worlds* and was hooked [28]. He began his investigations into rocket performance as early as 1903 and by 1909, completely unaware that Tsiolkovsky had already reached the same conclusion, had rejected other more fanciful notions and embraced the rocket as the answer to space travel.

Having experimented with available solid rockets and found them lacking, he built his own metal-cased rockets filled with double-base solid propellants,[2] themselves now considered to be limited in power and not widely used in space applications. To the exhaust end of his rockets he added funnel-shaped, divergent nozzles, developed in the 1880s by Swedish engineer Carl Gustaf De Laval to produce steam at maximum pressure from turbines. Thus Goddard improved the rocket's efficiency (in converting heat energy to kinetic energy) from about 2.5 per cent to an impressive 64.5 per cent, while increasing the exhaust velocity from 300 m/s to some 2400 m/s [29] (the product of the exhaust velocity and the mass of gas ejected per unit time gives the thrust of a rocket). In 1916, using an evacuated chamber, he also proved once and for all that rockets would work in a vacuum, thereby confirming the validity of Newton's Third Law of Motion [29]. At last, it could be said with certainty that the rocket did not require anything to push against!

Goddard reported his experiments in a renowned paper of 1919 entitled 'A Method of Reaching Extreme Altitudes', which was published by one of his sponsors, the Smithsonian Institution of Washington [30].[3] Based on work carried out between 1912 and 1916, it is the earliest technical, non-fiction publication in the English language on the use of rockets for space travel.

---

[2] Also known as homogeneous propellants (typically nitrocellulose plasticised with nitroglycerine plus stabilising products). The alternative heterogeneous or composite type consist of a mixture of fuel and oxidiser.

[3] This famous work appeared in Smithsonian Miscellaneous Collections, Vol.71, No.2, Publication #2540.

Figure 1.7   Robert Goddard [US Space and Rocket Center]

From 1920 onwards, Goddard devoted himself to liquid rockets, specifically those using gasoline as a fuel and liquid oxygen (LOX) as an oxidiser. Like Tsiolkovsky before him, enamoured with the idea of sending a spacecraft to distant planets, he was convinced that liquid propulsion was the key. The problems included overcoming the evaporation of LOX (which boils at a temperature of $-182.98°C$ and has to be stored in well-insulated tanks); cooling the engines (which operate at very high temperatures); feeding propellants to the combustion chamber in a carefully controlled manner; and precisely timing their ignition. He also had to address the more mechanical issues of stabilising the rockets in flight and recovering them for later analysis. Thus Goddard was the first to conduct a rocket development programme of the sort we would recognise today.

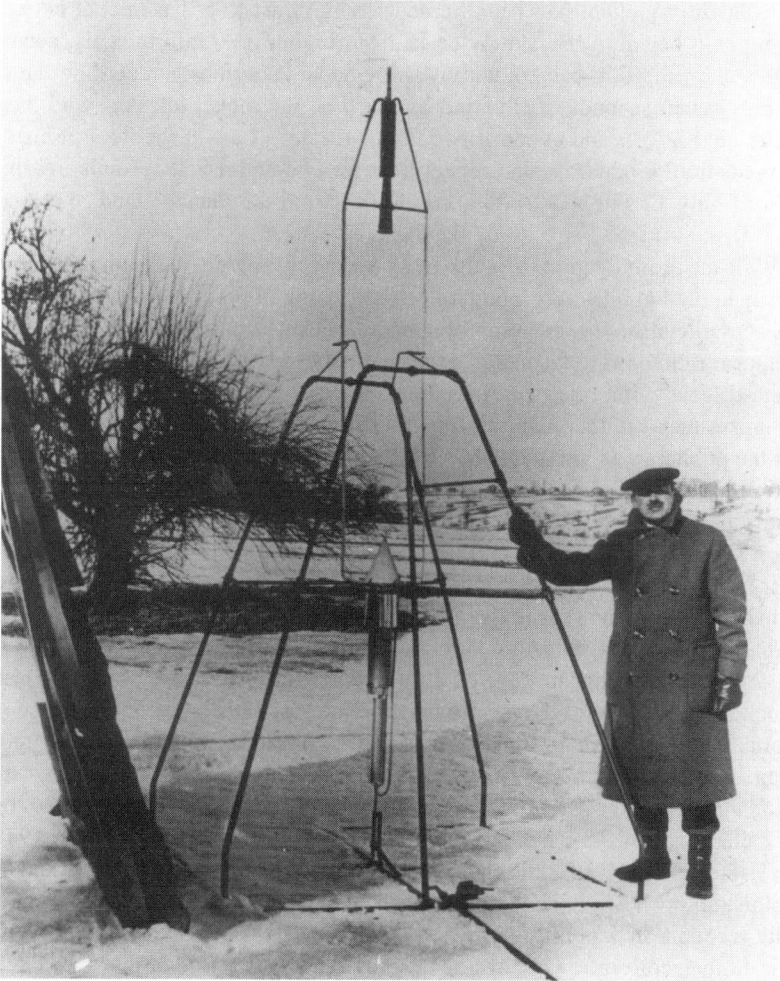

Figure 1.8   Goddard's first rocket [NASA]

In 1923, in a culmination of his early developments, Goddard made the first successful static firing of a pump-fed gasoline and LOX engine and, on 16 March 1926, conducted his first rocket launch from a site on his Aunt Effie's farm at Auburn, Massachusetts [31] (Figure 1.8). The rocket weighed almost 4.8 kg at ignition, but since its engine produced only 4 kg of thrust,[4] it remained within its launch stand

---

[4] Although thrust is measured in newtons (N) under the SI system, it is common to find 'kilograms/tonnes of thrust' or 'pounds/tons of thrust' quoted in historical documents. For reasons of comparison and easy conversion, this system is used here. As a rule of thumb, kg-thrust can be converted to N by multiplying by 10 (9.81 to be more exact).

until enough propellant had been burned for its thrust to at least equal its weight, whereupon it began to rise slowly on its historic journey. Unfortunately, according to one author, the film camera with which Goddard's wife was recording the event held only seven seconds of film and by the time the rocket left the stand the film had run out [32] – a sound confirmation of Murphy's Law in the field of historical documentation! Goddard's first rocket flew for two-and-a-half seconds, reaching a height of only 12.5 m and landing some 56 m from the launch stand, but it was a start [33].

While the promulgation of Tsiolkovsky's ideas was hampered by the State and his inability to fund publication, Goddard was the victim of mass misunderstanding. The press sensationalised the mention of a 'moon rocket' in his paper of 1919, making him appear ridiculous to its unenlightened readership, while professionals treated his advanced ideas with a degree of scepticism. There were also fears concerning the loud noises heard in the vicinity of his rocket tests. As a result, Goddard withdrew from the public arena and published little of his later work. In fact, the first public mention of Goddard's historic 1926 rocket launch did not appear until 1936, when the Smithsonian Institution published his 10-page monograph entitled 'Liquid-Propellant Rocket Development' [30].

In 1930, Goddard moved to Roswell, New Mexico, partly to avoid publicity but also to benefit from the year-round good weather. With the support of celebrated aviator Charles Lindburgh and with funding from the Guggenheim Foundation [34], Goddard continued his work in Roswell until 1941. With successive, improved designs, his rockets went on to reach a record height of almost 2300 m in 1935. He also made important developments in flight control and stabilisation, by incorporating moving air vanes linked to a gyroscope (Figure 1.9), and by 1940 developed an engine capable of producing around 400 kg of thrust [35], one hundred times that of his 1926 rocket.

Goddard gained 48 patents in his lifetime and following his death in 1945 his wife, Esther, obtained an additional 131; in 1960, the National Aeronautics and Space Administration (NASA) acquired the use of the patents at a cost of $1 million [36]. Public recognition was only bestowed posthumously, largely thanks to the efforts of his former colleague G. Edward Pendray and Esther Goddard, who edited his research notes into a book published in 1948 (*Rocket Development: Liquid-Fueled Rocket Research, 1929–1941*) [34]. This was, however, too late for Goddard's work to have a significant influence on those who followed.

Partly because of his reclusive nature but also because he failed to publish his results in the recognised way, Goddard's innovations were largely ignored by the US government and military officials who, by the early 1940s, were preoccupied with the development of the atom bomb ... seemingly uninterested in the concept of the rocket-based delivery system.

Meanwhile in Germany, the seeds for such a delivery system were being sown. Working independently from both Tsiolkovsky and Goddard, a third rocket pioneer, Hermann Oberth, had also decided on the liquid-propellant rocket as the ideal space booster (Figure 1.10).

Oberth was born in 1894 in Transylvania (a part of Austria–Hungary that became Romania in 1918) and later became a German citizen [37]. In the winter of 1905,

Figure 1.9   Gyroscope in Goddard rocket [NASA]

when Tsiolkovsky was refining his first liquid rocket design and Goddard was filling his notebooks with ideas about spaceflight, the 11-year-old Oberth read and began to memorise Jules Verne's *From the Earth to the Moon*. He later checked Verne's calculations and concluded that a cannon was 'not good for spaceflight', since his travellers would be 'flattened into pancakes' [38].

Oberth's theories were first published in Germany in 1923 in a small book called *Die Rakete zu den Planetenraumen* (By Rocket into Interplanetary Space). It was mathematical in nature and intended for scientists and engineers, who are said largely to have ignored or misunderstood it. Among other things, the book included the earliest known description of propellant injectors (which he called 'diffusors'), the use of multistage rockets, inertial guidance and ideas for manned orbiting stations [39].

This was followed in 1929 by a much-expanded version titled *Wege zur Raumschiffahrt* ('Ways to Spaceflight' or 'Paths to Space Travel', depending on the translator). In it, Oberth developed the Model E manned rocket he had proposed in *Die Rakete* and described a fictional circumlunar flight in the Model E *Luna*: it was 35 m tall and shaped like an artillery shell on four large fins. The book also contained discussions of space telescopes, lunar mining, colonies on Mars, solar power satellites and many other concepts which are still under discussion today (a fact that

Figure 1.10   Hermann Oberth [US Space and Rocket Center]

tends to support the cliché that there's nothing new under the sun!). However, even Oberth was not the first to come up with a spacecraft incorporating artificial grav- ity: remarkably, Hermann Ganswindt had anticipated Oberth's design by some 40 years [18].

In a sense, Oberth was a cross between Goddard and Tsiolkovsky, in that he conducted experiments with solid and liquid propellants as well as developing the theory. He wrote in a postscript to the 1923 edition of *Die Rakete* that only after typesetting did he become aware of Goddard's 1919 report, and later said that he heard of Tsiolkovsky only when the Russian sent him a copy of one of his publications in 1924. The fact that neither man's work was well disseminated and remained largely unknown to the public, restricted their influence.

Some have suggested that Oberth deserves the title 'Father of the Space Age' [40] because his work led others to follow directly in his footsteps. Indeed, his works were taken up enthusiastically by a group of disciples, including rocket-car builder Max Valier and author Willy Ley, who popularised Oberth's ideas and, in 1927, was a founding member of Verein fur Raumschiffahrt (literally the 'Society for Spaceship Travel'). The VfR, as it is better known, attracted almost 500 members in its first year [41]. Its journal *Die Rakete* was the first magazine devoted to spaceflight, although publication ceased in 1929 to allow the VfR, under its first president Johannes Winkler, to fund its own rocket experiments [42].

In fact, the VfR was not the first society for space travel to be formed. That title goes to the Soviet Union's Society for the Study of Interplanetary Travel, a division of the Moscow Society of Amateur Astronomers. It was founded in 1924 by Fridrikh A. Tsander, who became a leader in Soviet spaceflight circles after the First World War. Unfortunately, the society was short-lived and was disbanded the following year. The first in the West was the Austrian Scientific Society for High-Altitude Exploration, which was formed in 1926 [43]; perhaps its greatest call to fame was in providing a model for the VfR which was founded the following year.

Suddenly, it seemed, the spaceflight genie was out of the bottle. In Germany, the largely amateur membership of the VfR began experimenting, lecturing and writing articles on the subject, while Max Valier and Fritz von Opel experimented with rocket-propelled cars (experiments that would lead to Valier's death in 1930). Arguably more useful than the experiments were Valier's popular publications on spaceflight from 1924–30, including *The Thrust into Space* of 1924 (written in German), which ran to second, third and fourth editions in each of the subsequent years [44].

In 1928, Willy Ley edited a collection of essays by VfR members under the title *The Possibilities of Space Travel* (also in German) to help publicise the work of the society. Authors included Hermann Oberth and Walter Hohmann, an engineer from Essen who had published *The Attainability of Celestial Bodies* in 1925. The latter was a technically significant contribution to the literature, since it detailed the theoretical problems of space navigation and propellant requirements for planetary exploration. In it, he developed the famous Hohmann ellipses, or minimum-energy transfer orbits, which have since been followed by many interplanetary spacecraft designed to use the minimum amount of propellant while carrying the maximum payload.

Then in 1929, an early VfR member known as Hermann Noordung (the pseudonym of Herman Potočnic, a captain in the Austrian Army) wrote, in German, *The Problem of Navigating in Space*. It was the first comprehensive technical discussion of space stations [45] and was so comprehensive that nothing similar appeared until the mid-1940s [46]; indeed an English language translation was published by NASA in 1995 as *The Problem of Space Travel: The Rocket Motor* (SP-4026). The book describes a 30 m-diameter wheel-shaped space station (*Wohnrad* or living wheel), which features huge mirrors designed to collect solar energy for the station. It orbits in constellation with a separate solar mirror spacecraft and an observatory, all linked by pressure hoses for air and electrical connections (Figure 1.11).

Figure 1.11   Three elements of Hermann Noordung's space station seen through the window of a spaceship arriving in geostationary orbit [NASA]

Although some of the technical solutions may be questionable, an interesting aspect of the station was its position: 36,000 km above the Earth in what we now call geostationary orbit. Sixteen years later, Arthur C. Clarke would write an article for *Wireless World* magazine pointing out how three satellites placed in this orbit could provide almost worldwide communications coverage, though whether or not he was influenced by Noordung's book remains unclear. Clarke seems unsure on the point, but believes he may have been influenced by a series of science fiction stories by

George O. Smith known as the *Venus Equilateral* series about a manned radio station at one of the Sun–Venus Lagrangian points [47].

A science/fiction symbiosis

Science fiction became increasingly popular and more widespread as the 1920s progressed, and various aspects of space technology began to appear with great regularity as the genre developed in book and magazine form. For one thing, budding astronauts were relinquishing the fashions of Victorian England in favour of the spacesuit, which of course allowed the exciting literary device of the spacewalk. Thus, in the 1920s, astronauts began to appear in spacesuits on magazine covers, among the first of which was the August 1923 edition of *Science and Invention* (formerly *The Electrical Experimenter*) edited by the American publisher Hugo Gernsback (Figure 1.12). Labelled 'Science Fiction Number', this was Gernsback's first serious attempt to publish a science fiction magazine, a foray that would later produce such classics as *Amazing Stories* (from April 1927) and *Science Wonder Stories* (from June 1929).

In fact, Gernsback had begun his publishing career in 1903 with the much more down to earth, but no less forward looking, monthly magazine *Modern Electrics*, America's first radio magazine. Perhaps not surprisingly, given the immaturity of the field, Gernsback found some of his issues difficult to fill and began to include his own fictional stories. The story he wrote for the April 1911 issue featured a Martian spaceship tracked by what we now call radar, a technique actually used in the 1960s by the lunar-orbiting Apollo spacecraft. Apparently, the descriptions and diagrams that accompanied the story were so good that the US Patent Office later refused to consider Robert Watson-Watt's radio detection and ranging (radar) system as a new invention [48].

This is certainly not to suggest that Watson-Watt was influenced by Gernsback's story, but it does indicate the potential for science fiction to influence the increasing number of people who were open to its far-reaching suggestions. As the century progressed, an intimate link, or symbiosis, between space technology and science fiction began to develop.

It reached a significant peak in 1929 with Fritz Lang's seminal film *Frau im Mond* (Woman in the Moon), for which Hermann Oberth and Willy Ley acted as technical advisers. The film featured the multistage rocket *Friede*, which was based on the Model E in Oberth's book and considered by some to be the first realistic spaceship in movie history (Figure 1.13) [49]. Interestingly, the film is also credited with the first instance of 'regressive counting', now known as a countdown, which was considered more exciting than counting upwards to a preset figure. The German rocket engineer Krafft Ehricke, then aged twelve, saw the film a dozen times when it first opened, which prompted him to ask for several books on space and astronomy for Christmas [50]. Around 13 years later, in 1942, the tail fin of the first V-2 rocket launched from Peenemünde bore a drawing of a girl (a scantily clad witch riding a V-2 across the backdrop of a crescent moon) as a credit to the film that inspired Ehricke and the other members of his team [51].

Figure 1.12   Early science fiction pulp magazine [US Space and Rocket Center]

The 1920s was a period of development in other countries too. For example, Frenchman Robert Esnault-Pelterie (known to the cognoscenti as REP) is considered by some to be on a par with Tsiolkovsky *et al* because of his comprehensive publications and their influence on the work of others. In contrast with Goddard, Esnault-Pelterie was an active communicator. He began theorising on spaceflight as early as 1907 [52], calculated rocket flight times to the planets in 1912, and published his first major work, *L'Exploration par fusées* (Exploration by Rockets) in

Figure 1.13   Model of the Friede multi-stage rocket from the film Frau im Mond
[Ron Miller]

1928. This was followed by *L'Astronautique* in 1930, which among other things
was significant in coining a new word, 'astronautics' (the science of spaceflight,
otherwise known as cosmonautics) [46]. Among a multitude of contributions to *aeronautics*, Esnault-Pelterie was also the inventor of the familiar 'joy-stick' control for
aircraft [53].

Meanwhile, in Italy, Luigi Gussalli was experimenting with rocket motors and, in
1923, published a book which developed the concept of using a multistage rocket to
transport a pair of men to the Moon and back (another unwitting prediction of Apollo).

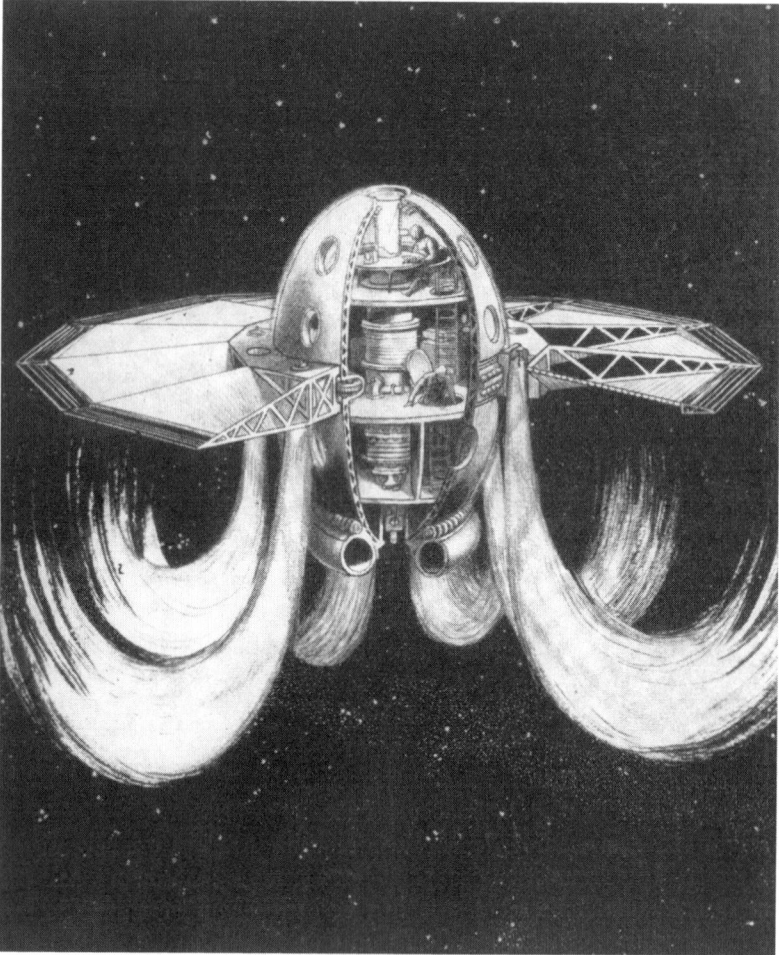

Figure 1.14   Ulinski's 'electron-powered' spaceship of 1927 [US Space and Rocket Center]

Gussalli's later publications, in the 1940s, expanded on the theme and included using the solar wind for interplanetary propulsion [34].

Austrian engineer Franz Abdon Ulinski was also active in the theory of space propulsion and, in 1927, presented a spaceship powered by an electron stream which he mistakenly believed to be recyclable (Figure 1.14) [54]. Though not as learned as Tsiolkovsky's thinking on a similar subject, this was effectively a precursor to the modern-day ion engine.

The 1930s continued the symbiosis between fact and fiction with the foundation of the American Interplanetary Society (AIS) in New York City. It was formed in 1930, mainly by writers of the early science fiction pulp magazines (so-called because of the cheap, poor quality paper on which they were printed). In 1931, its co-founder and first president David Lasser published *The Conquest of Space*, the first book on space travel and rocketry for a non-technical audience and highly regarded by rocket society members around the world. AIS vice-president G. Edward Pendray visited Germany in the spring of 1931 where he witnessed some of the VfR's rocket flights. As a result, the AIS began its own experiments that summer [55].

The society's new emphasis on rocket experimentation led to a change of name to American Rocket Society (ARS), as part of an attempt to attract professional engineers and scientists. At one point, Robert Goddard was asked to join the group but declined because he considered them amateurs. However, he is said to have regarded the ARS with more respect once a number of engineers had joined, so it seems that its policy was successful [56]. The Society survives to this day in the guise of the highly regarded American Institute of Aeronautics and Astronautics (AIAA).

Important developments from ARS members included, from 1933, Harry W. Bull's regeneratively cooled rocket nozzle and, from 1938, James H. Wyld's extension of the technique to all parts of the engine [57]; it seems they were unaware that both Tsiolkovsky and Oberth had written about it earlier and that Goddard and the VfR had actually tested it [56]. In December 1941, Wyld and three other ARS members formed America's first commercial rocket company, Reaction Motors Inc, and used the regenerative principle in a variety of rockets including, most notably, the Bell X-1 series of rocket-planes which broke the 'sound barrier' in 1947 [35]. As a result of Goddard's intransigence, therefore, the continuing thread of rocket development in the US was due more to the members of the AIS/ARS, and companies such as Reaction Motors, than to Goddard himself.

Another society formed in the 1930s, which survives to this day, is the British Interplanetary Society (BIS). Founded in 1933 by Philip E. Cleator and a group of like-minded individuals including Arthur Clarke, it set out to popularise space travel in the time-honoured fashion of publishing a journal and a number of books. The first British book on space travel, published in 1935, was *Stratosphere and Rocket Flight (Astronautics)* by Charles G. Philp, a science writer apparently unconnected with the BIS [58], followed closely in 1936 by Phil Cleator's *Rockets Through Space*.

Between 1937 and 1939, the BIS technical committee, composed of a number of engineers and scientists, put together a spaceflight plan which included the iconic BIS Spaceship, a design concept for a lunar lander that foresaw the Apollo lunar module. Its launcher was unusual in comprising six stages, each filled with a honeycomb cluster of individual solid rockets (168, 5 m-long, 35 cm-diameter motors in each of the first five stages and over a thousand smaller ones in the upper stage). The report was published in the Society's own journal (JBIS) in January 1939 and additional details were promised; however, World War II intervened.

Interestingly, the BIS rocket was designed to be launched from a floating platform, ideally on a high-altitude lake not far from the equator, an idea now experiencing reality in the form of the Sea Launch platform which operates from the equatorial Pacific. The BIS design specified liquid propellants for fine control, while steam jets were included for steering [59]. Although the use of steam has probably seemed archaic to most of the report's more modern readers, the practicality of a steam propulsion system was proved by the UK's Surrey Satellite Technology Ltd in 2003/04 using an experimental system on the UK-DMC microsatellite [60]. Once again, it is clear that many of today's expressions of space technology had their precursors.

The war years

In common with many technologies, the development of the rocket was driven by the needs of warfare; thus the Second World War produced a quantum leap in rocket technology which culminated in the terrifying V-2 ballistic missile.

The V-2 was the brainchild of Walter Dornberger, the head of rocket research in the German Army's Department of Ballistics and Munitions, and Wernher von Braun, a young VfR member recruited by Dornberger in 1932 as his first technical assistant [61]. In an attempt to develop a rocket that could carry a heavier, and thus more destructive, payload further than any existing artillery piece, Dornberger gradually assembled a competent technical team.

Those who accepted the military offer were relocated to the Berlin suburb of Kummersdorf, where the German Army, the Wehrmacht, had established an experimental rocket research station. Early tests at Kummersdorf were crude, to say the least, especially regarding the rocket's ignition system. In one test, von Braun apparently held 'a long stick with a can of gasoline fastened to the end, and shoved [it] under the exhaust nozzle as the propellants were let out' [62]. The engine was destroyed in the explosion, but amazingly no one was hurt (had von Braun been killed at this formative stage of his career, one wonders at the knock-on effect on German and, later, American rocket development).

By mid-1933, the first flight-model rocket, the A-1 (Aggregrate-1), was ready: it was just 1.4 m long and 30 cm in diameter. Both it and the similar A-2 used liquid propellants and were gyroscopically stabilised (the latter was an advance on the straightforward spin-stabilisation of simpler rockets). Success with the A-1 and A-2 led to the larger and more advanced A-3: it was 6.5 m tall, about 70 cm in diameter and produced 1500 kg of thrust for 45 s. The A-3 sported three-dimensional gyro control, molybdenum exhaust vanes, servo-valves actuated by electric motors, liquid nitrogen pressurisation and a radio link to cut the fuel flow to the engine [63].

In April 1937, the Army's rocket research team was moved to Peenemünde at the northern tip of the island of Usedom on Germany's Baltic coast. All four test-flights of the A-3, in December 1937, were failures: despite a good lift-off, the gyro system proved too weak and the vane movements too slow to counteract perturbing wind currents. However, the basic parameters of the A-4 (the original engineering designation of the V-2) had been set (Figures 1.15 and 1.16).

The A-4 itself was 14 m long and 1.7 m in diameter. Its take-off weight was about 12.6 tonnes and it could carry a warhead containing about a tonne of amatol high explosive. It was propelled by an engine burning a mixture of ethyl alcohol and water as the fuel and liquid oxygen as the oxidiser, producing a thrust of some 26.5 tonnes for 68 s. This gave it a maximum speed of about 5600 kph, a range just over 300 km and a peak altitude of close to 100 km [64].

Figure 1.15　Preparing a V-2 (A-4) rocket [NASA]

The development of the A-4 missile was fraught with problems and when war broke out on 3 September 1939, it was not even considered an option by most military officials. One significant problem was the difficulty in monitoring the operation of the vehicle in flight. Today's aerospace engineers take radio telemetry largely for granted, but the number of telemetered measurements made on early A-4 development flights was limited to four. For comparison, on the inaugural flight of the legendary Saturn V Moon rocket in 1967, a total of 3552 parameters were monitored continuously.

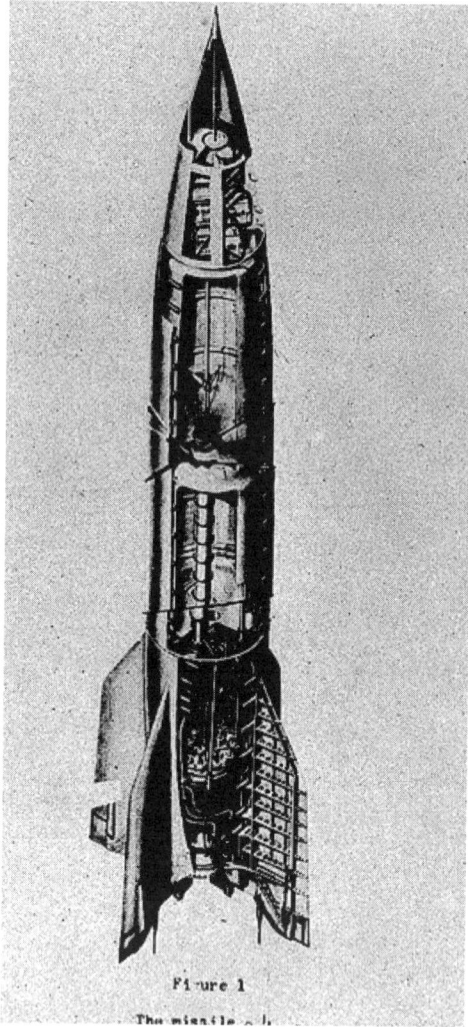

Figure 1.16   V-2 cutaway showing structure [NASA]

The lack of telemetry for the A-4 made it necessary to launch a large number of rockets simply to obtain enough data to perfect the design. For example, the development of the rocket's propellant shut-off valves required some 20 launches.

The A-4's propellant pumps were driven by a turbine, itself powered by the exhaust products from a gas generator. The resultant 'turbopumps' had to be light, yet capable of delivering fuel and oxidiser to the combustion chamber at a pressure of 300 psi (2070 kPa) at a rate of 50 gallons (190 l) or more per second. When von Braun approached pump manufacturers, he was surprised to discover that this was similar to the specification for centrifugal pumps used by fire-fighters, and they subsequently became the basis for those in the A-4 [65].

Another area of difficulty was the guidance system, for which there were two contenders. The first used a simple autopilot in the rocket and a large, expensive radar system on the ground as part of a radio-control system. Despite its accuracy, the system was vulnerable to jamming and made the already complex ground support equipment needed to launch the rocket even more complicated. The alternative was an inertial guidance system in which everything required to ensure the accuracy of the rocket would be carried on board. It comprised a number of gyroscopes, accelerometers and an analogue computer. Signals derived from the guidance system were used to activate hydraulically-powered thrust-vector control vanes, mounted in the exhaust like those of the A-3 but made from graphite.

Following a couple of launch failures, on 13 June and 16 August 1942, the A-4 performed its first successful flight on 3 October 1942. Far exceeding the performance of any previous rocket, it reached an altitude of about 90 km, travelled some 192 km from the launch site and landed within about 4 km of the target. According to a member of von Braun's rocket team, Ernst Stuhlinger, when the rocket launched, Walter Dornberger said 'today the space ship has been born! But I warn you: our headaches are only just beginning!' [66].

On 7 July 1943, Dornberger and von Braun, by then technical director of the Peenemünde rocket centre, showed the film of the first successful flight to Adolf Hitler, who until then had been somewhat dismissive of rocket technology. From that point on the A-4 was given the highest priority, and renamed 'Vergeltungswaffe (Vengeance Weapon) zwei', or V-2. Its forerunner, the V-1, known in England as the Doodlebug, was a rail-launched missile powered by an air-breathing pulse-jet engine. It travelled at subsonic speeds, could be shot down by aircraft and gave a distinctive audible warning to its human targets. By contrast, the V-2 was supersonic, impossible to shoot down and gave no convenient warning of its approach; indeed, it was said that if you heard a V-2, it had missed you!

The first two launches of the V-2 offensive were made on 6 September 1944 from a mobile battery stationed near Vielsalm, on the eastern fringe of Belgium, but both rockets failed when the fuel supply cut off prematurely. Two days later, having relocated to a point near Houffalize, the battery made its first successful launch towards Paris, which had fallen to the Allies on 25 August 1944. The world's first long-range combat rocket, now under the control of the notorious SS, made the 290 km flight in just a few minutes and impacted close to Port d'Italie, producing 'modest damage'. On the evening of 8 September 1944, other batteries located between The Hague and Wassenaar in Holland began their launch campaign towards London, aiming at a point near Waterloo station [67].

Launching as many as ten per day, a total of 1115 V-2s had reached England (of about 1400 targeted there) and an estimated 1775 of the 1852 successfully launched had reached targets in continental Europe by the end of the war [68]. The targeting of the V-2 was accurate to within a few kilometres, which was sufficient to justify its description as a 'terror weapon' when targeted at major cities the size of Paris or London, but it was never as accurate as its designers had hoped. Moreover, in Dornberger's own analysis, it was introduced too late to have a major effect on the war, which in Europe ended on 7 May 1945. Nevertheless, the V-2 had made an indelible mark on the history of rocket development.

While military rocket research was primarily concerned with lofting ever larger warheads on ballistic trajectories with greater ranges, it was obvious to engineers with more peaceful interests that range and payload mass could be traded for altitude. Although the V-2 was not designed as a space launch vehicle, its capability of delivering a payload to an altitude of 100km meant, by common definition, that it was capable of launching a payload into space, albeit briefly. In fact, while planned A-9 and A-10 versions of the rocket were designed to target more distance cities, such as New York, a future three-stage A-11 or its derivatives could have placed a satellite in orbit.

As a weapon of war, the V-2 killed many people in the Allied Nations and many more on German soil, employed as 'slave labour' to build the missiles. Much has been written about the underlying wishes of the V-2 rocket team to develop a satellite launcher and the apparent intentional delaying of V-2 development in horror of its destructive potential, while others have alleged outright complicity with Hitler's Nazi regime. It is difficult to see clearly through the fog of war and the full truth may never be known, but Albert Speer, the German armaments and munitions minister who encouraged Hitler to accord Peenemünde top priority, said of von Braun: 'For him and his team this was not the development of a weapon, but a step into the future of technology' [69].

An age of opportunity

The post-war transportation of Wernher von Braun and key members of his team to the US provided them the opportunity to continue their V-2 research and development efforts, albeit under a different Army command. Interestingly, the academic isolation and limited cross-fertilisation of ideas among early rocket pioneers does not appear to have extended to von Braun. Among the documents acquired by the Allies at Peenemünde was a German edition of a book by Tsiolkovsky, with 'almost every page ... embellished by von Braun's comments and notes' [70]. Von Braun was evidently not averse to standing on the shoulders of giants.

Without doubt, the V-2 proved crucial in the development of space technology by forming the technical basis for both Russian and American rocket programmes in the post-war period (as described in Chapter 2). Meanwhile, as part of a widespread desire to shake off the deprivations of war and look forward to an exciting future promised by the development of new technologies, a growing wave of interest began to grow in spaceflight beyond the field of rocket engineering.

In the professional arena, this new interest was highlighted by the formation of the International Astronautical Federation (IAF), a non-governmental association established to encourage the advancement of knowledge about space and the development and application of space assets [71]. Its origins are rooted in 1949, when the Gesellschaft für Weltraumforschung (GfW) of Stuttgart, Germany, proposed to other astronautical societies that they should hold a conference to develop cooperation. One of the first to support the idea was the British Interplanetary Society, which agreed to organise the conference in London in 1951.

In fact, the meeting was preceded by a large public gathering at the Sorbonne, Paris, in September 1950, which was designated the 'Premier Congrès International d'Astronautique'. Despite the generally unfavourable views on spaceflight common at the time, over a thousand people from many countries attended the congress. By default, this made the London conference, held in September 1951, the Second International Astronautical Congress (IAC). Astronautical and rocket societies from ten countries were represented and became signatories to the agreement founding the IAF, which continues to organise the IACs to this day.

Beyond the profession, too, the development of rocket technology continued to make its mark. For example, 1950 saw the release of George Pal's film *Destination Moon*, which was based on the novel *Rocketship Galileo* by Robert A. Heinlein. The lunar landscapes were designed by American space artist Chesley Bonestell and Hermann Oberth was the technical consultant. The film's slim, pointed rocket with tailfins, now accepted as science fiction's classical rocket shape, exhibits a striking similarity to the V-2 (Figure 1.17).

Once Wernher von Braun had settled in the United States, he began a new sub-career of space proselytisation, never missing an opportunity to try and interest people in space exploration. The best known example of this was a series of articles written by von Braun and his colleagues for *Collier's* magazine in the early 1950s.[5] Under the editorship of Cornelius Ryan, Willy Ley, Fred Whipple and von Braun wrote articles on rockets, wheeled space stations, lunar shuttles and journeys to Mars, illustrated by space artists Chesley Bonestell, Rolf Klep and Fred Freeman (Figure 1.18).

Collier's first space issue was published on 22 March 1952 under the title 'Man Will Conquer Space Soon'. The opening article, 'What Are We Waiting For?', began: 'On the following pages *Collier's* presents what may be one of the most important scientific symposiums ever published by a national magazine. It is the story of the inevitability of man's conquest of space. What you will read here is not science fiction. It is serious fact' [72]. Publicity for the issue included TV appearances by von Braun, window displays in selected American Express offices and the widespread distribution of press kits to the media and to schools and colleges [73].

Although many of von Braun's colleagues are said to have looked upon the *Collier's* series as 'speculative hoopla that served only the purposes of those who participated', many who went on to work in the US space programme attributed their initial interest in the subject to the series [74]. Spin offs from the series included two books edited by Ryan – *Across the Space Frontier* and *Conquest of the Moon*, published in 1952 and 1953 respectively – and a 1955 film, *Conquest of Space*, which used station designs by von Braun and sets by Bonestell (Figure 1.19).

---

[5] The series was published in eight issues of *Collier's* magazine, dated March 22, 1952; October 18 and 25, 1952; February 28, 1953; March 7 and 14, 1953; June 27, 1953; April 30, 1954.

Would you let <u>Your</u> Man take the first flight to the Moon?

COLOR BY **TECHNICOLOR**

IT'S CLOSER THAN YOU THINK! Rocket experts say that in our lifetime the moon-trip will be made exactly as you see it in this tense, believable picture! Will you have to say woman's most heart-breaking good-bye? Will your man take off on man's adventure into tomorrow? (*2 years in the making—the picture you've been reading about.*)

DESTINATION MOON

Produced by GEORGE PAL.   Directed by IRVING PICHEL.   Screenplay by RIP VAN RONKEL, ROBERT HEINLEIN and JAMES O'HANLON

Figure 1.17   Advertising George Pal's film Destination Moon *[Ron Miller]*

Figure 1.18 *The principals in the* Collier's *space series: (left to right) Willy Ley, Fred Whipple, Wernher von Braun, Chesley Bonestell, Rolf Klep, Fred Freeman and Cornelius Ryan [US Space and Rocket Center]*

The series also inspired a trio of Walt Disney TV films made to support the Tomorrowland section of the Disneyland theme park, which, incidentally, spread the word to a much wider audience. The first film, *Man in Space*, aired on 9 March 1955 to an audience of nearly 100 million [75], compared to an estimated 12–15 million readers of *Collier's* [76]. One viewer was US President Dwight D. Eisenhower, who was so impressed that he called Walt Disney personally and asked to borrow a copy, which he showed to Pentagon officials, among others [75]. Perhaps not entirely coincidentally, on 29 July of that year, Eisenhower's administration announced that as part of the International Geophysical Year (IGY), the United States would attempt to launch a satellite.

The second Disney film, *Man and the Moon*, also aired in 1955, while the third, *Mars and Beyond*, was originally scheduled for spring of 1956. Its completion was delayed when the Disney company was asked to produce a film on the IGY's project Vanguard. When Sputnik 1 was launched, in October 1957, the Vanguard film was about to go into production, but it was cancelled immediately, allowing *Mars and Beyond* to be completed in time for transmission on 4 December 1957 [77].

Two days later, a Vanguard rocket exploded on the launch pad and the launch of America's first satellite was left in the hands of von Braun's team, a line of rocket development that led eventually to the Saturn V which launched men to the Moon. Soon after the launch of that first satellite, Explorer 1, von Braun wrote a fictional story of the first manned flight to the Moon for *This Week* magazine: 'First Men to the Moon', published on 5 and 12 October 1958. It was so well received that he wrote two more: 'Five Days on the Moon' and 'Blast-Off from the Moon', published on 8 March and 12 April 1959, respectively [78].

In retrospect, it is almost as if he had written a draft plan for the following decade of space technology development, which would culminate in the Apollo missions to our nearest celestial neighbour. In parallel, other groups of engineers around the

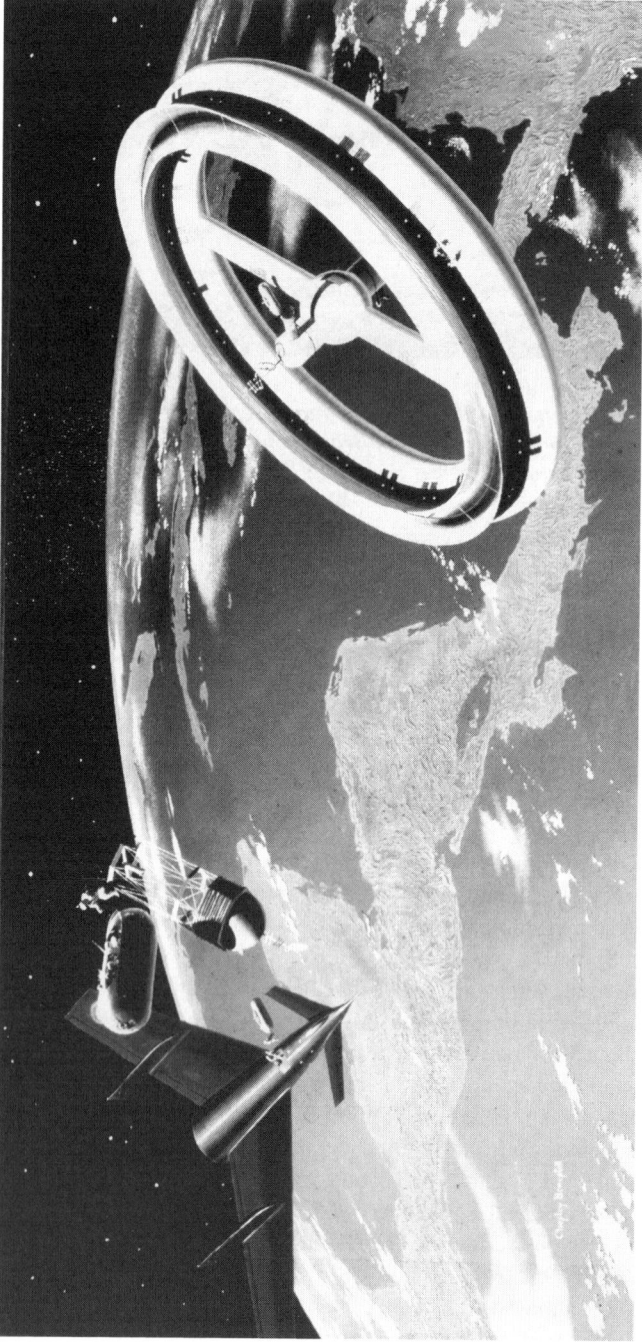

Figure 1.19 A vision of the future by Chesley Bonestell: the winged reusable third stage of von Braun's rocket arrives at a space station over Central America; a space telescope and attendant 'space taxis' orbit nearby [US Space and Rocket Center]

world would develop satellites and other unmanned spacecraft for civil, military, scientific and commercial applications. This book is the story of those technological developments.

References

1 Mackay, Alan L.: 'The Harvest of a Quiet Eye' (The Institute of Physics, Bristol, 1977) p. 164
2 Winter, Frank H.: 'Rockets into Space' (Harvard University Press, Cambridge, 1990) p. 2
3 Ordway, Frederick I.: 'Visions of Spaceflight' (Four Walls Eight Windows, New York, 2001) p. 30
4 ibid., p. 32
5 Miller, Ron: 'The Dream Machines' (Krieger, Malabar, 1993) p. 6
6 Ordway, Frederick I.: 'The Rocket from Earliest Times through World War I', in Ordway, Frederick I. and Liebermann, Randy (Eds): 'Blueprint for Space' (Smithsonian Institution, Washington, USA, 1992) p. 85
7 Canby, Courtlandt: 'A History of Rockets and Space' (Leisure Arts Ltd, London, 1964) p. 11
8 Winter, Frank H.: 'The First Golden Age of Rocketry' (Smithsonian Institution, Washington, USA, 1990) p. 1
9 Miller, Ron: 'The Dream Machines' (Krieger, Malabar, 1993) p. 8
10 Ordway, Frederick I.: 'Visions of Spaceflight' (Four Walls Eight Windows, New York, 2001) p. 33
11 ibid., p. 36
12 ibid., p. 39
13 Ordway, Frederick I.: 'Dreams of Space Travel from Antiquity to Verne', in Ordway, Frederick I. and Liebermann, Randy (Eds): 'Blueprint for Space' (Smithsonian Institution, Washington, USA, 1992) p. 43
14 Winter, Frank H.: 'The First Golden Age of Rocketry' (Smithsonian Institution, Washington, USA, 1990) p. 26
15 Miller, Ron: 'The Spaceship as Icon: Designs from Verne to the Early 1950s' in Ordway, Frederick I. and Liebermann, Randy (Eds): 'Blueprint for Space' (Smithsonian Institution, Washington, USA, 1992) p. 51
16 Ordway, Frederick I.: 'Visions of Spaceflight' (Four Walls Eight Windows, New York, 2001) p. 69
17 Winter, Frank H.: 'Rockets into Space' (Harvard University Press, Cambridge, 1990) p. 3
18 Miller, Ron: 'The Spaceship as Icon: Designs from Verne to the Early 1950s' in Ordway, Frederick I. and Liebermann, Randy (Eds): 'Blueprint for Space' (Smithsonian Institution, Washington, USA, 1992) p. 53
19 Miller, Ron: 'The Dream Machines' (Krieger, Malabar, 1993) p. 55
20 Winter, Frank H.: 'Rockets into Space' (Harvard University Press, Cambridge, 1990) p. 5–6

21 ibid., p. 9
22 Stoiko, Michael: 'Pioneers of Rocketry' (Hawthorn Books, New York, 1974) pp. 37–8
23 Winter, Frank H.: 'Planning for Spaceflight: 1880s to 1930s' in Ordway, Frederick, I. and Liebermann, Randy (Eds): 'Blueprint for Space' (Smithsonian Institution, Washington, USA, 1992) p. 105
24 Winter, Frank H.: 'Rockets into Space' (Harvard University Press, Cambridge, 1990) p. 10
25 Miller, Ron: 'The Dream Machines' (Krieger, Malabar, 1993) pp. 95–6
26 Booklet bound by 'Leonard Gustafssons Bokbinderi Stockholm': 'Konstantin Tsiolkovsky: Founder of Rocketry, Cosmonautics and theory of Interplanetary Flight' (Novosti Press Agency Publishing House, undated) pp. 12–13
27 Ordway, Frederick I.: 'Visions of Spaceflight' (Four Walls Eight Windows, New York, 2001) p. 89
28 Winter, Frank H.: 'Planning for Spaceflight: 1880s to 1930s' in Ordway, Frederick, I. and Liebermann, Randy (Eds): 'Blueprint for Space' (Smithsonian Institution, Washington, USA, 1992) p. 107
29 Winter, Frank H.: 'Rockets into Space' (Harvard University Press, Cambridge, 1990) p. 27
30 Ciancone, Michael L.: 'The Literary Legacy of the Space Age: An Annotated Bibliography of pre-1958 Books on Rocketry and Space Travel' (Amorea Press, Houston, 1998) p. 23
31 Stoiko, Michael: 'Pioneers of Rocketry' (Hawthorn Books, New York, 1974) p. 71
32 ibid., pp. 72–3
33 Winter, Frank H.: 'Foundations of Modern Rocketry: 1920s and 1930s' in Ordway, Frederick I. and Liebermann, Randy (Eds): 'Blueprint for Space' (Smithsonian Institution, Washington, USA, 1992) p. 97
34 Ciancone, Michael L.: 'The Literary Legacy of the Space Age: An Annotated Bibliography of pre-1958 Books on Rocketry and Space Travel' (Amorea Press, Houston, 1998) p. xvi
35 Winter, Frank H.: 'Foundations of Modern Rocketry: 1920s and 1930s' in Ordway, Frederick I. and Liebermann, Randy (Eds): 'Blueprint for Space' (Smithsonian Institution, Washington, USA, 1992) p. 101
36 Winter, Frank H.: 'Rockets into Space' (Harvard University Press, Cambridge, 1990) p. 34
37 ibid., p. 19
38 ibid., pp. 18–19
39 ibid., p. 20
40 ibid., p.25
41 Canby, Courtlandt: 'A History of Rockets and Space' (Leisure Arts Ltd, London, 1964) p. 59
42 Winter, Frank H.: 'Rockets into Space' (Harvard University Press, Cambridge, 1990) p. 37
43 ibid., p. 35

44 Ciancone, Michael L.: 'The Literary Legacy of the Space Age: An Annotated Bibliography of pre-1958 Books on Rocketry and Space Travel' (Amorea Press, Houston, 1998) p. 61

45 ibid., p. 46

46 Winter, Frank H.: 'Rockets into Space' (Harvard University Press, Cambridge, 1990) p. 26

47 Clarke, Arthur C.: 'How the World Was One: Beyond the Global Village' (Victor Gollancz Ltd, London, 1992) p. 164

48 Moscowitz, Sam: 'The Growth of Science Fiction from 1900 to the Early 1950s' in Ordway, Frederick I. and Liebermann, Randy (Eds): 'Blueprint for Space' (Smithsonian Institution, Washington, USA, 1992) p. 72

49 Miller, Ron: 'The Spaceship as Icon: Designs from Verne to the Early 1950s' in Ordway, Frederick I. and Liebermann, Randy (Eds): 'Blueprint for Space' (Smithsonian Institution, Washington, USA, 1992) p. 62

50 Chapman, John L.: 'Atlas: The Story of a Missile' (Harper and Brothers, New York, 1960) p. 169

51 Zukowsky, John (Ed.): '2001: Building for Space Travel' (Harry N Abrams Inc. and The Art Institute of Chicago, 2001) p. 13

52 Winter, Frank H.: 'Planning for Spaceflight: 1880s to 1930s' in Ordway, Frederick, I. and Liebermann, Randy (Eds): 'Blueprint for Space' (Smithsonian Institution, Washington, USA, 1992) p. 106

53 Wyld, James H.: 'History of the Rocket Engine' in Logan, Jeffrey: 'The Complete Book of Outer Space' (Maco Magazine Corporation, New York, 1953) p. 53

54 Ordway, Frederick I.: 'Visions of Spaceflight' (Four Walls Eight Windows, New York, 2001) p. 115

55 Winter, Frank H.: 'Foundations of Modern Rocketry: 1920s and 1930s' in Ordway, Frederick I. and Liebermann, Randy (Eds): 'Blueprint for Space' (Smithsonian Institution, Washington, USA, 1992) p. 100

56 Winter, Frank H.: 'Rockets into Space' (Harvard University Press, Cambridge, 1990) pp. 39–40

57 Wyld, James H.: 'History of the Rocket Engine' in Logan, Jeffrey (Ed.): 'The Complete Book of Outer Space' (Maco Magazine Corporation, New York, 1953) p. 65

58 Ciancone, Michael L.: 'The Literary Legacy of the Space Age: An Annotated Bibliography of pre-1958 Books on Rocketry and Space Travel' (Amorea Press, Houston, 1998) p. 49

59 Miller, Ron: 'The Dream Machines' (Krieger, Malabar, 1993) p. 242

60 Nice, Audrey: 'Surrey Successfully Demonstrate Steam Micro-Propulsion in-Orbit' (Surrey Satellite Technology Ltd press release, 17 March 2004)

61 Winter, Frank H.: 'Rockets into Space' (Harvard University Press, Cambridge, 1990) p. 46

62 ibid., p. 47

63 ibid., p. 48

64 ibid., p. 50

65 ibid., p. 49

66 Stuhlinger, Ernst: 'Gathering Momentum: Von Braun's Work in the 1940s and 1950s' in Ordway, Frederick I. and Liebermann, Randy (Eds): 'Blueprint for Space' (Smithsonian Institution, Washington, USA, 1992) p. 113,115

67 Ordway, Frederick I. and Sharpe, Mitchell R.; 'The Rocket Team' (William Heinemann, London, 1979) pp. 195–6

68 ibid., p. 251

69 Stuhlinger, Ernst: 'Gathering Momentum: Von Braun's Work in the 1940s and 1950s' in Ordway, Frederick I. and Liebermann, Randy (Eds): 'Blueprint for Space' (Smithsonian Institution, Washington, USA, 1992) p. 115 (quoting Speer, Albert: 'Inside the Third Reich: Memoirs' (Macmillan, New York, 1970) p. 367)

70 Siddiqi, Asif A.: 'Challenge to Apollo: The Soviet Union and the Space Race, 1945-1974' (NASA SP-2000-4408, Washington, USA, 2000) p. 27

71 Godai, Tomifumi: '50 Years of the International Astronautical Federation (IAF)', *Earth Space Review*, 2000, **9**(1), pp. 36–41

72 Ryan, Cornelius: 'What Are We Waiting For?', *Collier's*, March 22, 1952, p. 23

73 Liebermann, Randy: 'The Collier's and Disney Series' in Ordway, Frederick, I. and Liebermann, Randy (Eds): 'Blueprint for Space' (Smithsonian Institution, Washington, USA, 1992) p. 141

74 ibid., p. 142,144

75 ibid., p. 145

76 ibid., p. 137

77 ibid., p. 146

78 Ordway, Frederick I.: 'Visions of Spaceflight' (Four Walls Eight Windows, New York, 2001) p. 156

Chapter 2

Highway to space – the development of the space launch vehicle

The crux of the matter is the rocket engine

Sergei Korolev, 1934

The history of technology is replete with descriptions of the development of various types of vehicle – the steam locomotive, the motor car, the aircraft – all designed to transport passengers, goods or other cargo from one point to another. Well before rocketry was an accepted branch of engineering, successive generations of engineers had succeeded in developing transport systems that spanned the globe on land, on sea and in the air. It was therefore only a matter of time before they set their sights a little higher, towards space. Indeed, as the previous chapter has shown, in the minds of certain far-sighted individuals, space was just another destination in need of a transport system.

Although there is no legal or rigorous scientific definition of where space begins, the notional boundary between the Earth's atmosphere and space is generally accepted to be at an altitude of 100 km. While sustained orbital flight is not possible at this altitude, because of the frictional effect of the upper atmosphere, a payload on a suborbital trajectory will experience microgravity, much as it would in orbit; in addition, the atmosphere is thin and stars are visible in a black sky. In the early years of the Space Age, 50 miles (80.5 km) was a convenient figure to use, and it was this definition that allowed many of the X-15 rocket plane pilots to earn their astronaut wings (at least in the eyes of the US Air Force). In 2004, however, the legitimacy of the 100 km limit was supported by the completion of a contest – the Ansari X-Prize competition – to launch a privately-built spacecraft above this altitude twice in succession. As a result, it seems likely that this figure, which is recognised by the Fédération Aéronautique Internationale (FAI) for record-certification purposes, will become the *de facto* boundary to the space environment.

In terms of the nomenclature introduced by British mathematician and geophysicist Sydney Chapman, this places the transition to space within the thermosphere, some 15 km above the boundary with the mesosphere [1], at about ten times the cruising height of the average modern passenger jet. As discussed in Chapter 1, it was realised more than a century ago that the only vehicle capable of carrying a useful payload to such extreme altitudes was the rocket. The key to space exploration, therefore, was the development of a vehicle with this capability.

This chapter describes the evolution of the spacecraft launch vehicle from its origins in post-war rocketry experiments and ballistic missile development to the technical achievements that would take men to the Moon. Partly as a result of political aspirations, but also because the financial and other resources were made available, this was manifested predominantly as a 'space race' between the United States and the USSR.

US wartime rocket research

Despite the obvious potential of the rocket in reaching extreme altitudes, the exploration of space was not high on the list of national priorities in the immediate post-war world. A more immediate goal was the development of the rocket as a weapon delivery system, as demonstrated during the war by the German V-2.

Despite the work of Robert Goddard and other individuals, American wartime rocket research had been on a small scale compared with German efforts. Nevertheless, some significant theoretical and experimental work had been conducted under the GALCIT rocket project, named after the Guggenheim Aeronautical Laboratory, California Institute of Technology. It was established in 1936 by Frank J. Malina, a graduate student of the renowned aerodynamicist Theodore von Kármán, and a number of enthusiastic colleagues (Figure 2.1).

Robert Millikan, the physicist who headed Caltech, invited Goddard to cooperate with GALCIT in 1936, but Goddard declined. Malina visited the lone experimenter in New Mexico that summer, again regarding cooperation, but Goddard remained suspicious of other rocket developers and preferred to work alone. Even Harry Guggenheim, who had helped to finance both GALCIT and Goddard, failed to bring them together. Although Goddard's influence on the budding group could have been significant, his personality would not allow it [2] and US rocket research continued more despite him than because of him.

GALCIT's members soon found out for themselves how difficult it was to produce rockets that were sufficiently capable and reliable to, first, convince the sceptics that rockets were not simply noisy, smelly and dangerous and, second, to raise sufficient funding to progress their work. Things began to look up, however, when the commanding general of the Army Air Corps, H. H. 'Hap' Arnold, paid a surprise visit to the GALCIT lab in 1938. His fascination with the work, along with proposals from Malina, led to funding for work on jet-assisted take-off (JATO) rockets for aircraft, the first demonstration of which was in 1941 [3]. In March 1942, the key members of the team formed a private company called Aerojet, which foundered until 1943 when the US Navy awarded it a large contract for JATO systems for carrier-based aircraft [4]. Thus, it was Aerojet and America's first commercial rocket company,

Figure 2.1 *JPL dignitaries (left to right: William H. Pickering, former Director, Theodore von Kármán, co-founder and Frank J. Malina, co-founder and first Director) [NASA]*

Reaction Motors Inc (formed by members of the American Rocket Society in 1941), which shaped the nascent US rocket industry.

However, the true interest of the GALCIT founders had lain, not in designing rockets to help heavily laden aircraft struggle into the air, but in developing a sounding rocket to explore the extremities of the atmosphere, on the way to paving the highway to space. Their chance came in July 1944, when GALCIT was reorganised, expanded and renamed the Jet Propulsion Laboratory (JPL). Although, since then, JPL has entered the history books for its space-related achievements through its relationship with NASA, its name illustrates the scepticism which surrounded rocketry in the 1940s: jet was a broader term than rocket and avoided any stigma attached to the word [5].

Despite the initial uncertainties, the influence of JPL was to prove even wider than most people could have imagined, as shown by the story of the Chinese applied mathematician Tsien Hsue-Shen. Born in 1911, Tsien won a scholarship to the United States in 1935, spent a year at the Massachusetts Institute of Technology (MIT), then moved to Caltech to study under von Kármán. He wrote his first work on rocketry in 1937, gained his doctorate in 1939 and went on to become one of the co-founders of JPL. Despite his contributions to American rocketry, the post-war scrutiny of suspected communist subversives led the US Immigration and Naturalization Service to label him 'an undesirable alien who should be deported' [6], but his knowledge was considered too sensitive and he was forced to remain in California, effectively under house arrest.

Tsien continued to teach at Caltech until 1955, when he was returned to China with 93 fellow scientists in exchange for 76 American prisoners of war taken in Korea; it seems his knowledge was by then considered out of date. Once back in China, however, Tsien began to work on updating China's missile programme. On 8 October 1956, the Central Committee of the Communist Party of China established the Fifth Research Academy of the Ministry of National Defence, which included the Rocket Research Institute with Tsien as its first director. That date also marks the official foundation of the Chinese space programme [7]. It is deeply ironic that the nation which now fears and spends vast sums to guard against the spectre of the Chinese ICBM provided the academic training and professional development of the individual responsible for that rocket programme.

Winds of change

Although rocketry was by no means part of mainstream technology development in the United States, it was during the 1940s that American thinking on defence underwent a fundamental change: having discovered that its bordering oceans were no longer a significant buffer zone against potential invaders, an imperative for long range weapons began to crystallise. Thus JPL, operated by Caltech for the Army Ordnance Corps, became the official home for US work on guided missiles, specifically the nation's first tactical nuclear missiles, designated Corporal and Sergeant.

In essence, the overall goal was to produce a lighter and more accurate version of the V-2. Development began with a short-range solid propellant rocket called Private, and the Private A launches, in 1944, marked the Lab's first successful rocket programme.

Solid propellants, comprising a mixture of fuel and oxidiser, were chosen for many of the early rocket programmes because of their simplicity: they are generally easy to handle and store, and do not corrode their containers as some liquid propellants do. The solid propellant motor is also simpler in its construction, since it avoids the complex pipework, pumps and pressurisation systems of the liquid rocket engine, and the need to store fuel and oxidiser in separate tanks. It does, however, have to be designed to give precisely the required amount of thrust and total impulse, since, once ignited, it burns until its propellant is exhausted.

The Private A was 2.3 m tall and 26 cm in diameter. It was powered by an Aerojet-designed motor, using a solid propellant based on asphalt, which produced a thrust of 450 kg. It demonstrated a range of up to 18.3 km and a peak altitude of just 4.4 km [8]. It was no space rocket, but it marked a successful beginning to JPL's rocket programme.

Before moving on to the more complex Corporal, which used liquid propellants, JPL developed a smaller precursor missile dubbed WAC Corporal. Its name was derived from Women's Auxiliary Corps, because JPL's male engineers considered it the 'little sister' of the Corporal, but the acronym also stood for 'without attitude control' [9]. Amusingly, regarding the names of the missiles, von Kármán had suggested that the missile 'ranks' should rise only as far as Colonel, because that was the highest rank that worked [10].

The WAC Corporal's engine was a modified version of Aerojet's JATO engine, using aniline, which had been developed as a rocket fuel by GALCIT in 1942 [11], and red fuming nitric acid (RFNA) as the oxidiser. This propellant combination is termed hypergolic, since the components ignite spontaneously on contact, and is beneficial in that it allows designers to forgo the complexity of an ignition system. Thus, for the WAC Corporal, it was simply necessary to inject the propellants into the combustion chamber, engineer an efficient mixing process and let chemistry do the rest. The move to liquid propellants was, itself, an important step, because it allowed the development of more efficient rockets, as recognised by Konstantin Tsiolkovsky more than 40 years earlier and practised by Goddard in the 1920s.

The rocket itself was constructed mainly in steel and aluminium, with some magnesium components, and fabricated by Douglas Aircraft Company in one of its early forays into space technology [12]. It was inevitable that there should be a transfer of technology from the established aircraft industry to the nascent rocket industry, since aircraft manufacturers were already experienced in handling the materials and forming the structures required. Although there were obvious differences, the requirement for strong yet lightweight structures was similar, as was the need for quality and reliability. This was the beginning of a trend that would see most of the major American and European aircraft manufacturers developing space divisions in the first decade of the Space Age.

The WAC Corporal was small in comparison with the V-2, standing only 4.9 m high with a diameter of 0.3 m (against 14 m and 1.7 m for the V-2), but it set a new US altitude record on its first flight. On 11 October 1945, a perfect launch from the White Sands Proving Ground in New Mexico (not far from where Goddard had conducted his experiments) took it and its 10 kg payload to an altitude of around 70 km [9]. Its trajectory had been so vertical that after a flight of some seven and a half minutes it crashed back to Earth just over 1000 m from its launch point, perilously close to the launch team [13]. Indeed, it provided an excellent example of the benefits of trading range for altitude!

It was around this time that the focus began to shift, at least in some people's minds, from range to altitude. Some sources describe the WAC Corporal as 'the first man-made object to escape the Earth's atmosphere', but this of course depends on one's definition of where the atmosphere ends. And since reaching an altitude of 70 km does not make it a 'space rocket' under current definitions, this brings us back to the point of such a rocket, which is not simply to reach the threshold to space, but to deliver a payload, spacecraft or satellite beyond that threshold to space itself.

Moreover, simply reaching an altitude defined as 'space' is not sufficient to make a payload or spacecraft a 'satellite', since by definition a satellite is an object which orbits a parent body. In addition, if a satellite's orbit is too low, friction with the upper atmosphere will cause a gradual reduction in its altitude and it will re-enter.

For example, a typical spacecraft in a circular 250 km orbit will re-enter the Earth's atmosphere within about two months, while from 600 km it would take about 15 years. It is only for altitudes above about 850 km, where a large proportion of satellites orbit, that the decay time is measured in centuries [14].

Nevertheless, the 70 km achieved by the WAC Corporal flight of October 1945 (though following in the footsteps of the V-2, which had reached an altitude of 90 km in 1942) had set a new record for US rocketry (see Table 2.1). In addition, one of its descendents, the Aerobee, constructed by Aerojet for upper atmosphere research in the late 1940s, marked the eventual consummation of GALCIT's desire to develop a sounding rocket. However, as far as placing an American satellite in orbit was concerned, the lead came not from JPL, but from a group of transplanted German engineers.

American V-2 derivatives

In September 1945, just one month before the WAC-Corporal's inaugural flight, Wernher von Braun and many of his colleagues had been transported to the United States, under Project Paperclip, to begin a new career as conscripts of the US Army (or 'prisoners of peace' as von Braun called the team [15]). The transition was a difficult one, partly because of the language and cultural barriers, but also because they were accorded a fairly low status within the Army bureaucracy. They had only a 6-month contract with the Americans and, initially, were given little to do.

In recent times, historians have often concluded that, even while developing the V-2, von Braun and his colleagues were more interested in space travel than war and that von Braun organised the group's surrender to the Americans because he thought the US could provide the resources, ultimately, to make interplanetary travel a reality [16]. At first, however, they were to remain frustrated by an acute lack of resources and a US military which, post-war, was as interested in developing a long-range weapon system as the Nazis had been towards the end of World War II.

By February 1946, some 127 Peenemünde scientists and engineers had been transferred to Fort Bliss, near El Paso in Texas, where they would continue the development of the German A-series rockets under the US Army's direction. In parallel, the US military transported V-2 rockets, ground support equipment and as many spare parts as they could find to the White Sands Proving Ground. In fact, most of the V-2s would have to be made in the United States from parts acquired in Europe, because the missile was always intended to be launched as soon as possible after being manufactured, and significant stocks of completed V-2s did not exist [17].

The émigrés' initial role was to assist the American contractor General Electric in the pursuance of Project Hermes, the development of a series of long-range guided missiles (Figure 2.2). In part, this involved launching 64 liberated V-2s from White Sands between April 1946 and September 1952, while another was launched from an aircraft carrier at sea and two more from Florida as part of the Bumper programme [18].

The first launch, on 16 April 1946, failed when a fin came off at an altitude of some 5 km and the rocket had to be destroyed. Although wayward missiles were rare,

Table 2.1 *Performance of selected early rockets and launch vehicles in reaching 'space'*

Vehicle (Nation)	First launch success	Dimensions height/ diameter (m)	Launch mass (tonnes)	Thrust[a] (tonnes)	Payload capability (kg)	Altitude (range) (km)
First Goddard Rocket (USA)	16/3/26	3/-	0.0048	0.004	None	0.0125 (0.056)
A-4 (V-2) (Germany)	3/10/42	14/1.7	12.6	26.5	~1000	90 (192)
Private A (USA)	1/12/44	2.3/0.26	0.25	0.45	27	4.4 (18.3)
WAC-Corporal (USA)	11/10/45	4.9/0.3	0.3	22	11.3	70 (~zero)
US V-2[b] (Project Hermes, USA)	10/5/46 17/12/46 29/10/51	—	—	—	—	113 186.7 213.2
Bumper #1 (USA)	13/5/48	18.9/1.7	13	27	10	127
R-1 (USSR)	10/10/48	14.6/1.6	13.5	25	—	— (290)
Bumper #5 (USA)	24/2/49	18.9/1.7	13	27	10	393
R-5 (USSR)	2/4/53	21/1.7	28.5	51	1300	512 (800)
Redstone (USA)	20/8/53	19.2/1.7	27.8	35	2800	LEO (370)
Jupiter-C (USA)	20/9/56	21/2	29.5	38	18 to LEO	1097 (5472)
Jupiter IRBM (USA)	1/3/57	18.4/2.7	50	68	—	LEO (2800)
R-7 (USSR)	21/8/57	29/3+ Boosters (10 across fins)	270	504	~2000 (As A-1: 4800 to LEO)	LEO (>8000)

[a] Thrust: to convert to kN multiply by 9.8.

[b] Table shows altitudes reached by Rocket numbers 3, 17 and 60 (V-2s fired vertically).

LEO = Low Earth Orbit.

when they did go astray they did so with a certain flair. In May 1947, for example, a V-2 suffered a gyroscope malfunction and headed south from White Sands for El Paso on the Mexican border. Passing over the city, it crashed into a rocky hillside just outside Tepeyac Cemetery, 2 km south of the Mexican city of Juárez, barely missing a construction company's explosives store. Although unarmed, the missile blasted a crater 10 m deep and 15 m across . . . and marked America's first firing of a

Figure 2.2   Hermes A-1 (modified V-2) launch from White Sands proving ground [NASA]

guided missile against a foreign country. According to the range safety officer, within 10 minutes of the impact, food stalls in the vicinity of the cemetery were selling V-2 souvenirs, still warm! [19].

As far as breaking the notional 100 km boundary to space was concerned, the first rocket to do this appears to have been rocket number 3, which reached an altitude of about 113 km on 10 May 1946, while rocket no. 17 peaked at 187 km on 17 December of that year [20]. On 29 October 1951, an American V-2 (no. 60) reached an altitude of 213.2 km, which appears to be the record for a single-stage V-2 [21].

As well as assisting American endeavours in rocket development, the captured V-2s were put to good use as sounding rockets since they could deliver a payload of about a tonne to some five times the altitude of contemporary sounding balloons [22]. The V-2s carried their payloads to the upper atmosphere, making measurements of layers in the ionosphere, discovering solar X-rays and characterising the Sun's far-ultraviolet spectrum, which was impossible from the ground because of the absorption properties of the atmosphere [23].

Payload experiments were coordinated by the V-2 Upper Atmosphere Panel and provided by scientists including James Van Allen, who would later be instrumental in the discovery of radiation belts around the Earth. Another member of the panel was Fred L. Whipple, a scientist who had been stationed with the US Army Corps in the United Kingdom from late 1943 to work on the development of 'chaff', aluminium foil strips dropped from aircraft to confuse enemy radar. As a result, he was known unofficially as the 'Chief of Chaff' [24]. In 1946, Whipple made some initial calculations on the presumed hazards of space debris to spacecraft, designing the 'meteor bumper shield', which later became known as the Whipple Shield. It has since featured on spacecraft as diverse as the Giotto Comet Halley interceptor and the International Space Station.

The acquisition of V-2 technology, and of course leading members of the original 'rocket team', provided a significant boost to American rocket research. The V-2 incorporated the key technologies of efficient liquid propellants (including cryogenic liquid oxygen); turbopump propellant feeding; inertial guidance; and a rudimentary telemetry system. Although the American V-2 launches did little to advance rocket technology *per se*, they were crucial in bringing America 'up to speed' in rocket research and development. After all, having a flight-proven rocket to tweak and modify beats starting from scratch.

However, with Project Bumper, conducted between 1948 and 1950 by von Braun's team in cooperation with JPL, America began to make real advances. The project involved modifying a series of eight V-2s to accept a second stage in the shape of a WAC-Corporal (Figure 2.3). Bumper, or Bumper–WAC, was America's first two-stage missile and important for its role in developing the technology to separate a warhead from its booster.

This so-called staging technology was also directly applicable to space exploration, both in separating a spacecraft from its launch vehicle and, fundamentally, to allow a useful payload to reach orbit. The main advantage of the multi-stage rocket is that empty tanks and associated structure do not have to be carried all the way to space, with the result that a larger payload can be delivered to a given altitude. Of course, the chief disadvantage of any expendable (as opposed to reusable) rocket remains, in that most, if not all, of the vehicle is discarded on every mission, which is rather like scrapping an airliner after a single flight. Nevertheless, the development of the multi-stage rocket, or step-rocket as they were called in the early days, was a crucial prerequisite for space exploration – and one which would probably have been developed much later were it not for the military requirement of warhead separation.

The first launch of the two-stage Bumper–WAC, conducted from White Sands on 13 May 1948, reached a peak altitude of 127 km, which was lower than intended because the WAC-Corporal cut-off early. The programme suffered several failures, but the fifth flight, on 24 February 1949, broke the altitude record by a good margin, reaching 393 km [25]. The little known Bumper 5 is sometimes described as the 'first recorded man-made object to reach extraterrestrial space' [26], but, again, this begs the question as to where outer space begins, since the single-stage V-2 had already exceeded the 100 km ceiling.

The last two Bumper flights, conducted in July 1950, were historically interesting for an entirely different reason: they were launched from the Long Range Proving Ground on Florida's Cape Canaveral, which later became the US Air Force Eastern

Figure 2.3   Bumper–WAC launch from White Sands [NASA]

Test Range and Kennedy Space Center, the point of departure for the Apollo lunar missions. For the Bumper flights, a make-do gantry was constructed from painters' scaffolding poles and a tarpaper shack served as a blockhouse – a far cry from the facilities that would dominate the Florida landscape in decades to come (Figure 2.4).

Meanwhile, the development of rocketry was continuing in parallel in the Soviet Union.

Soviet V-2 derivatives

After the Second World War, as America was organising the removal of V-2 engineers and hardware to the United States, the Soviet Union too was putting together its own German rocket team. Thus the V-2 was not only instrumental in the development of American post-war rocketry, but also assisted progress in the Soviet Union.

Again, as in America, it is important to note that there was indigenous rocket development prior to the acquisition of the German experts. Indeed, the modern era of Soviet rocketry had opened on 15 May 1929, when research began into both liquid rocket engines and electric propulsion at the Gas Dynamics Laboratory (GDL) in Leningrad. The first working model of a liquid propellant rocket, called ORM-1, was completed in early 1931 [27], at about the same time the VfR and AIS rocket societies

Figure 2.4   Missile Row, Cape Canaveral, Florida in November 1964: just visible at the top of the picture is the Apollo–Saturn Vehicle Assembly Building (VAB) and the crawlerway to the two launch-pads [NASA]

were conducting their initial experiments in Germany and America, respectively (see Chapter 1) (Figure 2.5).

The Soviet plan for its German rocket team was different from that of the Americans, at least initially, and from late 1945 it began to re-establish V-2 production facilities *inside* Germany. As a result, by the beginning of September 1946, a pilot production line had manufactured 30 flight-worthy V-2s and some 5000 scientists and technicians had been assigned to the programme [28].

Some of the Soviet V-2s were fired in a static test stand at Lehesten under the direction of Russian Colonel Valentin P. Glushko [28], who already had considerable experience in liquid propellant rocketry dating back to the mid-1930s, when he and his colleagues had built and tested rockets in Moscow. Glushko was one of a band of like-minded people around the world who had been influenced in his youth by stories of space travel and, in 1926, at the age of 18, had himself published an article called 'Station Beyond Earth', which described artificial satellites and space stations [29]. He would later become a key figure in the Soviet space programme.

Russian engineers learnt from the Germans, who were encouraged to suggest improvements to the V-2 design. This led, among other things, to pressurised

Figure 2.5   Members of Russian amateur rocket group GIRD work on the first Soviet liquid propellant rocket (left to right): Sergei Korolev, Nikolai Yefremov and Yuriy Pobedonostsev. The rocket reached an altitude of 400 m on 17 August 1933 [NASA/Asif Siddiqi]

propellant tanks, the relocation of all control equipment in one place behind the tanks, and turbopumps driven by exhaust gases from the engine rather than a separate gas generator [28]. However, Soviet developments were restricted by the relatively small number of V-2s available for flight testing, the Americans having taken most of the better hardware. As a result, engineers were ordered to start work on a copy of the rocket, designated R-1, the R standing for 'raketa' or missile [30].

However, the Soviet engineers – including Glushko's deputy, Sergei P. Korolev (Figure 2.6), who was destined to become the Soviet Union's mysterious

Figure 2.6   Sergei Korolev in July 1954, with a dog that had just returned to Earth following a suborbital hop on an R-1D scientific rocket [NASA/Asif Siddiqi]

'Chief Designer' of rockets and spacecraft – had little enthusiasm for copying a rocket they felt had limited capabilities and was cumbersome in terms of operations. So, in early 1946, they began work on an improved, 'stretched' version of the V-2, designated R-2 and known to British intelligence as the K-1 (Figure 2.7) [31]. Its propellant tanks were some 3 m longer and the thrust developed by its engine was increased from 25 to 32 tonnes, giving it twice the range of its predecessor [32].

The British, incidentally, were also thinking about improving the V-2 and were among the first to suggest modifications that would allow a piloted version to perform a suborbital flight. In 1946, H. E. Ross and R. A. Smith of the British Interplanetary Society proposed a design for a V-2 with a capsule called Megaroc; unsurprisingly, the British government was not interested in funding it [33].

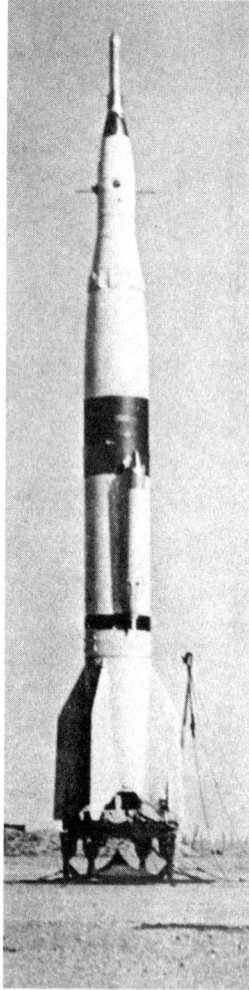

Figure 2.7   Soviet R-2A missile [NASA/Asif Siddiqi]

Sergei Korolev's rise to fame – at least within the Soviet political system – gives an indication of the fluidity of political opinion behind the Iron Curtain: a Soviet source reports that, back in 1939, Korolev had been unable to witness the first flight of a rocket prototype 'for reasons beyond his control'; in fact he had been accused of treason, arrested and sent to a forced labour camp in Siberia [29].

Once back in favour, Korolev was keen to do things his own way, rather than accepting suggestions from the German engineers [32]. His wish for Russian independence was echoed by the Kremlin and the Soviet military, which, in October 1946, brought Stalin's decree to the attention of the German rocket workers in no uncertain terms: they and their families were to be transferred to the USSR, where they would receive a new contract based on conditions applied to skilled workers in the Soviet Union. Thus from 12–16 October 1946, according to some sources, some 6000 German engineers and technicians, together with their wives, children and other relatives (some 20,000 people in all), were loaded onto 92 trains bound for the Soviet Union [34,35].[1]

As it turned out, in a strange parallel with von Braun's team in Texas, the Russian rocket team, under Helmut Gröttrup, an engineer with experience of the V-2's control electronics [36], found themselves evacuated to an unfamiliar country where they were underpaid and undervalued. It was only in the early 1950s that 'Russia's Germans' were no longer considered useful and were allowed to return to their homeland [37].

The first 'Russian V-2', designated T-1, was launched on 18 October 1947 from a base in the steppes, some 200 km east of Stalingrad (renamed Volgograd), which would later be known as the Kapustin Yar space launch facility. The rocket travelled some 200 km from the launch site, but came down about 30 km to one side of the target because of a guidance system failure. A second launch, on 20 October, was even more of a disappointment since the rocket landed about 180 km from the intended target [38].

Like their American counterparts, Soviet scientists wanted to capitalise on the availability of the V-2 for high altitude research and revived their Commission for the Study of the Stratosphere, whose work had been in limbo during the war. As a result, the nose sections of a few rockets were equipped with an ionization chamber and a gas discharge counter, while cosmic ray detectors were placed between the rocket's fins. The total mass of the science payload was about 500 kg. The rocket's main equipment section was modified to include a science telemetry system, whose antenna was mounted at the rear of the vehicle. The first of these scientific launches, on 2 November 1947, reached an altitude of 80 km, providing about 3 minutes of data [39].

However, the Soviet V-2 'campaign' was short-lived compared with America's Project Hermes, because officials were able to obtain only 18 rockets from Germany, 11 of which were actually launched (rockets 10 and 11 were both fired on 13 November 1947). As a result, the Soviet rocket engineers were obliged to pursue the development of their own variant, the R-1 (referred to in all technical documentation as 'product 8K11', a number-letter-number style of nomenclature that would continue into the space era) [40]. The first successful launch, following several failures, occurred on 10 October 1948; the final R-1 was launched on 5 November, marking nine successes out of 12 attempts. Success was a relative term, however, since only one missile fell in the 16 km × 8 km target area [41].

Subsequent Russian rockets followed a complicated designation system, in that there were many variants of the R-1 (R-1A, R-1B, etc.), as well as further numerical variants (R-2, R-3, R-5) and their sub-variants. One of the most interesting from a technical perspective was the R-10, also rather confusingly known as the G-1 or 'German rocket'. In common with the Americans, the Soviets had used the German

---

[1] It should be noted that, as with many 'historical facts', these figures are disputed by some commentators, who believe the numbers are exaggerated.

expertise and workforce to make improvements to the basic V-2, which they did more or less in parallel with Korolev's efforts to design the R-2, which was slow to come to fruition.

The R-10 was similar in size to the V-2, but the important difference lay in its mass ratio, the ratio of its fully fuelled mass at lift-off to the mass of the vehicle without propellant (its 'dry mass'); the quantity is also known as the 'propellant mass ratio' or 'propellant mass fraction'. The V-2's launch mass was about 12.8 tonnes, while its dry mass was 4 tonnes, giving a mass ratio of 3.2. By contrast, the figures for the R-10 were 18.4 tonnes and a little under 2 tonnes, respectively, giving a mass ratio of around 9 (showing that 90 per cent of the launch weight was propellant) [42].

It takes little technical expertise to comprehend the importance of a threefold improvement in mass-efficiency, but it was this type of improvement that was to prove crucial, for later vehicles, to the capability of placing a useful satellite in orbit. Expressed simply, any reduction in the structural mass of the launcher would increase the mass of payload it could carry to a given altitude.

The R-10's improved mass ratio was achieved partly by replacing the propellant tanks mounted inside an aerodynamic fairing with a monocoque structure, in which the pressurised tanks formed an integral, load-bearing part of the rocket. It also featured a modified propulsion system, which increased the engine's thrust from the V-2's 25–26 tonnes to some 32 tonnes. This was achieved by enhancing the turbopumps to increase propellant flow rates, thus raising the combustion chamber pressure. The V-2's gas generator, which powered the engine's turbopumps, was replaced by a system using engine exhaust gases, thus saving about 180 kg in weight. The pumps were brought up to speed before launch using compressed nitrogen [43].

A further key to the development was being able to separate the warhead from the rocket, as opposed to carrying the deadweight of the entire rocket to the target [43]. As in America, the importance of staging was clearly understood. Although the R-10 itself does not appear to have been produced in quantity, the technology developments provided the basis of many Soviet missiles of the 1950s [44]. Among other things, the Soviet V-2 derivatives were used to carry instruments and animals to altitudes of up to 109 km [45].

Ballistic missiles

Much has been written about the political uses of technology, not least space technology, and the role of the rocket in Cold-War politics is a prime example.

The priority, at least initially, was to develop a long-range delivery system for nuclear armaments – the intercontinental ballistic missile (ICBM) – but the technology was directly transferable to the satellite launch vehicle and the two applications eventually followed parallel tracks.

Building on its experience with V-2 technology, the Soviet Union continued to develop rockets and their engines throughout the late 1940s and early 1950s, with the ultimate goal of developing a missile capable of reaching the United States. Indeed, as early as 1947, Stalin had proposed an ICBM based on the V-2, believing it would be 'an effective straitjacket for that noisy shopkeeper, Harry Truman' [46].

Despite its work on the Bumper–WAC programme, the need for an ICBM was not universally recognised in the United States in the immediate post-war period. America was concentrating on the delivery system it knew best, the long-range bomber aircraft, in a belief that the bombs under development were far too large to be carried on any conceivable rocket. In a classic example of the short-sightedness that pervades the history of technology, Dr. Vannevar Bush (co-inventor of the computer and proponent of the atom bomb) told a congressional committee in 1946 that 'a 3000-mile high-angle rocket . . . is impossible today and will be for many years'. Such testimony did little to close the technology gap that was widening between America and the Soviet Union in the field of rocket research.

Soviet ICBM

Soviet developments in the 1950s produced a range of missiles known in the West by their NATO codes: SS-1, SS-2, and so on. The SS-1 (also known as Scunner) and SS-2 (Sibling) represented the furthest extent of the Soviet Union's exploitation of German technology; Soviet designers knew these missiles as the R-1 and R-2. The SS-3 (Shyster), or R-5, and subsequent missiles were based primarily on domestic developments.

The R-5 was a medium range ballistic missile (MRBM) with a specified range of about 800 km and was similar in appearance to the US Redstone (described later). It used liquid oxygen (LOX) and kerosene in its propulsion system, developed a thrust of some 51 tonnes and made its first flight in 1953. By the mid-1950s, it was also being used as a sounding rocket, capable of carrying a 1.3 tonne payload to a height of 512 km [47].

Soviet and American ICBM developments of the 1950s followed opposing design philosophies. While the Americans concentrated on reducing the size and mass of their nuclear payloads to match the lift capabilities of their rockets, the Soviets concentrated on building rockets capable of lofting their heavy nuclear payloads [48].

Part of the solution in designing a large and capable rocket is staging, but there was little experience with staging technology in the 1950s and it was still a risky and complex procedure. First, it was necessary to arrange for the remote firing of a separation system that would split the stages cleanly, without introducing aerodynamic instabilities to throw the missile off course. And second, the upper stage engines had to be ignited at the correct time, typically after stage separation but before too much velocity was lost. In effect, the upper stage ignition was a separate launch from a moving platform at high altitude, a far from trivial exercise.

The Soviet design solution for a large vehicle was to cluster together a number of propulsion units, each housing several identical engines, and ignite all engines on the ground. The breakthrough came with the R-7 (or 'product 8K71'), known in the West as the SS-6 (Sapwood). It comprised four identical boosters grouped around a sustainer, or core stage. All five segments were ignited together to launch the rocket,

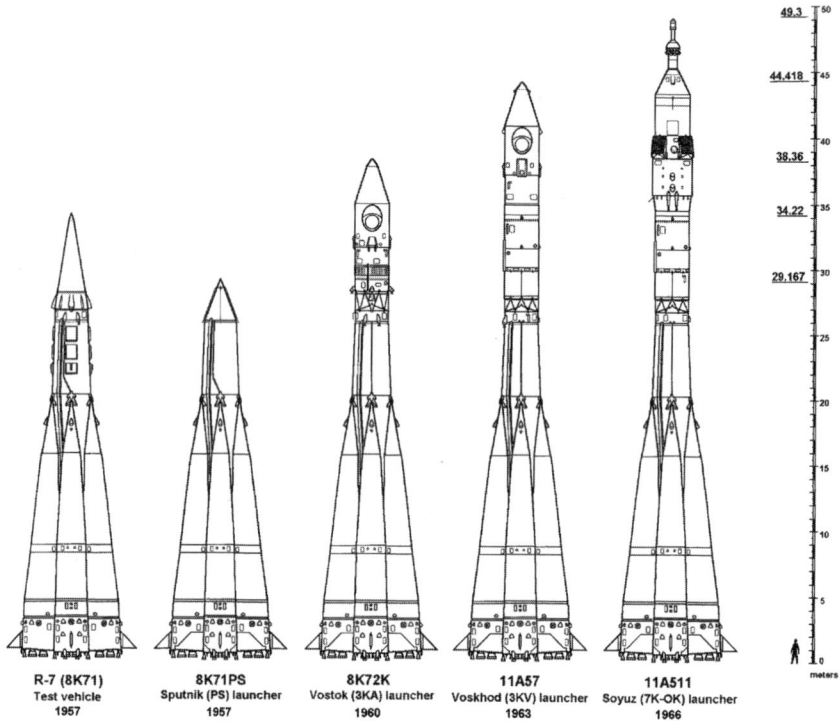

R-7 (8K71)	8K71PS	8K72K	11A57	11A511
Test vehicle	Sputnik (PS) launcher	Vostok (3KA) launcher	Voskhod (3KV) launcher	Soyuz (7K-OK) launcher
1957	1957	1960	1963	1966

*Figure 2.8   Evolution of Soviet space launch vehicles (left to right) R-7 ICBM,
            Sputnik, Vostok, Voskhod and Soyuz launchers [NASA/Asif Siddiqi]*

with the four boosters being jettisoned when their propellant was exhausted (after
120 s of flight), leaving the core stage to sustain the thrust of the vehicle for an
additional 150 s (Figure 2.8).

The R-7 was designed and developed by Sergei Korolev's Special Design Bureau
No. 1 (OKB-1), which had been formed in April 1950. Its propulsion units, developed
by the bureau between 1954 and 1957, represented a quantum leap in Soviet rocketry.
The 28 m-long, 3 m-diameter sustainer, know as the RD-108, was powered by LOX
and kerosene propellants supplied by a central turbopump to a cluster of four rocket
engines which together produced a thrust of about 96 tonnes. The four 19 m-long
RD-107 boosters had similar, self-contained propulsion systems, each producing a
thrust of 102 tonnes, thus giving a total 504 tonnes of thrust [49]. The R-7 was, by
any means of accounting, an extremely powerful rocket.

In addition to the main engines, each segment carried a number of smaller, gimballed engines for steering, thus dispensing with the graphite control vanes of the
V-2 and its derivatives; there were four of these engines on the sustainer and two on
each of the boosters. Engine gimballing – usually of the main engine itself – was to
become the most common method of launch vehicle steering.

The existence of the R-7 ICBM was announced by Premier Khrushchev in 1957 [50]. Its first flight on 15 May 1957 was a failure, but a test conducted on 21 August of the same year showed the vehicle to have a range of more than 8000 km. However, its use of liquid propellant was a disadvantage for an operational weapon, since it could not be stored fully fuelled, and only about a dozen were deployed up to the early 1960s [49].

Nevertheless, it was its secondary role as a spacecraft launch vehicle that really proved its capability, as well as its use as a Cold-War political tool. Designated 'A' under the widely used American code for Soviet launchers, it launched the early Sputniks (Figure 2.9) and, as the A-1 variant, also the early lunar probes, such as Luna 1

Figure 2.9   R-7 launch of Sputnik 2 [NASA/Asif Siddiqi]

which it launched on 2 January 1959. By that time, having developed some experience with staging operations, it was possible to contemplate a more complex launch vehicle. Thus the A-1 incorporated an upper stage with a single engine, mounted on top of the sustainer unit, giving the vehicle the capability of delivering a 4800 kg payload to low Earth orbit or 400 kg on a lunar trajectory.

The launch site for these space vehicles, and the ICBM tests that preceded them, was established in 1955 near the town of Tyuratam, in Kazakhstan, to the east of the Aral Sea. Although the Soviets referred to it as the Baikonur Cosmodrome, it is about 300 km from the town of Baykonyr (alternative spelling) and only about 20 km from Tyuratam; thus the naming policy has since been characterised as a Cold War misinformation tactic.

Apart from the R-7/A-1's ability to lift large payload masses, it also possessed the advantage of physical size, specifically the 3 m-diameter of its sustainer section. This meant that a payload fairing of the same diameter could readily be incorporated with the rocket, allowing large satellites and even manned spacecraft to be carried.

So it was that several other variants of the basic rocket were developed to support the Soviet manned space programme. The baseline two-stage space launcher was known as the RNV (*Rakyeta Nosityel Vostok*), a powerful rocket developing a thrust of some 570 tonnes, while the three-stage RNV-2 (Voskhod) and RNS (Soyuz) produced 650 tonnes and 700 tonnes of thrust, respectively [51]. The rocket designations mirrored those of the spacecraft they were primarily designed to launch: the Vostok ('East'), Voskhod ('Sunrise') and Soyuz ('Union') spacecraft. All of these vehicles – most notably Vostok 1, which carried Yuri Gagarin into orbit on 12 April 1961 – were launched from the Baikonur Cosmodrome, which has been used ever since for both manned and unmanned launches. Interestingly, following the dissolution of the Soviet Union, Russia was obliged to lease the facilities from Kazakhstan.

One of the keys to success for Soviet rocketry was the clustering of a number of thoroughly tested, highly reliable and relatively simple propulsion units. Indeed, the arrangement has proved so reliable that it has become a signature technology of the Soviet space programme. For example, the impressive array of 20 large engine nozzles flanked by 12 smaller thrusters was used to good effect when the vehicle was first shown publicly at the 1967 Paris Air Show, the 38 m long, gleaming white rocket with its gold-painted engines causing quite a stir.

To western experts, however, the rocket's public debut was a relief. Until then, practically everyone had believed in the Cold-War myth of Russian 'super rockets' and even 'super fuels', which were thought to have been developed in secrecy behind the infamous Iron Curtain [51].

Despite the relief, it was clear that the Soviet Union had a rocket capable of delivering large payloads to orbit, or on a ballistic trajectory to the United States. America's contemporary developments were on a much smaller scale.

American MRBM

In 1950, Wernher von Braun's 'rocket team' was moved to the Redstone Arsenal in Huntsville, Alabama (now the site of the NASA Marshall Space Flight Center).

Figure 2.10   Redstone in gantry at Cape Canaveral [Mark Williamson]

Its main role was to continue the development of ballistic missiles, based heavily on experience with the V-2. The development of the Redstone booster, named after the team's new home, began on 1 May 1951, with a first firing from Cape Canaveral on 20 August 1953 (Figure 2.10).

The Redstone was slightly smaller than the Soviet R-5 MRBM: both were 1.7 m in diameter and the Redstone was 19.2 m long, compared to the R-5's 21 m. The R-5 was the more powerful, however, and had a greater range. The two vehicles also used a similar steering system, based on the V-2 [47].

As previous events had shown, guidance was at least as important in rocket development as the operation of the engine itself and, in the context of a tactical missile, the aiming accuracy was of paramount importance. The Redstone received a significant

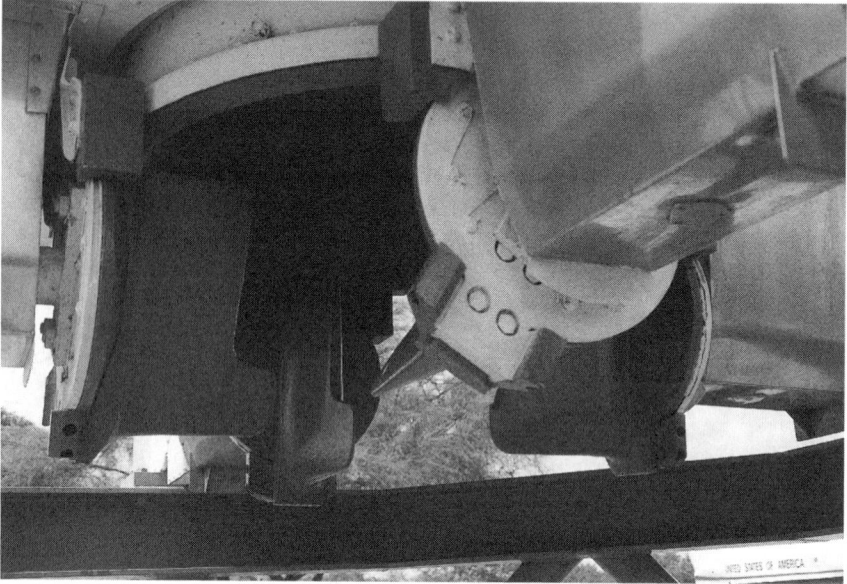

Figure 2.11   Steering vanes in Redstone exhaust [Mark Williamson]

inheritance from the V-2, not least in its LEV-3 control system which provided guidance for both the booster and the warhead section after separation. During the boost phase, while the rocket was travelling through the denser layers of the atmosphere, correction signals were sent to four air vanes which acted much like aircraft rudders. Each vane was connected, by means of a chain drive, to a graphite rudder in the exhaust, upon which the rocket relied for steering outside the atmosphere (Figure 2.11) [52].

Using the LEV-3 allowed the qualification of the missile structure, propulsion system, and so on, while its new ST-80 inertial guidance platform was under design and development. In common with other inertial platforms, it was designed to use a set of gyroscopes to define a frame of reference, with respect to which the vehicle's attitude could be determined and controlled. A typical platform comprises three gyros and three accelerometers mounted on gimbals to isolate them from the vehicle. The gyros are set spinning with their axes in the three orthogonal directions which, however the vehicle moves, remain fixed in inertial space. The accelerometers detect any acceleration imposed on the platform, and their outputs are processed to calculate the vehicle's velocity and position. The vehicle's autopilot then corrects any errors.

Although the Army had intended to follow its traditional approach in weapons development and design, build and test the Redstone in-house, it soon realised that a project of this scale was beyond it and hired a number of large industrial companies to provide major assemblies and components. This departure was an early example of the collaboration between the military services and the aerospace industry which gave rise to the term military–industrial complex, a cooperative bonding that often makes it difficult to separate the two.

Thus the prime contractor for all but the early development models of the Redstone was the Chrysler Corporation; the engine was built by the Rocketdyne Division of North American Aviation; and the ST-80 guidance platform was manufactured by Ford Instrument Co. with gyroscopes from Sperry Rand Corp. The designation of conventional aerospace contractors for all major hardware was a sign that rocket building had entered a more professional era: by contrast, the contractor chosen to manufacture the Redstone's LEV-3 autopilot control system to Army design specifications had been Waste King Corp of Los Angeles, a manufacturer of household appliances such as garbage disposers, dish washers and waste incinerators [53].

In May 1955, the US Air Force decided to develop an intermediate range ballistic missile (IRBM)[2] called Thor, in parallel with its Atlas and Titan ICBM programmes (discussed below with reference to space missions). The Thor became the basis for a series of satellite launchers, including the Thor-Able, Thor-Able Star and Thor-Agena, each with a different upper stage. NASA later matched Thor with a re-engineered upper stage, known as the Delta since it was the fourth variant, to form the Thor-Delta (a name subsequently simplified to Delta). The Delta became an important launch vehicle, in several variants, for NASA's satellite and interplanetary spacecraft programmes and continued to be used beyond the turn of the century.

In terms of launch vehicle technology, the Thor was perhaps most notable for its introduction of solid propellant strap-on boosters, which are used to augment the thrust of a basic or core vehicle. The Thrust Augmented Thor (TAT) was formed by the addition of three Castor I solid motors, each 6 m long and 80 cm in diameter, which increased the lift-off thrust from 78 tonnes to nearly 152 tonnes (Figure 2.12). This new variant was first launched on 28 February 1963 with an Agena D upper stage, but failed to reach orbit; the first successful launch occurred on 18 May 1963.

US space launcher

Despite America's apparent, post-war reluctance to embrace the ICBM as a weapon delivery system, by the early 1950s ballistic missiles were *de rigeur*, and America's military services were all developing ICBMs that could be converted into space launch vehicles. The Army, under the guidance of Wernher von Braun, was working on the Redstone and its derivatives; the Air Force was intending to convert the Atlas, an ICBM programme begun in the early 1950s; and the Navy had proposed an essentially civilian programme called Project Vanguard, using the Viking sounding rocket it had developed with Glenn L. Martin Co. in the 1940s [54].

If America had been intending simply to develop a satellite launch vehicle, this dilution of national technical expertise could be perceived as a contributory factor in America's failure to orbit the first satellite, but the primary use of the technology was a military one. Nevertheless, the administrative procedure which eventually led to the first American satellite could have been more efficient. On 29 July 1955, President

---

[2] Although definitions are imprecise, an IRBM tends to have a range greater than an MRBM and less than an ICBM.

Figure 2.12   Delta rocket with strap-on solid boosters (prior to failed ITOS-B flight in October 1971) [NASA]

Eisenhower's press secretary, James Hagerty, announced that America would launch a satellite as part of US participation in the International Geophysical Year (IGY) of 1957–8 [55]. It was another 6 weeks, however, before the Pentagon announced – on 9 September – that the Navy was to take charge of the launching programme [56]; and it was 6 October before the Naval Research Laboratory was officially notified to take charge [57]. It appeared that the Navy's Vanguard would launch America's first satellite.

Despite the implicit rejection of the proposal to develop the Redstone as a space launch vehicle, von Braun remained with the Army and continued his work towards this goal. Meanwhile, as planning for the Army's next-generation ballistic missile, the Jupiter IRBM, got underway, it became apparent that a new organisation was needed for its development, so on 1 February 1956 the US Army Ballistic Missile Agency (ABMA), directed by Brigadier General John B. Medaris, was established at the Redstone Arsenal [58] (Figure 2.13). In pursuance of the Army's requirement for a new

Figure 2.13   ABMA officials, including Hermann Oberth (centre) and Wernher von Braun (second from right) [NASA]

missile, 25 Redstones were modified internally to act as a testbed for components that would be incorporated into the new rocket. They were flown as the Jupiter A series.

The Jupiter C (C-for-Composite re-entry test vehicle) was a more extensively modified Redstone, with two upper stages, developed to test the ablative nosecone of the Jupiter's warhead. In what was becoming a common ploy in rocket design, the Jupiter C was an amalgam of a number of existing military missiles. Its upper stages, contained within a tapered fairing mounted above the Redstone, used the Army's Sergeant solid propellant guided missile, scaled to one third of its standard size. The second and third stages were made from groups of 11 and 3 of these 'Baby Sergeants', respectively [59].

On 20 September 1956, von Braun's team successfully fired the first Jupiter C to an altitude of 1097 km and a range of 5472 km. The Jupiter's potential as a satellite launcher was obvious but, as a result of President Eisenhower's commitment to Project Vanguard, the Pentagon prohibited any attempt by von Braun and the Army to launch a satellite.

In an interesting historical parallel, by 1956 the Russian rocket designer Sergei Korolev had brought the development of his ICBM to a stage where he too wanted to test his creation by launching an artificial satellite. He was told, however, that this was not a valuable project and to continue to develop the booster as a military launch vehicle. Thus, in August 1957, his team achieved their successful test-firing of the R-7 rocket with a range of some 8000 km and the capability to carry a 2 tonne warhead.

This succeeded in capturing the interest of Khrushchev, who suddenly seemed to realise that Korolev's new missile could now be used as an international, Cold-War bargaining tool. To his credit, however, he also realised that, to be believed, a public demonstration would be needed. He ordered Korolev to launch a satellite . . . and quickly. It took only six weeks to assemble the multi-stage rocket and its simple payload, Sputnik 1.

The launch of the Sputnik, on 4 October 1957, provided the ultimate spur to the development of American launch vehicles; yet, still, official hopes remained with the Vanguard. It took a second Soviet triumph – the launch of the half-tonne Sputnik 2 the following month – to shift attention to von Braun's team. Five days later, on 8 November, it was given the go-ahead to convert the Jupiter C into a satellite launcher called Juno I [59]. This was done simply by adding a fourth stage in the shape of a Baby Sergeant, already used in clusters for the Jupiter's second and third stages, a combination capable of placing up to 200 kg in a low altitude orbit.

Before Juno I could be brought into service, however, there was more embarrassment in store for US rocketry. The US government's insistence on billing Vanguard as America's answer to Sputnik, and the ensuing publicity, forced a launch attempt on 6 December 1957: the rocket rose hesitantly from the pad, fell back and exploded (Figure 2.14). Thus it was Juno I, not Vanguard, which launched America's first satellite, Explorer 1, into orbit on 31 January 1958. At last the Americans could claim a modicum of success.

A second attempt to launch a Vanguard failed on 5 February 1958 when the rocket exploded at a height of 7 km, but the project finally succeeded in placing a satellite in orbit on 17 March 1958 when it launched the Vanguard 1 satellite (see Chapter 3). The programme as a whole lacked momentum and only 3 out of a total

Figure 2.14   Vanguard launch failure [NASA]

of 11 Vanguard launch attempts were successful, a record which contributed to the inescapable impression that rockets were a relatively unreliable form of transport. And since US rocket launches were aired in full sight of the public, via the print and television media, this impression was destined to last. Indeed, Juno I itself was successful in only three of its six launches before it was abandoned, thus becoming the only space launch vehicle to retire from service before NASA came into existence in October 1958.[3]

[3] The National Aeronautics and Space Administration (NASA) was formed on 1 October 1958 to coordinate United States research and development into aeronautics, space science and space technology. The organisation was based on the structure of its predecessor, the National Advisory Committee for Aeronautics (NACA).

Nevertheless, the basis of the Juno and the Jupiter – the Redstone which formed their first stages – could be deemed a success. Its seven-and-a-half year engineering phase, which ended with the successful launching of the last R&D missile on 3 November 1958, produced not only a highly reliable short-range tactical missile but the launch vehicle for America's first artificial satellite.[4]

By May 1958, however, this was largely academic: the Soviets had orbited Sputnik 3, weighing in at 1327 kg, and America feared that the USSR was preparing to launch a man into space. A different race had begun.

Atlas, Titan and Saturn

As indicated earlier, the heritage of rocket programmes, names and hardware is a complicated web of interrelationships, born of both conflicting and mutually supportive applications addressed by multiple design teams. Although it is not possible to do them justice in a limited space, as far as the United States was concerned, three important programmes should be mentioned in the context of their support of manned spaceflight in the 1960s: the Atlas, Titan and Saturn vehicles, which launched the Mercury, Gemini and Apollo spacecraft, respectively.

As far as space exploration was concerned, launch vehicle development was no longer simply a matter of 'breaking the altitude barrier'; it had evolved into an effort to deliver larger and heavier payloads into orbit in a reliable fashion. If the manned vehicles had retained the early rockets' unreliability, there would have been no manned space programme.

The Atlas was America's first true ICBM, initially capable of delivering a two megaton warhead across a range of some 8000 km, and even further for later versions. The name Atlas was selected in 1951 by Karel J. Bossart, head of Consolidated Vultee Aircraft (Convair) Corp, which later became a division of General Dynamics; it was an obvious nod to the Greek god of the same name, but also recognised Convair's owner, the Atlas Corporation. At that stage, the Atlas was little more than a concept, which Convair continued to develop throughout 1953 and 1954, finally securing a contract from the Department of Defense in January 1955 [61].

The Atlas ICBM made its first flight in June 1957, but this and its second flight in September were failures; its first successful flight occurred in December 1957 (some four months behind the Soviet R-7) and it became an operational weapon system in September 1959.

The vehicle was ground-breaking in many ways, not least in its use of so-called 'balloon tanks' conceived by Bossart in 1946 and used for the Atlas and the associated Centaur upper stage. The pressure-stabilised, thin steel, monocoque structure formed a 'steel balloon', which served as part of the primary structure of the stage

---

[4] Aficionados of rocket numbering systems might like to note that the Army used a system for its Redstone rockets which substituted the letters of the word HUNTSVILLE (minus the second L) for the numerals 1 to 9 (H1, U2, N3, T4, S5, V6, I7, L8, E9) while zero was represented by an X. For example, as the 29th Redstone produced, the first Juno I carried the letters UE [60].

as well as the skin of the propellant tank. The thickness of the material varied from just 0.25 mm to about 1.3 mm, so thin that the tank could not support its own weight without being pressurised. The pressurant maintained the tank's rigidity without the need for heavy structural members, thereby improving the vehicle's mass ratio.

The original Atlas was considered a 'one-and-a-half' stage vehicle. Its three main engines were ignited simultaneously at launch, but only one, the sustainer, remained attached to the propellant tanks for the full duration of the flight. The other two, called boosters, were jettisoned along with their fairings and other structural components, part way through the flight (in a similar manner to the Soviet ICBM, except that the Atlas engines were fed from common propellant tanks). A key advantage of the one-and-a-half stage configuration was that it avoided having to start a large engine at high altitude (in vacuum conditions), something with which designers had little experience at the time.

This configuration became the first stage of successive versions of the Atlas launch vehicle, all of which used the propellants liquid oxygen and kerosene. The Atlas-D variant, for instance, was used to orbit the Mercury capsule; the Atlas-Able and Atlas-Agena two-stage vehicles launched the Ranger probes to the Moon and the early Mariners to Mars and Venus; and the Atlas-Agena D carried the Agena target docking vehicle for the Gemini programme.

The Centaur stage used on the Atlas-Centaur was America's first liquid oxygen/liquid hydrogen stage: it first flew successfully in November 1963, and became operational in May 1966 with the launch of the lunar probe Surveyor (Figure 2.15). It could lift more than 4.5 tonnes to low Earth orbit (LEO) or place somewhat less than a tonne on a planetary trajectory. The Atlas itself was ultimately so successful as a space launch vehicle that its later variants were still in use beyond the turn of the century, albeit using Russian-built engines (an ironic twist for a product of the Cold War).

As the development of the Atlas was nearing completion, a second ICBM called Titan began its development as an insurance policy against potentially significant failures and delays in the Atlas programme. As with most rocket vehicles, several variants ensued: the two-stage Titan I used the same propellants as the Atlas (LOX and kerosene), while the two-stage Titan II used the storable liquid propellants nitrogen tetroxide and aerozine-50.

Instead of steel, the Titan was constructed from light copper-aluminium alloys and did not use the Atlas 'balloon tank' design. Again, unlike the Atlas, it utilised a standard two-stage design, which was tested successfully on its fourth flight (albeit with a dummy second stage) on 4 May 1959. Interestingly, the Titan's first flight – on 6 February of that year – had also been a success, the first time that the maiden flight of a large ballistic missile had been successful [62].

The Titan II met the strategic requirements of the day, specifically the need to maintain the possibility of a retaliatory strike under the most severe bombardment. Where the Atlas and the Titan I had to be fuelled in a vertical position on the launch pad just prior to launch (since LOX boils at $-182.98°C$ and cannot be stored onboard), the Titan II could remain fuelled in hardened underground silos, ready to launch at a moment's notice. In addition, the Titan II had fewer moving parts in its propulsion

Figure 2.15   Atlas launch of Surveyor 1 [NASA]

system and benefited from improvements in electronic components made in the early 1960s. It was this version which, being capable of placing some 3.7 tonnes into LEO, later launched the manned Gemini spacecraft (Figure 2.16).

A further development was the more powerful, three-stage Titan III, which incorporated two solid propellant strap-on boosters. Its later addition of the Centaur typified the mix-and-match philosophy of launch vehicle designs, which often utilised relatively high performance liquid propellants in the core vehicle, while adding solid strap-ons to improve the payload capability of a given variant. The Titan III was used to launch military payloads and planetary spacecraft: in the 1970s, for example, it boosted the Viking spacecraft to Mars and the Voyagers to Jupiter and Saturn. A later Titan IV variant was developed chiefly as a military heavy-lift launcher.

Figure 2.16   Titan launch of Gemini 11 [NASA]

So by the early 1960s, there were two more launch vehicles – the Atlas and the Titan – available to the nascent American space programme. However, neither of these new vehicles were capable of performing the task of transporting the Apollo spacecraft to the Moon. For this, Wernher von Braun and his team, now under the auspices of NASA, developed the Saturn V.

The process began with a smaller vehicle, the two-stage Saturn I, which itself was based partly on earlier, single-stage rockets: its first stage was effectively eight Redstones clustered round a central tank based on the Jupiter. It would ultimately be capable of lifting a 10 tonne payload into LEO, which was almost unimaginable in 1957, when it was first studied as a concept: at the time, the diminutive Vanguard launcher was still six months from its first flight and could place little more than 10 kg in orbit.

In recognition of its heritage, the rocket was formerly known as the Juno V, but in October 1958 Wernher von Braun proposed that this be dropped in favour of Saturn, which was both a Roman god (in the Army tradition of naming rockets) and also the next planet out beyond Jupiter. Initially known as the Saturn C-1, because it was the Saturn C concepts which included the chosen LOX/liquid hydrogen upper stages, the designation was later simplified to Saturn I [63]. Despite fears that it would complicate the design, liquid hydrogen (LH_2) was chosen as a fuel because of its high specific impulse (a measure of its mass-efficiency and performance). Today, liquid hydrogen technology is recognised as one of the key enabling technologies for the manned lunar landing missions.

Figure 2.17   Saturn V launch [NASA]

An uprated version of the rocket, the Saturn IB, comprised an S-IB stage, which used LOX and kerosene, and an S-IVB LOX/LH$_2$ upper stage. This combination gave it a payload capability of some 18 tonnes to LEO, allowing it to launch either an Apollo command and service module (CSM) or lunar module (LM) for in-orbit tests. It was first launched in February 1966 with an unmanned CSM, which it placed on a ballistic trajectory to test propulsion systems and the re-entry of the command module. It was also used in October 1968 to launch the first manned mission, Apollo 7; in 1973 to launch three crews to the Skylab space station; and in 1975 for the Apollo-Soyuz Test Project (ASTP).

The crowning glory of American rocket development of the 1960s was the Saturn V itself (Figures 2.17 and 2.18). Formerly known as the Saturn C-5, it comprised three stages: the S-1C first stage, which burnt LOX and kerosene, and the S-II and S-IVB LOX/LH$_2$ stages. The Saturn V was first launched on 9 November 1967 for the unmanned Apollo 4 test-launch, and last used on 14 May 1973 to launch Skylab (itself a converted S-IVB stage).

In total, complete with Apollo spacecraft and launch escape tower, the Saturn V was more than 110 m in height, 10 m in diameter at the base, and weighed over 2800 tonnes fully fuelled. The five, clustered F-1 engines of its first stage produced a lift-off thrust of about 3400 tonnes by burning propellant at the staggering rate of some 12.9 tonnes per second (Figure 2.19) [64].

Figure 2.18   Saturn V first stage in VAB [NASA]

Figure 2.19   Increasing size and complexity of US rocket engines: (a) Redstone, (b) Centaur, (c) Saturn V F-1 [NASA]

(c)

Figure 2.19   Continued

At last America had a rocket to beat the Soviet Union's best: a single F-1 engine, developing a specified 680 tonnes of thrust, was more powerful than an entire Soviet RNV at 570 tonnes. Moreover, the Saturn V as a whole could deliver 95 tonnes to Earth orbit and 48 tonnes to a lunar trajectory, compared with about 5 tonnes and half a tonne, respectively, for the RNV. By any measure and from any angle, the Saturn V was a colossus of the Space Age.

Although details were not publicly known at the time, it was later revealed that the Soviet Union had attempted to develop its own Moon rocket known as the N-1 (N stood for 'nositel' or 'carrier'), as part of an ill-fated manned lunar programme. In fact, the programme's existence was not admitted until the coming of *glasnost* and an unretouched photo of the N-1 was not published until 1992 (Figure 2.20) [65].

Towards the end of 1961, Korolev's experimental design bureau had begun work on a booster intended to deliver between 40 and 80 tonnes (in different versions) to LEO. By mid-1964, it was decided that the N-1 should be part of the Soviet effort to send a man to the Moon and it was planned to increase the payload capability to about 95 tonnes for this purpose (the same as the Saturn V).

Whereas the Saturn V was transported to the launch pad in a vertical orientation, using a heavily modified construction vehicle known as the Crawler, the N-1 was transported in a horizontal position by two twin diesel locomotives running on parallel railway lines and erected at the launch pad (the method common to all large Soviet launchers).

The N-1 had two main disadvantages compared with the Saturn V: its propellants were a less energetic combination than the Saturn's, and its launch site was further from the equator. Rockets launched from equatorial latitudes benefit most from the inherent rotation of the Earth which adds to their orbital velocity. Thus the

Figure 2.20   Two Soviet N-1 launch vehicles on their pads at Baikonur Cosmodrome in July 1969 (one in background is ground test model) [NASA/Asif Siddiqi]

Saturn V was some 10 per cent more efficient simply by virtue of its launch site being in Florida. The Saturn also had one less rocket stage and many fewer engines than the N-1, which made it simpler and, in principle, more reliable [66].

Indeed, it was the N-1's reliability which let it down: while all Saturn V flights were successful, all four N-1 launch attempts, made between February 1969 and November 1972, were failures and the programme was cancelled.

Beyond the superpowers

Although the space race had been conducted between the two major superpowers, this is not to say that other technologically capable nations were absent from rocket development. This was particularly true for Europe, whose major nations

Figure 2.21   Blue Streak engine test at Spadeadam Rocket Establishment, Cumbria (UK), March 1963 [EADS, formerly De Havilland Aircraft Co. Ltd.]

conducted largely separate lines of development for many years, in parallel with both the United States and the USSR, before cooperating to produce what could have been an independent European satellite launcher by the late 1960s. While recognising that other nations, such as China and Japan, were also active during this period, Europe provides a good illustration of the international nature of launch vehicle development.

In the 1950s, like America, Britain made the decision to develop its own intermediate range ballistic missile (IRBM) capable of delivering a nuclear warhead. It was called Blue Streak (Figure 2.21). Following a UK Government operational requirement issued in June 1953, contracts were awarded in great secrecy, in 1955, to De Havilland Propellers (later Hawker Siddeley Dynamics) for the airframe of the missile and to Rolls Royce for the engines [67]. Outside of the United States and USSR, Blue Streak was then the largest rocket under development, which meant that Britain, in cooperation with Australia, was the world's leading non-US/USSR participant in space research and technology [68].

The Australian connection dated back to Britain's need for a test range from which to fire its rockets without endangering the densely populated land masses of Europe. Despite its distance from the United Kingdom, the wide open outback of Australia seemed ideal, and so, through an Anglo-Australian agreement of 1946, a test centre for guided weapons and the town of Woomera came into being [69].

Blue Streak was widely regarded as a very capable vehicle. Its size and thrust – and, by extension, launch capability – approached that of the American Atlas and it used the same propellants, kerosene and LOX. In addition, it had a good reliability record and flew for the intended duration on most flights (the exception being flight F4, which was terminated after 140 s for safety reasons when a problem was detected). This shows that whether a launch vehicle is man-rated (intended to carry people) or not, good design engineering, attention to detail and 'learning from history' can reap rewards. In the case of Blue Streak, it could be said that its designers learnt from the mistakes of the Atlas, which was unsuccessful in 15 of its first 25 flights [70].

As a result of a technology transfer agreement made with the US Government, many of the basic techniques and assembly methods were common to both vehicles; there was, after all, no point in 'reinventing the wheel'. However, there were also substantial differences driven by their different operational requirements. For example, Blue Streak had two engines in a single-stage design, whereas the Atlas had three in a one-and-a-half stage configuration. In addition, the Blue Streak's kerosene tank was designed to be self-supporting in the vertical position by adding stringers to the outside, giving the lower part of the stage a ribbed appearance. The balloon tank design of the Atlas was confined to Blue Streak's LOX tank, which was housed in the upper section of the rocket. Such was the structural efficiency of Blue Streak that its dry mass was only about 6 per cent of its fully fuelled mass. In other words, in its standard configuration, of a total launch mass of some 95 tonnes, more than 88 tonnes was propellant.

When the United Kingdom abandoned the concept of an independent nuclear deterrent in April 1960, the Blue Streak programme was cancelled. In an attempt to salvage something from the years of development, it was then proposed as the first stage of a European three-stage satellite launcher. The new vehicle was originally called ELDO-A after the newly created European Launcher Development Organisation, the convention for which had been signed in London on 29 March 1962 [71]. Appropriately enough, it was later re-christened Europa.

This was not exactly the favourite plan among UK industrialists, who would have preferred to use Blue Streak as a first stage, modify the highly successful Black Knight sounding rocket as a second stage and add a small solid third stage, 'thus producing a very commendable all-British space launcher' [72]. But the government could not see its way to supplying the further £60–70 million that would be necessary and, as a compromise, the Europa plan was devised.

The founding countries of ELDO, in order of their financial contributions, were Britain, Australia, West Germany, France, Italy, Belgium and the Netherlands. Its goal was to develop a three-stage launch vehicle capable of launching a satellite weighing 1000 kg into a 500 km-high orbit by early 1967. Britain was to supply the first stage, in the shape of Blue Streak, France would build the second stage, Germany the third, Italy the satellite payloads, Belgium the guidance and tracking stations and the Netherlands the telemetry facilities. Australia, of course, was responsible for the launch facilities at Woomera.

The flight testing of Europa was divided into three phases to qualify, respectively, the first stage alone, the first and second stages together and, finally, the complete system. Although Europa was a shadow of the three-stage Saturn V, being developed by America at the same time, an overview of the test programme provides a stark illustration of the difficulties of integrating three distinct rocket stages – in Europa's case from three separate manufacturers in three different countries.

Blue Streak's first launch, as Europa 1-F1 (flight no.1), was conducted from Woomera on 5 June 1964: the rocket reached a height of 177 km and a speed of 10,300 km/h before impacting 966 km from the test stand [73]. Phase 1 of the programme was completed with another two tests of the Blue Streak, both of which – apart from the minor anomalies expected in a test programme – were considered successful.

Meanwhile, the French second stage, known as Coralie, was undergoing tests of its own. Before its debut as Europa's second stage, Coralie flew as the first stage of a test rocket called Cora, which carried a mock-up of Europa's third stage and a dummy satellite. The name Coralie was in fact a contraction of Cora-Australie, signifying that it would be launched from Woomera, Australia; Cora itself was launched from the French national test site at Hammaguir, Algeria. Cora was launched twice in 1966: the first vehicle had to be destroyed after 62 s due to a guidance failure, but the second flight of 176 s was a success [74].

Phase 2 of the Europa flight-tests began in 1966 with two successful flights of the Blue Streak carrying mock-ups of the second and third stages, but it was when the time came to add a 'live' second stage that the problems began. On the sixth Europa flight, Blue Streak performed well, but the Coralie, which used the propellants nitrogen tetroxide and UDMH (unsymmetrical dimethylhydrazine), failed to ignite. It is believed that an electrostatic discharge which occurred at stage separation caused the Coralie's ignition sequencer to fail. The next flight was a virtual carbon copy, as the sequencer failed again to initiate ignition: this time the explanation was that some electrical connector pins had not disconnected simultaneously. Things did not look good for Coralie, especially since, in separate tests, it had suffered another guidance problem due to a wiring failure [75].

Europa's final test phase began on 30 November 1968, with a flawless performance from both Blue Streak and Coralie; unfortunately the German third stage, Astris, exploded shortly after ignition when a bulkhead broke. The following flight was a shattering repeat performance. Then finally, on Europa's tenth test-flight, all three stages worked; unfortunately, the payload fairings failed to jettison and the satellite, trapped within, was unable to reach orbit. The following flight was cancelled due to budgetary constraints [75].

Europa had one last chance to prove itself, in an upgraded version sanctioned by a conference of ministers, which met in 1966 while the second phase tests were underway. Europa 2 would incorporate a fourth stage, or perigee–apogee system (PAS), to boost a satellite to the high-altitude geostationary orbit. It would be launched from a new facility near Kourou in French Guiana, where the French national space agency (CNES) had established a launch range for its small satellite launcher, the Diamant. The idea was that Europe would be able to launch its own scientific payloads and would have its own experimental communications satellite in geostationary orbit by 1970. For the first time, it would be completely independent of the United States.

The first flight model of Europa 2, designated F11, was launched from Kourou on 5 November 1971 and exploded 150 s into flight while the first stage was still burning. The problem was traced to electrostatics once again: friction with the air had charged the fairing, which was not electrically grounded; this led to electrostatic discharges which caused the guidance system computers, housed in the upper stage, to fail. Since the first stage was no longer receiving guidance signals, the vehicle banked, and increased aerodynamic loads caused it to break up. Subsequent flights were cancelled in April 1973 and the Europa 2 programme was abandoned [76].

ELDO itself was disbanded in 1973 and its activities were amalgamated with those of the European Space Research Organisation (ESRO) to form the European Space Agency (ESA) in May 1975. In contrast to Britain, France had invested heavily in its national space programme, which had produced a proposal for a new three-stage launch vehicle called L3S. Thus when ESA initiated the Ariane launch vehicle programme, based on the L3S, it delegated its management to CNES, giving France a leading role in the project. This led, in due course, to the Ariane 4, which by the early 1990s had captured more than 50 per cent of the world's commercial satellite launch market.

In the years since Robert Goddard's first liquid propellant rocket made its two-and-a-half second flight from a field in Massachusetts, the subject of rocketry has developed from an obscure scientific pursuit to a globally competitive industrial undertaking. Launch vehicles based on the principles expounded by Goddard and Oberth, and Tsiolkovsky before them, have boosted satellites into orbit, astronauts to the Moon and spaceprobes to the distant planets of the solar system.

Vertically, space may only be 100 km away, but it has proved to be one of the most challenging journeys technology has ever been required to make.

References

1 Perek, L.: 'Deep Space at WARC-ORB 88' (IISL-89-061 1989): 32nd International Colloquium on the Law of Outer Space, 40th IAF Congress, Malaga, Spain, 1989

2 Koppes, C. R.: 'JPL and the American Space Program' (Yale University Press, New Haven, CT, 1982) p. 4

3 ibid., p. 8,11

4 ibid., p. 16

5 ibid., p. 20

6 ibid., p. 30

7 Harvey, Brian: 'The Chinese Space Programme: from Conception to Future Capabilities' (Wiley-Praxis, Chichester, 1998) pp. 4–5

8 Baker, David: 'The Rocket' (Crown Publishers Inc, New York, 1978) pp. 73–4

9 Koppes, C. R.: 'JPL and the American Space Program' (Yale University Press, New Haven, CT, 1982) p. 23

10 ibid., p. 22

11 ibid., p. 14, 23

12 Ley, Willy: 'Rockets and Space Travel' (The Viking Press, New York, 1948) p. 356

13 Koppes, C. R.: 'JPL and the American Space Program' (Yale University Press, New Haven, CT, 1982) p. 24

14 'In the matter of Mitigation of Orbital Debris' (Federal Communications Commission (FCC), Notice of Proposed Rule Making, IB Docket No. 02-54; Adopted: March 14, 2002; Released: March 18, 2002)

15 Ordway, Frederick I., and Sharpe, Mitchell R.: 'The Rocket Team' (William Heinemann Ltd, London, 1979) p. 347

16 Miller, Ron: 'The History of Rockets' (Grolier Publishing, New York, 1999) p. 61

17 Ordway, Frederick I., and Sharpe, Mitchell R.: 'The Rocket Team' (William Heinemann Ltd, London, 1979) p. 278

18 ibid., pp. 353–4

19 ibid., pp. 354–5

20 'Final Report, Project Hermes, V-2 Missile Program' (General Electric Report No. R52A0510, September 1952); Michael Ciancone, personal communication 9/2/05

21 Jung, Philippe, personal communication 22/4/05

22 Ordway, Frederick I., and Sharpe, Mitchell R.: 'The Rocket Team' (William Heinemann Ltd, London, 1979) p. 354

23 Stuhlinger, Ernst: 'Gathering Momentum: Von Braun's Work in the 1940s and 1950s' in Ordway, Frederick, I., and Liebermann, Randy (Eds): 'Blueprint for Space' (Smithsonian Institution, Washington, USA, 1992) p. 119

24 Whipple, Fred L.: 'Recollections of Pre-Sputnik Days' in Ordway, Frederick, I., and Liebermann, Randy (Eds): 'Blueprint for Space' (Smithsonian Institution, Washington, USA, 1992) p. 127

25 Baker, David: 'The Rocket' (Crown Publishers Inc, New York, 1978) p. 234

26 Koppes, C. R.: 'JPL and the American Space Program' (Yale University Press, New Haven, CT, 1982) p. 41

27 Shelton, William: 'Soviet Space Exploration: The First Decade' (Arthur Barker Ltd, London, 1969) pp. 24–5

28 Ordway, Frederick I., and Sharpe, Mitchell R.: 'The Rocket Team' (William Heinemann Ltd, London, 1979) p. 321

29 Winter, Frank H.: 'Rockets into Space' (Harvard University Press, Cambridge, 1990) p. 41

30 Siddiqi, Asif A.: 'Challenge to Apollo: The Soviet Union and the Space Race, 1945–1974' (NASA SP-2000-4408, Washington, USA, 2000) p. 41

31 ibid., p. 42

32 Ordway, Frederick I., and Sharpe, Mitchell R.: 'The Rocket Team' (William Heinemann Ltd, London, 1979) p. 322

33 Miller, Ron: 'The History of Rockets' (Grolier Publishing, New York, 1999) p. 64

34 Ordway, Frederick I., and Sharpe, Mitchell R.: 'The Rocket Team' (William Heinemann Ltd, London, 1979) p. 324

35 Smolders, Peter L.: 'Soviets in Space' (Lutterworth Press, Guildford, 1973) pp. 56–7

36 Baker, David: 'The Rocket' (Crown Publishers Inc, New York, 1978) p. 113

37 Ordway, Frederick I., and Sharpe, Mitchell R.: 'The Rocket Team' (William Heinemann Ltd, London, 1979) p. 326

38 Siddiqi, Asif A.: 'Challenge to Apollo: The Soviet Union and the Space Race, 1945-1974' (NASA SP-2000-4408, Washington, USA, 2000) pp. 55–6

39 ibid., p. 56

40 ibid., p. 61

41 ibid., p. 62

42 Ordway, Frederick I., and Sharpe, Mitchell R.: 'The Rocket Team' (William Heinemann Ltd, London, 1979) p. 329

43 ibid., p. 330

44 Ordway, Frederick I., and Sharpe, Mitchell R.: 'The Rocket Team' (William Heinemann Ltd, London, 1979) p. 331

45 Miller, Ron: 'The History of Rockets' (Grolier Publishing, New York, 1999) p. 66

46 Osman, Tony: 'Space History' (Michael Joseph, London, 1983) pp. 36–7

47 Baker, David: 'The Rocket' (Crown Publishers Inc, New York, 1978) p. 117

48 ibid., p. 118

49 ibid., p. 227

50 Smolders, Peter L.: 'Soviets in Space' (Lutterworth Press, Guildford, 1973) p. 63

51 ibid., p. 65

52 Sharpe, M. R., and Burkhalter, B. B.: 'Mercury-Redstone: the First American Man-rated Space Launch Vehicle' (IAA-89-740): 40th IAF Congress, Malaga, Spain, 1989, p. 10

53 ibid., p. 9

54 Miller, Ron: 'The History of Rockets' (Grolier Publishing, New York, 1999) p. 67

55 McLaughlin, Constance, and Lomask, Milton: 'Vanguard: A History' (Smithsonian Institution Press, Washington, USA, 1971) p. 37

56 ibid., p. 1

57 ibid., p. 57

58 Ordway, Frederick I., and Sharpe, Mitchell, R: 'The Rocket Team' (William Heinemann Ltd, London, 1979) p. 379

59 Ulanoff, Stanley: 'Illustrated Guide to US Missiles and Rockets' (Doubleday & Co Inc, New York, 1959) p. 116

60 Wilson, Andrew: 'Jupiter C/Juno I – America's First Satellite Launcher', Spaceflight, 1981, **23** (1), pp. 12–17

61 Baker, David: 'The Rocket' (Crown Publishers Inc, New York, 1978) pp. 231–2

62 Burgess, Eric: 'Long Range Ballistic Missiles' (Chapman & Hall, London, 1961) pp. 48–9

63 Baker, David: 'The Rocket' (Crown Publishers Inc, New York, 1978) p. 243

64 ibid., p 245

65 'First Untouched N-1 Picture', *Spaceflight*, 1992, **34** (3), p. 79
66 Mishin, V. P.: 'The Role of Academician S P Korolev in the Development of Space Rocket Vehicles for the Lunar Exploration with the help of Manned Spaceships' (IAA-91-674): 42nd IAF Congress, Montreal, Canada, 1991, p. 7
67 Martin, Charles H.: 'De Havilland Blue Streak' (British Interplanetary Society, London, 2002) p. 1
68 Allward, Maurice (Ed.): 'The Encyclopedia of Space' (The Hamlyn Publishing Group Ltd, Feltham, 1968–9) p. 118
69 ibid., p. 82
70 General Dynamics (Convair Aerospace Division) document 574-0-67-102: 'Chronological Order of Atlas Flights', dated 17 November 1967
71 Allward, Maurice (Ed.): 'The Encyclopedia of Space' (The Hamlyn Publishing Group Ltd, Feltham, 1968–9) p. 778
72 Pardoe, G. K. C.: 'The Challenge of Space' (Chatto & Windus, London, 1964) p. 121
73 Baker, David: 'The Rocket' (Crown Publishers Inc, New York, 1978) p. 221
74 Rothmund, Christophe: 'Coralie, The Forgotten Rocket' (IAA-91-678): 42nd IAF Congress, Montreal, Canada, 1991, p. 6
75 ibid., p. 8
76 ibid., p. 9

Chapter 3

Looking at space – the development of the space science satellite

From the rocket we can see the huge sphere of our planet. We can see how the sphere rotates and how within a few hours it shows all its sides successively. This picture is majestic, attractive and infinitely varied

Konstantin Tsiolkovsky, 1911

In the early years of the Space Age, the development of space technology was intimately linked with that of military missile and warhead technology, but it was a marriage of convenience rather than a love affair. Both partners in the arrangement were focused on their individual goals – the exploration of space and the protection of state, respectively – but each needed the other.

In the early days of rocket development, which predated the Space Age, many of the people developing the technologies for military applications also had aspirations towards spaceflight. Later, as the respective fields grew in complexity, these individuals began to specialise in one application or another, but they remained linked by the basic technologies: propulsion, power, electronics and so on. Despite the dual-use nature of the basic technologies, the military and civilian partners were not equal, the military side attracting much greater levels of government funding. Nevertheless, civilian space technologists were keen to push the technology towards their own goals and were certainly not above jumping on political bandwagons when they were going their way.

It is undeniable that the early developments of the Space Age were driven as much by politics as they were by science or engineering. Both the USSR and the United States of America were technically capable of launching the first satellite in the mid-1950s, but both Sergei Korolev and Wernher von Braun were held back by political apathy. Eventually, the Soviet Union won what became known as 'the space race' because of two complementary factors: a desire on the part of the Soviet leadership to prove superiority in at least one high-technology field; and an incapacity on the part of the American leadership to believe that the Soviet Union was capable of such a feat.

Indeed, it was the American public that was most outraged, and worried, by the Soviet triumph. They believed that if the USSR could place a satellite in orbit, it could just as easily target a nuclear warhead on an American city. Thus, it was in this climate of fear that the early development of spacecraft technology was pursued (Table 3.1).

Space science from Earth orbit

In the 1950s, there were no specific space applications as there are now: a description of the early spacecraft in today's terms might be 'technology demonstrators engaged in scientific measurement of the space environment'.

In effect, the spacecraft were conducting what we now call 'space science'. Although the term has been in use for many years, it has no widely agreed definition and some general dictionaries do not even recognise the term. Within the space community, however, space science is understood to be 'that branch of science concerned with the features and processes of the space environment [including] "space physics" and "space astronomy"' [1], in other words 'the science *of* space'. With developments in manned spaceflight, however, the definition has been extended to include topics such as space medicine and microgravity research ('science *in* space'), but they are beyond the scope of this book.

So while access to space was the remit of the rocket engineer, the development of the payload was largely down to the science community, specifically (since there were no space scientists) the geophysics community. As implied earlier, research was being pursued in parallel by both military and civilian teams, with some degree of cross-fertilisation, but we shall concentrate here on the civilian aspects.

International Geophysical Year

In a sense, the developing interest in the physics of planet Earth was fortuitous for space technology, since it provided a reason for the development of satellites. The two were brought together by the International Geophysical Year (IGY), an international scientific programme conducted jointly by 54 nations between July 1957 and December 1958.[1] Its principal task was the comprehensive study of the influence of solar activity on the various phenomena observed in the atmosphere, ionosphere and near-Earth space. Thus the time selected for the IGY corresponded to a period of maximum solar activity.

---

[1] IGY studies continued into 1958–59, which was referred to as the International Geophysical Cooperation Year (IGCY). A similar programme to the IGY, known as the International Quiet Sun Year (IQSY), was carried out during the 1964–65 solar-minimum period [2].

Table 3.1   Selected space science satellites

Launch date	Name	Nation	Comments
4 October 1957	Sputnik 1	USSR	First satellite in orbit
3 November 1957	Sputnik 2	USSR	First animal in orbit: Laika
31 January 1958	Explorer 1	USA	First American satellite in orbit; detected radiation belt
17 March 1958	Vanguard 1	USA	Satellite orbit used to measure shape of Earth
26 March 1958	Explorer 3	USA	First use of tape recorder in space
15 May 1958	Sputnik 3	USSR	Geophysical and astrophysical measurements
26 July 1958	Explorer 4	USA	ARPA Project Argus: mapped radiation
7 August 1959	Explorer 6	USA	First deployable solar arrays
19 August 1960	Sputnik 5	USSR	Recovery of dogs, Belka and Strelka, from orbit
7 March 1962	OSO 1	USA	First astronomical observatory in space (Orbiting Solar Observatory)
16 March 1962	Cosmos 1	USSR	First of long-running Cosmos (Kosmos) series
6 April 1962	Cosmos 2	USSR	Radiation belt and cosmic ray data
26 April 1962	Ariel 1	UK/USA	Ionospheric satellite; first satellite designed by one nation and launched by another
29 September 1962	Alouette 1	Canada	Ionospheric satellite; first Canadian satellite
5 September 1964	OGO-1	USA	First Orbiting Geophysical Observatory
15 December 1964	San Marco 1	Italy	First Italian satellite
26 November 1965	A1 (Asterix)	France	First French satellite
29 November 1965	Alouette 2	Canada	Ionospheric satellite
8 April 1966	OAO 1	USA	Orbiting Astronomical Observatory; battery failure on second day
5 May 1967	Ariel 3	UK	First all-UK satellite
29 November 1967	WRESAT 1	Australia	First Australian satellite
17 May 1968	ESRO 2B		First successful European Space Research Organisation (ESRO) satellite
7 December 1968	OAO 2	USA	Payload: 11 telescopes and two scanning spectrometers
15 December 1968	HEOS		Highly Eccentric Orbiting Satellite (2400 km × 223,000 km); studied magnetic fields (ESRO)
8 November 1969	Azur 1 (GRS A)	Germany	First German satellite
11 February 1970	Osumi	Japan	First successful Japanese launch of indigenous satellite
24 April 1970	DFH 1	China	First Chinese launch of indigenous satellite (Dong Fang Hong – 'The East is Red')
21 August 1972	OAO 3	USA	Also known as Copernicus

The idea for the IGY was spearheaded by renowned American geophysicist Lloyd Berkner in the spring of 1950, apparently during an informal gathering of fellow scientists at the Maryland home of James Van Allen [3]. Among the group was British geophysicist Sydney Chapman who, with Berkner, set the idea in motion. By 1953, an international Organising Committee for the IGY had been established, with Chapman and Berkner as president and vice-president, respectively.

Although one naturally thinks of geophysics as a predominantly ground-based activity, even in the 1950s there were some members of the geophysics community whose interests lay in the upper atmosphere, which was difficult to research because of its inaccessibility. Indeed, 'outer space' was seen by some as an exciting extension of the geophysical environment and, since the technology to launch a satellite was becoming available, it seemed logical to include this aspect of investigation in the IGY. Among those individuals was Fred Whipple, who had served on the V-2 Upper Atmosphere Panel, which coordinated the use of the V-2 for sounding rocket experiments in the 1940s; he chaired the IGY technical panel on rocketry [4].

As a result of the influence of those early space scientists, the IGY organising committee, which met in Rome in 1954, made the following recommendation:

> In view of the great importance of observations during extended periods of time of extraterrestrial radiations and geophysical phenomena in the upper atmosphere, and in view of the advanced state of present rocket techniques, the Special Committee for the International Geophysical Year recommends that thought be given to the launching of small satellite vehicles [5].

The phrase 'extended periods of time' is significant in that until satellites could be placed in stable, semi-permanent orbits, observation times were limited to the few minutes available from suborbital sounding rockets. So although space science could be said to have begun shortly after the Second World War, when America and Russia began to experiment with German V-2 rocket technology, experiments were extremely transitory in nature and far from easy to repeat – hardly the ideal prerequisites for good science!

Despite the enthusiastic and, without doubt, politically directed language of the recommendation in citing the 'advanced state of present rocket developments', a rocket capable of launching a satellite into orbit had yet to be developed. Moreover, although America thought of itself as the *de facto* leader in the field, it had yet to decide which satellite to launch. As described in Chapter 2, there were two candidate satellites and launchers on offer from competing teams: the US Navy's Vanguard satellite and identically named launcher; and the US Army's Explorer satellite and the Juno I launcher being built under von Braun's supervision. The satellite programme eventually chosen to contribute to the IGY was Vanguard, and contribute it did, but it was beaten into orbit by an external competitor, the Soviet Union.

Sputnik 1

The world's first artificial satellite, Sputnik 1 (Figure 3.1), was launched on 4 October 1957 at 6.00 am Moscow time and the event was announced by Radio Moscow

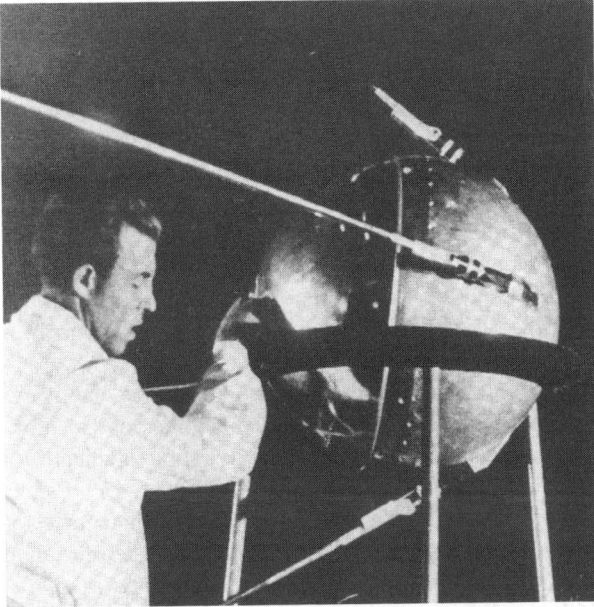

Figure 3.1 The first artificial satellite, Sputnik 1 [NASA]

that evening. Sputnik was the Russian word for satellite, but is often more lyrically translated as 'traveller' or 'fellow traveller' [6].

In Washington, at a reception for delegates of an IGY conference, Lloyd Berkner was able to congratulate the happy Russian contingent who, in typical style, pleaded ignorance of the details [7]. It soon became clear, however, that the launch of Sputnik 1 was a significant accomplishment: it was an aluminium-alloy sphere, measuring 58 cm in diameter and weighing 83.6 kg. Compared with anything America had planned, it was huge.

Many in the West found it hard to believe that the Soviet Union had managed to launch a satellite, but proof was provided by the Jodrell Bank radio telescope in England, which was brought quickly to operational status especially to track the mysterious object.

The launch took America in particular by surprise. Satisfied with its lead in many products of technology, from motor cars to the atom bomb, the nation found it difficult to believe that its superiority was threatened. Such was America's complacence that it had ignored a catalogue of announcements which could have prepared it for Sputnik:

- As early as 1948, Colonel Gregory Tokayev, a Russian rocket expert who had sought refuge in England, expressed the conviction more than once that the Soviets intended to place a satellite in orbit [8].
- On 27 November 1953, Prof Alexander Nesmeyanov, the president of the Academy of Sciences, declared at the World Peace Council in Vienna that science

had 'reached a stage where it makes sense to speak of sending a rocket to the moon and of creating an artificial earth satellite' [8].

- In May 1954, George P. Sutton of North American Aviation (the company which would later build the Apollo command module) told the audience of a space travel symposium at New York's Hayden Planetarium that the Soviets would soon be capable of launching a satellite which could 'bombard' the Earth with 'radio propaganda' [8].
- In July 1955, two days after America's intentions to launch a satellite for the IGY had been announced, Prof Leonid Sedov, the head of the Interdepartmental Commission for Interplanetary Communications, revealed during the sixth IAF Congress that the Soviet Union was planning a satellite which, in his opinion, could be launched within two years [8].
- In September 1956, at an IGY preparatory conference, another member of the Academy of Sciences, Ivan Bardin, repeated that the Soviet Union would launch a satellite within the IGY, even going as far as listing its scientific objectives [9].
- In October 1956, a Moscow news programme discussed a possible satellite [10].
- In May, June, July and August of 1957, Soviet radio amateurs were given instructions on how to build a suitable radio receiver 'to listen for an artificial moon which would broadcast on wavelengths of 7.5 and 15 m' [9].
- In June 1957, Nesmeyanov had announced in the Soviet press that the Russians would launch a satellite within the next few months [9], while the IGY committee had been told that the satellite was ready [10]; and in August 1957, the USSR had proudly reported the successful firing of its ICBM [10].

Although the earlier announcements may have come too early, considering the poor opinion America had of Soviet technology, and some may not have been disseminated beyond the academic environment in which they originated, one would have thought that sufficient uncertainty remained for the US president's advisors to err on the side of caution and recommend an acceleration of their nation's own satellite programme.

Since then, however, a possible explanation has emerged, which presents the American administration as well aware of the possibilities: if it allowed the Soviet Union to set a precedent by over-flying foreign nations with the Sputnik, there could be no objection to flights of American spy satellites over the Soviet Union. Since America's U-2 spyplanes were increasingly in danger of being shot down by Soviet surface-to-air missiles, the development of the spy satellite was a logical step. Of course, such satellites did not exist pre-Sputnik and were officially denied for many years after they were fielded, so it is difficult to be sure of the finer details of American policy at the time. The assertion that America allowed the Soviets to win the space race could represent a retrospective rewriting of history to save face. On the other hand, it could explain why von Braun's team was prohibited by the Pentagon from attempting a launch a satellite on a rocket that was ready to do just that in 1956.

Whatever the current uncertainties, the public announcements made in 1957 were clearly aimed at downplaying the Soviet achievement. Even after Sputnik's bleeping signal had been heard across the western world, President Eisenhower referred to it disparagingly as 'one small ball in the air, something that does not raise my appre-hensions, not one iota'. Leaping to his president's assistance, in what seemed like a lame attempt at damage limitation, White House aide Clarence Randall called Sputnik 'a silly bauble . . . a bubble in the sky'. Randall was reassured by Queen Elizabeth and Prince Philip, who visited Eisenhower shortly after Sputnik's flight and said, he reported, that 'People in London gave it one day of excitement, and then went about their business' [10].

But the American public had a less relaxed attitude: they wanted to know why America had failed to beat the Soviets into space. Undoubtedly, one of the reasons was the lack of political support to those who had been calling for a programme of space exploration for more than a decade. For America, Sputnik 1 was the ultimate 'wake-up call'.

In the context of the part of the American space programme that did exist – the launch vehicle – the size and mass of Sputnik 1 was a key marker, and American designers knew only too well that their relatively puny launch vehicles came nowhere near the benchmark set by the Soviets. Placing the mass of a fully loaded refrigerator in orbit was something America could only dream of at that point.

Korolev's design bureau, OKB-1, had begun the development of the satellite, officially designated Simple Satellite No. 1 (PS-1), in November 1956. It was originally intended that it, and its sister PS-2, would weigh about 100 kg and be launched in April or May 1957, after the R-7 ICBM had been flight-proven, but problems with the development of the R-7 delayed the launch of the first Sputnik to October.

Sputnik 1 is now recognised as both an historic achievement and a political tool: not only was its launch intended to beat America into space, its transmission fre-quencies had been chosen to match the receiving capabilities of existing professional and amateur radio receivers (20.005 MHz and 40.002 MHz, corresponding to wave-lengths of 15 m and 7.5 m, respectively) [7]. The characteristic rhythmic bleeping of Sputnik 1 became one of history's most evocative sounds, ranking alongside Morse's dots and dashes, and Edison's phonographic rendition of 'Mary had a little lamb'.

The satellite's four radio antennas, or 'whip aerials', were 2.4 m and 2.9 m in length, and with the spherical body completed an external design that would become a visual icon in the history of science and technology (Figure 3.2). Their technical function was to transmit radio signals on the two frequencies: a 0.3s-pulse on one frequency followed by a 0.3 s pause, then a similar pulse on the other frequency and a similar pause. The transmission power was about one watt [11].

Although Sputnik 1's radio transmitter payload was primarily designed to get the satellite noticed, it also made it possible to convey data on the internal temperature of the spacecraft by varying the duration of the signal pulses and the pauses between them. This was, in effect, the first attempt at spacecraft telemetry (a word – meaning 'measurement at a distance' – which is particularly apposite for spaceflight). More-over, it also allowed a degree of serious research into the propagation of radio signals through the atmosphere, which would earn it the honour of being the first orbiting space science experiment.

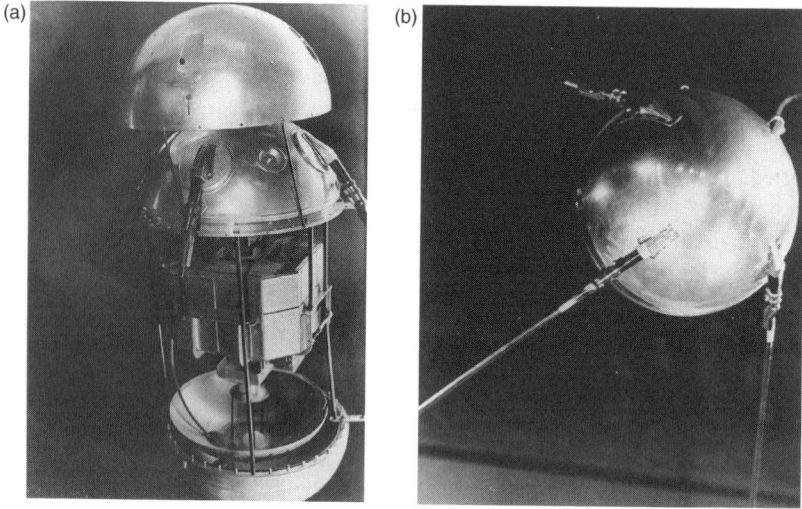

Figure 3.2   Sputnik 1, inside (a) and out (b) [NASA]

As an indication of how much information could be gained from so little hardware – at least in the days of relative ignorance of space and atmospheric science – Sputnik 1's carefully machined and highly polished spherical surface allowed an additional line of research. Measurements of the differences between the optical and radio 'rising and setting' of the satellite made it possible to determine the distortion of the radio beam by the ionosphere and to deduce its electron content [12,13].

Of course, the materials from which the satellite was constructed were not chosen solely for political purposes: the highly polished surface of the aluminium-alloy sphere was a key component of its passive thermal control subsystem by virtue of its efficiency in reflecting incident heat from the Sun. In addition, since the density of aluminium is about half that of steel, it allowed designers to keep the weight of the spacecraft – although significant – to a minimum. Thus its structure amounted to about 14 kg, some 17 per cent of the total [11].

Pure aluminium is a relatively soft metal, but as the German metallurgist Alfred Wilm discovered in 1906, it alloys easily with other metals to produce a more durable material. In fact, aluminium alloyed with copper and magnesium, known as 'Duralumin' after the Durener Metal Works which bought the patent from Wilm, has a long pedigree as a light-weight aerospace metal: it was used for the zeppelin airships that conducted bombing and reconnaissance missions in World War I [14].

Apart from its structure, a significant contributor to a spacecraft's mass budget is its power[2] subsystem. Sputnik 1's power source was a chemical battery – a silver–zinc accumulator – which, according to one writer, accounted for some 38 per cent of its

---

[2] To the physicist, power is the rate of doing work (measured in W or J/s) and energy is the capacity for doing work (measured in J). However, in engineering and everyday use the word 'power' (measured in watts) generally describes a commodity that can be generated, stored and used. It is used in this way here.

total mass [15]. Another quotes the mass of the 'power source unit' (which one might translate as power subsystem, of which batteries are a part) as 51 kg, which is over 60 per cent of the total [11]. Although Sputnik 1 was unique, the significant proportion of the mass budget that must be devoted to the power subsystem is common to all spacecraft, and even in more modern satellites could account for up to 25 per cent of the total spacecraft mass [16].

Since batteries are, by their nature, relatively heavy devices, they should produce as much power per unit mass as possible. The mass-efficiency of a battery is charac-terised by its specific energy, measured in watt-hours per kilogram (W-h/kg). Early spacecraft batteries, of the zinc–mercuric oxide or silver–zinc types, had specific energies of about 30 W-h/kg, which compared very favourably with the 7 W-h/kg available from the lead-acid batteries commonly used in terrestrial equipment [17]. Had it been necessary to use the lead-acid formulation in spacecraft, the mass of the battery system would have increased four-fold over the typical silver-zinc system, significantly reducing the mass available for the payload. Later spacecraft carried the even more mass-efficient rechargeable nickel-cadmium (NiCd) battery, which was rated at up to 35 W-h/kg.

Sputnik 1's non-rechargeable battery allowed transmissions to continue for just 21 days, but this was long enough for it to accomplish its mission, and it re-entered the Earth's atmosphere, as planned, following some thirteen weeks in orbit.

Sputnik 2

America had barely recovered from the launch of Sputnik 1 when, just 1 month later, on 3 November 1957, the Soviet Union did it again. The second Sputnik was a much larger vehicle: its total mass was 508.3 kg, six times that of Sputnik 1. What was not widely realised at the time, however, was that the Sputnik 2 spacecraft remained attached to the central core stage of its launch vehicle, the two being thrown into orbit together. This meant that the total mass injected into orbit was some six-and-a-half tonnes [18], a staggering capability to those who realised it. That, however, was only part of Sputnik 2's achievement.

To the amazement of a world already fascinated or, in many cases, frightened by the advent of Sputnik 1, Sputnik 2 delivered a dog called Laika to orbit, thus arguably opening the field of veterinary space science. If Sputnik 1 had conducted science by default, Sputnik 2 was a true space science platform. In addition to its radio transmitter and a pressurised cabin for Laika, it carried a package of solar UV and X-ray detectors, mounted on a frame within the nosecone, and a cosmic ray experiment mounted on the body of the rocket itself.

The solar radiation experiment consisted of three photomultipliers mounted 120° apart and equipped with filters to separate the X-ray bands and the hydrogen line in the solar UV spectrum [19]. Photons striking the photomultiplier tubes produced electrical signals which were amplified and transmitted to Earth. To save power – a scarce resource in the days of the non-rechargeable battery – a photomultiplier and its instrumentation was activated only when an associated photoresistor or 'light sensor' indicated that it was facing the Sun; thus when the satellite changed its orientation relative to the Sun or went into eclipse, the instruments were deactivated.

The two identical detectors in the cosmic ray experiment worked by registering the scintillation (flash of light in a phosphor) caused by electrically charged particles. The impulses were counted by a semiconductor (triode-based) circuit which was designed to produce a signal when a predetermined number of cosmic rays had been detected.

All the scientific data were transmitted to Earth by a preprogrammed telemetry system, which was activated periodically and operated on the same frequencies as Sputnik 1. It was housed in a spherical container along with temperature sensors, a thermal regulation system and a power supply [20]. In fact, the spherical container was a virtual replica of Sputnik 1, thus marking the first use of one of the tenets of spacecraft design, which is to use tried and tested equipment wherever possible. Today, we refer to this proven technology as a spacecraft's or subsystem's 'design heritage'.

Sputnik 2 transmitted for one week as planned: the 20 MHz signal was radiated with an average duration of 0.3 s, which was varied slightly to indicate the variation of temperature and pressure within the spacecraft; the 40 MHz signal was continuously modulated by data from the scientific experiments which was transmitted in a predetermined order. This included details of Laika's pulse, respiration, arterial pressure and heart function as well as the temperature and air pressure within her cabin [21]. The cabin was even equipped with a slow-scan TV system, transmitting 200 lines at ten frames-per-second, which was used to monitor Laika [22].

Sputnik 2 was a remarkably complex spacecraft for the time, not least because of the Soviet advantage of extremely potent launch vehicles, which made payload mass much less of a constraint. As a direct result, in stark contrast with America's early satellites, it was possible to consider payloads with relatively high power requirements, since the mass of batteries required could be easily accommodated. Of course, the power resources still had to be carefully managed. The power required for Sputnik 2 was several times that of the first Sputnik, not only because of the extra scientific equipment, but also because of the life support system for Laika. Among other things, this comprised an oxygen supply, a waste absorption system and a number of feeding troughs [21].

Sputnik 2 made more than 2300 orbits before re-entering the Earth's atmosphere on 14 April 1958. The fate of Laika, dubbed Muttnik by the American press, has provoked questions ever since: although it was intended to administer a lethal injection before the cabin air ran out, Laika is believed to have died from heat exhaustion (following a failure of the environmental control system) on the fourth day of the mission [23].

Explorer 1

As a result of delayed political commitment to a satellite launcher, and the failed attempt to launch a Vanguard in December 1957 (which the media characteristically dubbed 'Flopnik' or 'Kaputnik' [24]), America had to wait almost four months after the launch of Sputnik 1 to see its first satellite in orbit. That satellite, Explorer 1, was launched on 31 January 1958.

Since America's rockets were so much smaller and less powerful than the Soviet vehicles, its satellites too were significantly smaller and lighter. Thus Explorer 1 weighed just 8.2 kg, compared with Sputnik 1's 83.6 kg, although complete with the expended solid propellant rocket, the mass delivered to orbit by the Juno I launch vehicle was some 14 kg [25].

Interestingly, the seminal photograph showing William Pickering, James Van Allen and Wernher von Braun triumphantly holding aloft an engineering model of what is usually described as 'Explorer 1' in fact shows the satellite attached to its Baby Sergeant upper stage. A picture of this famous trio attempting to hold the 75 cm-long, 15.3 cm-diameter Explorer alone would have made a much less dramatic visual statement (Figure 3.3).

Despite its small size, Explorer 1 carried a reasonably advanced payload. Mounted at the tip of the pencil-shaped spacecraft was an external thermometer, which recorded temperatures varying between $+60°$ and $-30°$C, depending on the Sun's illumination of the satellite; another thermometer inside recorded a range of 10–40°C. The payload also included a microphone to detect micrometeoroid impacts (about one every 15 days), a cosmic ray detector and a Geiger counter experiment. The Geiger tubes were the most significant part of the payload, since they were instrumental in discovering what became known as the Van Allen radiation belts, but the impact results can now be seen in a new light. The same experiment repeated today would almost certainly record a greater number of impacts, because of the significant increase in man-made orbital debris since 1958 (the unfortunate result of unintended rocket stage explosions and stage separation debris, to name but two sources).

One can imagine that including all this payload, miniaturised as it was, in such a small cylinder left little room or mass margin for batteries. However, Explorer 1's zinc–mercuric oxide accumulators were sufficient to operate one of the two radio transmitters for 15 days and the other for almost four months; their radiated powers were just 60 mW and 10 mW, respectively. What the Americans lost against the Russian competition in the ability to launch large payloads, they gained in component miniaturisation and longevity of operation. Whereas Sputnik 1 had transmitted for 21 days and Sputnik 2 had expired within just seven, Explorer 1 continued to transmit data until 23 May 1958, almost four months after its launch [26].

Despite Explorer 1's successes, its mission began with a problem when, shortly after separation from the launch vehicle, it became unstable. The concept of stabilisation is of crucial importance to most spacecraft, since it allows all or part of the spacecraft to maintain its attitude with respect to the Earth or some other planetary or astronomical body. In an attempt to provide the gyroscopic stability of the classical spinning top, Explorer 1 was spun about its long axis before separation from the launch vehicle.

Although theoretical analyses of rigid-body motion had indicated that such spin stabilisation would maintain its stability in orbit, Explorer was not entirely rigid: its four flexible whip-like antennas could vibrate, thus dissipating kinetic energy. As

Figure 3.3   *Proudly holding a model of Explorer 1 and its upper stage: (left to right)*
William H. Pickering, former JPL director, James A. Van Allen of the
State University of Iowa and Wernher von Braun. A model of the launch
vehicle stands on the desk [NASA]

a result, within 90 minutes, the spacecraft's longitudinal spin had transformed into
rotation about a transverse axis – a 'flat spin'. This was deduced from a modulation,
or periodic 'fade-out' of the received signal. The event was, nevertheless, useful
in informing the field of spacecraft stability, since it showed that spacecraft spin-
stabilisation could only be sustained if the intended spin axis coincided with the axis
of maximum moment of inertia.

Vanguard 1

When the US Navy finally solved its problems with the Vanguard launcher programme, and succeeded in launching the Vanguard 1 satellite (on 17 March 1958,
some 5 months behind Sputnik 1), it was a cause for celebration in the homeland.

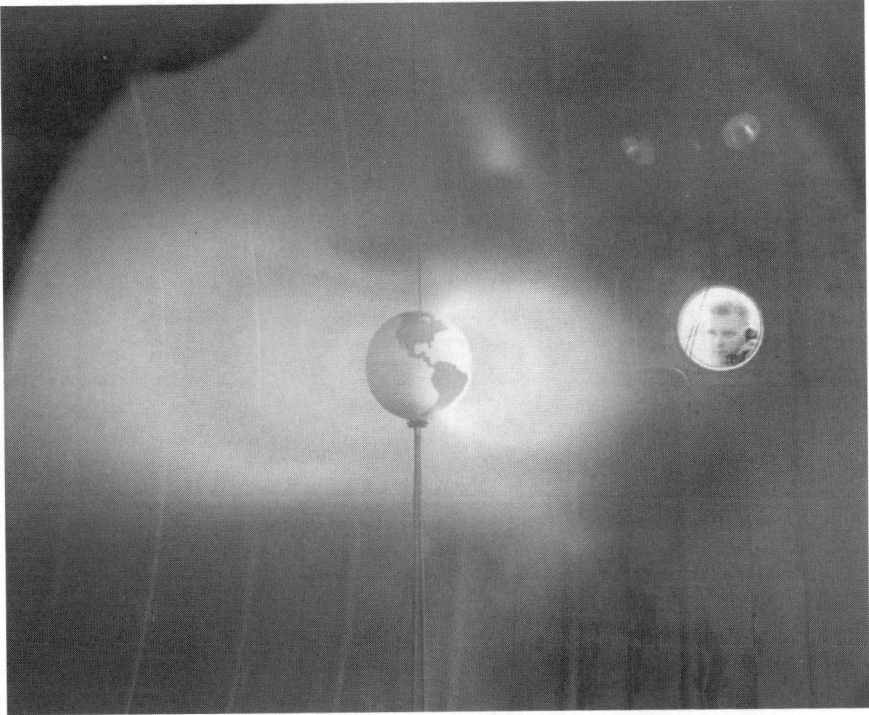

Figure 3.4   Simulation of Earth's Van Allen radiation belts generated by a plasma thruster at NASA's Lewis Research Center [NASA]

However, at 16 cm in diameter and 1.5 kg in weight, it was even smaller than and not one-fifth the mass of Explorer 1. Considering that Sputnik 2 had been more than 300 times heavier, the spherical Vanguard 1 – nicknamed the 'grapefruit' by Soviet Premier Khrushchev [27] – was a political gift to the Soviets (Figure 3.5).

Nevertheless, Vanguard 1 represented a significant advance from the engineering point of view, in being the first satellite to derive its power from solar cells. The first practical photovoltaic cell had been developed at AT&T's Bell Telephone Laboratories in 1954 and Vanguard 1 presented the first opportunity to fly it in space. The 34 cells were mounted behind six tiny protective windows spaced around the sphere, and produced just 10 mW of DC (direct current) power [28]. At the time, these groups of cells and their accompanying rechargeable batteries were termed 'solar batteries'. The term arose because, in the early days of the Space Age, the majority of launch vehicle and spacecraft power was derived from non-rechargeable batteries, and the novelty of the solar cell led people to distinguish between a battery charged on the ground and one continually recharged by an array of solar cells [29].

As far as its payload was concerned, Vanguard 1 carried two radio transmitters, one powered by mercury batteries, the other by the solar cells. The only data

Figure 3.5   Vanguard 1, 'the grapefruit', showing solar cells [NASA]

transmitted – a temperature reading which varied between 20°C and 90°C – were transmitted on a frequency of 108.012 MHz via four stub aerials on the surface of the sphere [28]. A problem was, however, that no cut-off switch could be provided, because of the severe mass constraints, and the satellite continued to broadcast its trivial data until May 1964, some six years after launch [26]. This prompted one critic to dub Vanguard 1 an 'inexhaustible and unwelcome babbler' [28]. Ironically, had the satellite been confined to the power from a non-rechargeable battery, rather than carrying the innovative solar cells, it would have ceased transmitting far sooner.

On the other hand, there was an advantage to be had from the satellite's 'babbling' which led to a significant scientific discovery. Tracking its transmissions throughout its extended lifetime permitted an excellent study of the changes and perturbations of the satellite's orbit, leading to important conclusions on the shape of the Earth. Among other things, the study of Vanguard 1's motions produced an accurate figure for the flattening of the poles and showed that the southern hemisphere was larger than the north. This led to such headline writers' gems as 'Grapefruit says Earth is Pear-shaped' [30]. More importantly, it was a justification of the foresight of those geophysicists who had proposed that satellites could be useful for studying the Earth.

Deep Space Network

The tiny amount of DC power available to the early satellites such as Explorer and Vanguard placed a limitation on the level of RF (radio frequency) power that could be radiated as a signal carrier; thus a satellite's signal was little more than a faint whisper from the cosmos. The solution was to build extremely large receiving dishes, or earth stations, which would act as a 'giant ear', collecting and amplifying the extremely low-power signals received from spacecraft.

In effect then, it was the mass and power constraints on spacecraft technology that led NASA to establish a network of tracking and communications stations around the world known as the Deep Space Network (DSN). Although 'deep space' is generally defined as space beyond the vicinity of the Earth–Moon system (the radius of the Moon's orbit being approximately 380,000 km), the DSN is also used for spacecraft in lunar orbit, highly elliptical Earth orbits, and for close-Earth fly-bys of interplanetary spacecraft using the planet's gravity to boost their velocity in a so-called 'gravitational slingshot' manoeuvre [31].

The DSN evolved from the tracking and data recovery experience of the Jet Propulsion Laboratory (JPL) gained in work for the US Army in the 1950s. Specifically, it had developed Microlock, a phase-locked loop tracking system developed from early guidance research for the Corporal missile programme that could 'lock onto' very low power signals.

For the Explorer 1 mission, JPL expanded its initial network of two Microlock tracking stations – in California and Florida – in cooperation with the British IGY committee and new stations were established in Nigeria and Singapore. Later in 1958 the Advanced Research Projects Agency (ARPA), which had been established within the Department of Defense to handle the military space programme and any eventual civilian space projects, authorised JPL to establish the DSN – at that stage known as the Deep Space Instrumentation Facility (DSIF) – beginning with a 26 m antenna at Goldstone, in California's Mojave Desert. The network did not become fully operational until it was used to track Mariner 2 to Venus in 1962.

The original contract called for the construction of three large earth station complexes, separated by approximately 120° of longitude: one at Goldstone (Figure 3.6); one at Robledo de Chavela, west of Madrid, Spain; and one near the Tidbinbilla Nature Reserve, south-west of Canberra, Australia. The geographical separation allowed respective stations to hand over control to the next as the spacecraft receded over the local horizon.

Each of the three complexes was situated in semi-mountainous, bowl-shaped terrain to shield against terrestrial radio interference and comprised at least four deep space stations, one with a 70m-diameter parabolic reflector, one with a 26 m reflector (mainly for Earth orbit operations), and two 34 m antennas (for both Earth orbit and deep space applications), all of the fully steerable type.

The physical size of the DSN antennas presented a number of engineering challenges above and beyond the RF hardware. For a start, in order to detect signals of such low magnitude an antenna needs not only to be large, but also to be extremely

*Figure 3.6   Goldstone Pioneer station antenna: (a) under construction; (b) com-
pleted [NASA]*

accurate in terms of its overall shape and surface profile relative to the wavelength
of the radiation. If not, incident radiation will not be brought to a single focus and
waves reflected from different parts of the dish will not be in phase. And since very
large parabolas will also tend to distort under gravity as they track from horizon to
horizon, they must have extremely rigid support structures.

In addition, antennas must be pointed precisely at the source and maintained by continuous monitoring of the signal strength. This is no mean feat, but in order to communicate with the early satellites, the techniques had to be developed quickly in parallel with the design of the space-based hardware.

The DSN was designed to operate in the two main bands used for spacecraft telemetry, tracking and command: S-band and X-band (around 2 GHz and 8 GHz respectively). Telemetry includes any spacecraft-systems data, or scientific instrument readings and images, modulated onto an RF carrier signal and transmitted by the spacecraft to Earth. Tracking of a spacecraft is typically accomplished by detecting and monitoring the carrier signal or a dedicated RF beacon and, in the case of deep space missions, allows the range and velocity of a spacecraft to be determined. Command data are modulated onto a carrier of a different frequency and transmitted from a DSN station to the spacecraft.

In terms of amplifying the very weak signals from spacecraft, the DSN's 70 m antennas were designed to have a receive-gain of some 63 dB at S-band and 74 dB at X-band (amplification factors exceeding a million and ten million, respectively). From the antennas, the signals were fed to low noise amplifiers (LNAs) which were cryogenically cooled, using liquid helium, to virtually eliminate thermal noise from the amplifier itself and thereby improve the signal-to-noise ratio. The technologies of large antennas and low noise amplifiers were also applicable to radio astronomy, where the object was to receive faint radio signals from distant astronomical sources, and it was this experience on which the DSN and other systems were built.

However, spacecraft communications systems also had to operate in the opposite sense – transmission – since not only were the early spacecraft incapable of transmitting high-power signals, they were also incapable of receiving low-power signals. This meant – and to a large extent still means – that the onus was on the earth station on the transmit or 'uplink' path as much as the receive or 'downlink' path.

As a result, for the transmission of commands to spacecraft, the earth stations' amplifiers had to be the most powerful available. So the 34 m DSN stations use high-power klystron amplifiers, specified at 4 kW for X-band and 20 kW for S-band, while the 70 m stations boast the Network's 'ghetto blasters', radiating 400 kW of radio-frequency power [32].

Sputnik 3

America had been forced to play a 'catch-up' role following the Soviet Union's pre-emptive strike with Sputniks 1 and 2, but this did not mean that the USSR was resting on its laurels. The West's habit of publishing scientific results in the open literature, coupled with America's wish to portray its capabilities in the emerging field of space science, produced immediate improvements in Soviet spacecraft.

An attempt was made on 27 April 1958 to launch what the Soviets called 'Object D', but resonant vibrations in the launch vehicle caused it to break up and the satellite was lost [26]. When the 'third artificial satellite', later named Sputnik 3,

Figure 3.7   The United Kingdom's Jodrell Bank radio telescope was used to track many of the early spacecraft, including the Sputniks, since it was larger and more sensitive than any other 'deep space' antenna of the time [Mark Williamson]

reached orbit on 15 May 1958, it incorporated two technological features which first appeared on American satellites. The first was a complement of instruments to detect the Van Allen belts discovered by Explorer 1; the second was a solar energy collection system [33]. Given the former aggressive launch policy of the Soviets, and the fact that six months had passed since the launch of Sputnik 2, it seems likely that Sputnik 3 was delayed to fit these instruments [34].

Continuing the Soviet trend towards ever larger spacecraft, Sputnik 3 was a 3m high aluminium-alloy cone measuring 1.73 m across its base and weighed 1327 kg, of which 968 kg was payload [35]. Panels of silicon photovoltaic cells, distributed about the outer surface of the spacecraft so that power could be generated regardless of the satellite's orientation, were used to recharge its silver–zinc batteries. Each cell produced 0.5 V, at a conversion efficiency of 9–10 per cent. The inclusion of the solar cell gave the spacecraft much greater longevity, allowing some of its equipment to function after two years in space.

The spacecraft also carried a simple electronic computer, which programmed the measurements, switching equipment on and off in a given order, stored the data in memory, and fed then to the transmitter at the appropriate moment [36]. It also controlled an innovative temperature regulation system subsequently used on Russian manned spacecraft. It comprised a number of louvres on the outside of the satellite which opened and closed in response to the internal heat balance, exposing or covering thermal radiators as required.

As with its forerunners, Sputnik 3 carried several scientific experiments designed to make both geophysical and astrophysical measurements, as well as transmitting spacecraft engineering data. Such was the haul of data from Sputnik 3 that memoranda were still being published in Soviet scientific journals some four years after its demise on 6 April 1960, when it re-entered the atmosphere [37].

Like Sputniks 1 and 2, telemetry was again transmitted on 20 and 40 MHz: the 0.25 W 40 MHz transmitter operated for only a month, but the 20 MHz transmitter worked for the full, two-year life of the spacecraft. The latter transmitter, named Maiak (for 'beacon'), marked an increase in the complexity of spacecraft telemetry systems, needed because of the increasingly sophisticated payloads satellites were required to carry. It radiated four separate signals: the first was 300 ms in duration and invariable; the second was either 50 ms or 150 ms depending whether the spacecraft was operating from batteries or solar cells; the third and fourth had durations of 50, 100 or 150 ms depending on the intensity of cosmic radiation or light flux [38].

This represents a good example of the iterative nature of spacecraft system design: Sputnik 1 transmitted a fixed duration signal on the standard frequencies; Sputnik 2 varied or modulated the signals slightly; and Sputnik 3 radiated a far more complex telemetry signal. Using this method, designers could build on what they knew would work in space and make step-by-step improvements, another example of design heritage.

Technology development

It is important to note that these early spacecraft were representative of the state of the art in just the first eight months of the Space Age. However, once the gateway to space had been pushed open, there was no stopping the development of ever more complex spacecraft, which were launched with increasing frequency. The recorded number of launches, including failures, in the first four years of the Space Age was as follows: three in 1957; 23 in 1958; 25 in 1959; and 40 in 1960 [39].

In addition to the trail-blazers discussed above, the count included, from America, the world's first weather satellite (Tiros 1), the first passive and active communications satellites (Echo 1 and Courier 1B respectively), and the first navigation satellite (Transit 1B). The USSR's achievements, in the same period, included the launch of the first spacecraft to fly beyond the Moon (Luna 1), the first to impact the Moon (Luna 2), and the first to image the far side (Luna 3); it also launched the first prototype manned spacecraft (Sputnik 4), which in its operational version, Vostok 1, would carry Yuri Gagarin into orbit in April 1961. These and other missions are described in the context of their applications in following chapters.

Although the story of spacecraft development is, by necessity, dominated by the 'space race' between the United States and the USSR, it is also important to note that, as with launch vehicles, other nations were interested in exploring the new frontier. Although the leaders had the field to themselves for the first few years of the Space

Figure 3.8   Ariel 1 [NASA]

Age, they were soon joined by the United Kingdom, which accepted an offer made in 1959 by the United States to COSPAR (the international Committee on Space Research established by the International Council of Scientific Unions (ICSU) in 1958). In effect, the US offered to launch scientific equipment or complete satellites designed and supplied by COSPAR member nations.

As a result, the British National Committee for Space Research (BNCSR) came to an arrangement with NASA to design and launch three ionospheric satellites designated UK-1, UK-2 and UK-3, the first two of which would be built in the United States and the third in Britain; the payloads themselves, which included X-ray and cosmic ray detectors, were to be designed and manufactured by British universities and industry. Thus it was that UK-1, renamed Ariel 1 in orbit, became the first 'international satellite', designed by one country and launched – on 26 April 1962 – by another (Figure 3.8) [40].

Less helpful in terms of international collaboration was the US Starfish high-altitude thermonuclear explosion test conducted in July 1962, which created artificial radiation belts through which Ariel 1 was obliged to pass; as a result, its p/n-type solar cells sustained severe damage and functioned only intermittently thereafter. UK-2 and 3 were subsequently equipped with n/p-type cells[3] which were found to be more resistant to radiation [41].

After the United Kingdom, the next country to develop its own satellite was Canada, which launched Alouette 1 on an American Thor-Agena vehicle on 29 September 1962. As with Ariel 1, Alouette was an ionospheric research satellite designed in collaboration with NASA's Goddard Space Flight Center in Maryland. Later in the decade, other countries such as Italy, France and Japan joined the fray, but the great majority of spacecraft were designed, manufactured and launched by the United States and the Soviet Union. As a result, the majority of the advances in space technology were made either by, or in collaboration with, those nations. This led to a number of standards in electronic components, measurement and communications devices and power subsystems, which are described below.

Electronic components

Electronics – which can be defined as 'the science and technology concerned with the development, behaviour and applications of electronic devices and circuits' [43] – was, and still is, fundamental to spacecraft development. Given the limitations of the early launch vehicles, the miniaturisation of devices to control spacecraft payloads and other equipment became a key driver of this development, and as electronic components were made smaller and lighter the complexity of spacecraft increased.

The application of electronics to spacecraft was, in essence, no different from its application to terrestrial devices, as far as they used the same types of components and circuits. In the early days of spaceflight, electronic circuits – in space and on the ground – used resistors, capacitors, diodes and transistors mounted on printed circuit boards. The differences lay in the way spacecraft electronics, as a discipline, was required to cope with the constraints placed on it by the space environment, and the applications which electronic components and subsystems were required to support in various types of spacecraft. Of course, these differences remain to this day.

The key design constraints on electronic components are led by the two different environments to which they are subjected: first, the launch vehicle environment, which subjects a payload to noise, vibration and acceleration forces, while also placing strict limits on the size and mass of the payload; and second, the space environment, with its inherent vacuum, microgravity, temperature and radiation characteristics.

Although components and subsystems must be designed to withstand the rigours of the launch environment, and particularly delicate components are ruled out for

---

[3] Silicon was the predominant semiconductor used in solar cells in the 1960s. The addition of phosphorus (a process known as 'doping') results in an n-type semiconductor; doping silicon with boron produces a p-type material ('n' stands for negative and 'p' for positive). Ingots of pure silicon doped with boron are cut into disks and phosphorus is diffused into the top surface to make a semiconductor (p/n) junction across the surface of the cell; an n/p junction results from the reverse process [42].

use on spacecraft, the far more serious constraints are those of size, mass and power consumption, as shown by the development of the early satellites.

As far as the space environment is concerned, the near-perfect vacuum is in one sense an advantage, since it improves component isolation, thereby reducing the separation required for high voltage components. Microgravity, or 'weightlessness', has no effect on electronic subsystems, although it may affect the mechanical systems to which they are connected. However, the inherent lack of convection in the microgravity environment means that excess heat can only be removed from components by conduction and eventual radiation to space. This makes temperature an important constraint in space systems design and the thermal engineer an important member of a spacecraft design team. Although the exterior of a spacecraft may vary between $-70°C$ and $+140°C$, depending on materials and illumination, the interior is typically required to remain at 'room temperature' (15–20°C on average), because the equipment inside is designed, built and tested in facilities maintained at this temperature.

In the early years of spaceflight, radiation was less of a problem than it is today, especially since it was a nascent subject of scientific investigation in itself. Later, as the feature size of microprocessor components decreased, they became more susceptible to radiation effects such as 'bit-flips', which can switch equipment on or off at random. As a result, solutions such as 'radiation hardening' and shielding became more important.

Although it is true to say that all of the early spacecraft required reliable electronic components that would survive the launch and be capable of operating in the relatively harsh environment of space, the basic components and circuits were no more exotic than those developed for harsh terrestrial environments. Then, as now, components were selected by bench-testing, to confirm the specification and weed out the 'weaker' samples, and then tested in environmental chambers to 'space qualify' individual flight-model components. What the space industry did, however, was provide a welcome stimulus to the process of component miniaturisation, which had already begun.

According to one engineer writing in the early 1960s, satellite equipment was already being produced with component densities of 300–1500 units per cubic inch (18–90 units per cubic centimetre), saving space, weight and power. Contrasting the packaging density of pre-war valves and similar components (at one or two per cubic inch) with later developments, he highlighted the density of 'transistors and sub-miniature components' at $20–30/in^3$, 'micro modules (components on wafers)' at $200–400/in^3$ and 'micro circuits (printed circuit functions on wafers)' at $1500–6000/in^3$. The holy grail at the time appeared to be 'solid circuits: fashioned circuits on semiconductor wafers' at a density of $20,000–60,000/in^3$ [44].

Another contemporary source states that in the 20 years between 1947 and 1967, the weight and bulk of electronics was reduced in a ratio of about 20,000:1, which supports the above. In 1964, for example, 110 circuits of five different types (representing 1100 transistors and 4200 resistors) could be assembled on a 1 inch silicon disc [45].

Considering the mass and volume limitations of spacecraft, space exploration as we know it would have been impossible without the transistor and other miniature components. This is not to say, however, that space exploration *per se* would have been impossible. Given the political impetus behind the early launches, it is possible that development would have continued to the point where large space stations stacked with valves and vacuum tubes were commonplace.

Indeed, Arthur C. Clarke and other space advocates, writing before the Space Age began, assumed that 'communications satellites' would be large manned space stations. 'The station could be provided with living quarters, laboratories and every-thing needed for the comfort of its crew...', wrote Clarke in his seminal *Wireless World* article entitled 'Extra-Terrestrial Relays' [46]. 'It could be provided with receiving and transmitting equipment... and could act as a repeater to relay trans-missions between any two points on the hemisphere beneath, using any frequency which will penetrate the ionosphere'. Although Clarke did not go into details of subsystem design, it was widely assumed that a crew would have to be on board to operate and maintain the spacecraft, mainly because the necessary levels of component reliability and control system autonomy were unimaginable before transistorisation.

Interestingly, had development proceeded in this manner, the Soviet Union would have been much better placed than America to implement such a station, because its rockets were far more capable. On the other hand, continuing the same speculative vein, had component miniaturisation not occurred, it seems likely that America would have developed its own heavy-lift rockets earlier than it did. What we can conclude from this is that it is always difficult to devise alternative histories with any degree of reliability.

Basic devices

As shown by the examples of early spacecraft, space technology relied on a num-ber of essential, basic electronic devices which had already been developed for use on Earth. These included sensors or 'probes', such as thermocouples (for converting temperature readings to voltages) and piezo-electric crystals (which produce a signal proportional to pressure); photoelectric cells sensitive to visi-ble, IR and/or UV wavelengths; and microphones for the detection of sounds or impacts.

The outputs of sensors and science payloads were transmitted to Earth using a specialised communications circuit – the telemetry subsystem. In common with any communications system, a spacecraft telemetry system is built from modules of filters, amplifiers and mixers. The key element of the mixer (used for frequency conversion) is the signal generator, which is also used to produce waveforms (sine wave, square wave, saw tooth, etc.) to characterise the information conveyed from sensors and other subsystem equipment.

A spacecraft telemetry system is usually combined with a chain of electronic components designed to receive commands from Earth-based mission controllers, the combination being known as a telemetry, tracking and command (TT & C) or telemetry, command and ranging (TC & R) subsystem. The tracking or ranging element is engineered by a ground station[4] which locks onto the spacecraft's telemetry transmissions or, in larger spacecraft, a dedicated radio beacon.

One of the main problems in communicating with the early spacecraft was the low altitude of their orbits – a function of launch vehicle capability – and the resultant limited visibility above the local horizon. In keeping with the laws of physics, satellites in low altitude orbits have shorter orbital periods than those in higher orbits, which means that they spend less time above the horizon from a given earth station's position. For example, Sputnik 1 was launched into an elliptical orbit with a high point (apogee) of 939 km and a low point (perigee) of 215 km; this gave it an orbital period of 96.2 min. The apogee and perigee of other early satellites varied considerably, but their periods were generally between 90 and 110 min (an exception was Vanguard 1, which because of its high apogee, 3897 km, had a period of 133.6 min) [26].

What this meant was that a typical satellite would only be within range of a given earth station for 15 or 20 minutes on each orbit and, until the DSN and similar systems became available, this placed the satellite out of contact. This led the Soviet Union to deploy special tracking ships in various ocean locations to keep tabs on its spacecraft.

Since command signals could only be transmitted to the early satellites during limited sections of their orbital periods, they often carried timers to supply control signals at predetermined times. Moreover, the fact that they could often not transmit their data to a ground station in 'real time' also drove the need for onboard tape recorders; they were inherently unreliable because of their complexity, but necessary nevertheless. The first use of a tape recorder in space was on Explorer 3, which was launched on 26 March 1958 [47], while Courier 1B, a US Army communications satellite launched in October 1960, carried five tape recorders, one for voice and four for Teletype.[5] It could transmit up to 800,000 words of text in the 14 minutes it was in view of the ground station, but unfortunately failed after 17 days [48].

Alfred Bester, in his quirky little book *The Life and Death of a Satellite*, describes the two tape recorders on NASA's Orbiting Solar Observatory (OSO), which was launched in 1962 (as described below). The recorders each contained some 250 m of tape in a continuous loop and were designed to record at slow speed for 90 minutes, then dump their data in 4 minutes when the spacecraft was over the ground station. 'But how do you get a tape recorder running on a single watt of power for six months, and only measuring eight inches by four inches?' asks one of Bester's sources rhetorically. Apparently the subcontractor made a fixed price bid of $33,000 for each recorder and 'lost twice as much as he bid' [49]. The story illustrates concisely that designing equipment for use in space is far from easy, as many contractors have since discovered to their cost.

---

[4] The term ground station is more or less synonymous with earth station, and is used that way here, but it is worth mentioning that the term for any future installation on another planetary body would probably be ground station.

[5] Teletype was a trademark for a make of teletypewriter (US) or teleprinter (UK), a telegraph apparatus consisting of a keyboard transmitter that converted a typed message into coded pulses, which were 'typed out' at the receiving end. A later international service using this technology was called telex (for teleprinter exchange).

Bester also discusses the extreme care with which manufacturing components for space use must be conducted, albeit with a somewhat unorthodox slant. The danger of mechanical stress and strain on circuit boards during manufacture, he writes, 'is so critical that women are not permitted to work on delicate components during their menstrual periods; engineers dare not run the risk of subjecting the components to the extra acidity of women's skin at those times of the month'. In classic contemporary style, he adds that women are well suited to delicate work, but 'display a very human weakness. After six or eight months on the job many of them become bored by the niggling work, get careless, and have to be fired' [49]. The case for men is not recorded!

Power

Once in orbit, many of the early satellites were expected to operate only for days or weeks before re-entering the Earth's atmosphere or suffering some sort of electrical failure. This minimal lifetime requirement meant that power could be derived from non-rechargeable storage batteries based on a technology already developed for aircraft and rockets. As a result, the first spacecraft were powered by batteries alone.

However, shortly after Vanguard 1's first flight of a solar cell in space, solar arrays featuring large numbers of electrically coupled photovoltaic cells became the primary power source for the vast majority of spacecraft, mainly because a battery would rapidly become discharged if relied upon for any significant period. For this reason, the battery subsystem is often still referred to as the secondary power supply.

Thus within less than a year of the dawning of the Space Age, the basic power subsystem technology for spacecraft had been established. Energy derived from solar cells was used to power both payload and support subsystems while the spacecraft was in sunlight, and to recharge a battery that would power at least some of the equipment when the spacecraft moved into the Earth's shadow (which for satellites in very low-altitude orbits was about half the time).

Apart from being relatively short-lived, the earliest spacecraft carried fairly simple payloads, which meant that their electrical subsystems were equally simple. Improvements in the capacity and reliability of power sources allowed an increase in spacecraft complexity and lifetime to the point where what was once a simple electrical supply became a subsystem in its own right.

However, as spacecraft designs began to feature ever more complex packages of scientific equipment, meteorological sensors and telecommunications payloads, demands on the power subsystem became greater and greater. Physical limitations on the size of solar arrays meant that severely limited energy supplies had to be managed efficiently – using the concept of the power budget – and distributed effectively to a fast-growing range of spacecraft hardware using different voltages. In the same way that domestic battery-operated equipment requires batteries of different voltages, spacecraft equipment must be designed to operate within a certain range of 'bus-voltages'.

Throughout the 1960s, spacecraft development in the West was largely dominated by the United States, which meant that almost any spacecraft subsystem developed by US companies became the *de facto* standard for the industry. Thus the power subsystems for the first satellites of the European Space Research Organisation (ESRO), designed and launched in the late 1960s, were derived directly from US concepts. The operating voltage, at 16 V DC, was low, but considering the relatively low power demands of the spacecraft, this was acceptable. The main distribution bus for ESRO I, as with most American satellites, was unregulated (which means that the output voltage was variable, dependent on the power source and electrical load), but since the subsystem was relatively simple (with about 1000 components compared to as many as 20,000 in a mid-1980s satellite) this was felt to be acceptable [50].

However, whereas American industry had set the initial standards, it was Europe that sought to push for further development. The launch of ESRO's TD-1 satellite in 1972 marked an interesting milestone in power subsystem design. Its design engineers recognised that a raw power source which varied between 19 V and 25 V was 'not particularly user-friendly' [50] and offered a *regulated* 16 V line. More significantly, they also proposed an alternating current (AC) supply of 18 V for the first time. Unfortunately the low frequency of the supply, at 2400 Hz sine wave, made it heavy and inefficient, and although the experimenters were satisfied, the concept of distributing AC power was shelved for a decade or so [50].

It was not until the early 1970s, with the development of the COS-B scientific satellite, that it was felt necessary to abandon the 16 V standard in favour of a system that would accommodate the continually increasing power demands of contemporary spacecraft. Thus 28 V, which had already been used on communications satellites, became the accepted standard for scientific spacecraft built on both sides of the Atlantic.

The only way spacecraft could become more complex, and more useful, was through the development of high-capacity power supplies featuring photovoltaic cell arrays and relatively high voltage distribution (later rising to 50 V and beyond). This allowed a rapid increase in the power available to spacecraft hardware and instruments, and the development of spacecraft for a wide range of applications, from high-power communications to radar imaging. Thus the development of spacecraft electrical systems has been fundamental not only to the development of spacecraft but to the development of the space industry itself.

Space astronomy

The rapid increase in the frequency of launches in the early 1960s mirrored the widening interest in a growing number of space applications. Far from limiting their interests to geophysics and space radiation, scientists and engineers were pushing forward on all fronts, from earthbound navigation to planetary exploration. Both civilian and military applications were being pursued and manned spaceflight was also under rapid development.

Many of these developments – especially those involving astronauts and cosmonauts – tended to push the scientific applications of space technology into the background. When men were floating outside capsules in low Earth orbit, few people (led, it must be said, by the whims of the media) had an interest in automated scientific recording instruments. As a result, the development of the next generation of satellites designed to investigate the space environment went largely unnoticed, a trend that was only reversed by the introduction of the Hubble Space Telescope in 1990.

The early space telescopes conducted astronomical observations in a field now known as space astronomy, that is astronomy performed from space rather than from astronomical observatories on the Earth. Its main advantage over terrestrial astronomy is its elimination of the deleterious effect of the Earth's atmosphere, which not only absorbs some wavelengths of radiation but also distorts the images of stars and other objects of interest. The distortion problem, the most obvious effect of which is the 'twinkling' of stars nearest the horizon, has been of note since the earliest days of astronomy, but the solution did not become available until the dawn of the Space Age.

One of the chief protagonists of the space telescope was Lyman Spitzer, a Princeton University astronomer and member of the IGY's technical panel on the 'Earth Satellite Program'. As early as 1946, Spitzer wrote a paper on the advantages of a space-based telescope, effectively laying the groundwork for his future career and those of many others. In fact, it is said that the origins of the planning for the much later Hubble Space Telescope can be traced back to this paper [51]. In tribute, NASA renamed its Space Infrared Telescope Facility (SIRTF) the Spitzer Space Telescope in December 2003, following a naming contest which produced more than 7000 entries [52].

Orbiting Solar Observatory

The first astronomical observatory in space was NASA's Orbiting Solar Observatory, OSO-1, which was launched on 7 March 1962, just two weeks after John Glenn had become the first American to orbit the Earth. It is hardly surprising that few outside the space astronomy community noticed the event.

OSO-1 was built under contract to NASA's Goddard Space Flight Center by Ball Brothers Research Corp, an example of a company that evolved to meet the challenges and opportunities of the Space Age. It was founded in 1880 by five brothers in Buffalo, New York, as a producer of wood-jacketed, tin oil cans and, in its first half century, developed into a manufacturer of glass containers, battery cans and other similar products, eventually becoming a modern-day packaging supplier for the food and beverage industries. In 1956, however, the company diversified into the high-technology market with the formation of its Research Corporation, based in Boulder, Colorado, and quickly became established as a spacecraft manufacturer.

The OSO satellites were designed to expand on the work of the long-running Explorer series. Previous science spacecraft had been designed as individual instrument carriers, with no attempt at standardisation: it had been largely a matter of

Figure 3.9   OSO 1 [NASA]

bolting detectors to a structural frame and adding solar cells and antennas to convenient attachment points. Often, the only similarity between successive spacecraft was
their name. OSO was NASA's first standardised observatory spacecraft, of which
nine similar models were constructed.

Structurally, it was more complex than most previous satellites, comprising a
nine-sided base section about a metre across and a fan-shaped upper section mounted
perpendicular to the base (Figure 3.9). The base section formed the standardised
platform on which a variety of payloads could be mounted. Thus, although OSO-1
weighed only 206 kg at launch (in 1962), by the time OSO-8 was launched in 1975,
the standardised platform (though adapted to the form of a cylinder) was supporting
a spacecraft weighing 1052 kg.

The distinction between the functions of platform and payload would later lead to
a modular concept for spacecraft design, in which the platform (containing the main
support subsystems) would be termed the service module and the section containing
the instruments the payload module. The distinction would also allow an innovation in spacecraft stabilisation known as the dual-spin design, a concept eventually
applied to a variety of different types of spacecraft, including communications and
weather satellites, which soon adopted a standardised cylindrical form covered in solar
cells.

Although a simple spacecraft could be stabilised by imparting a spin, this meant
that payloads requiring a fixed view of anything were impractical. Alternatively, if
a spacecraft was long and narrow, it could be gravity-gradient stabilised by making use of the decrease in gravitational field-strength with distance from the Earth's

Figure 3.10   OSO 7 [NASA]

centre; as long as its main axis pointed towards the Earth, the difference in grav-itational attraction between the two ends of the spacecraft acted to maintain its orientation. Again, however, the design concept was of limited use for more complex spacecraft.

Following a suggestion made by Fred Singer in 1954 [53], later spacecraft were spin-stabilised using an on-board propulsion system, either in their entirety (often with their spin axes aligned with the Earth's polar axis) or in part, with a spin-stabilised service module and de-spun payload module (the dual-spin design). The simplest propulsion systems used a gas – such as nitrogen, argon, freon or propane – as the propellant. It was stored at high pressure and fed, through a pressure-reducer, to an expansion nozzle to produce thrust. This type of thruster, termed a cold-gas thruster since no combustion or heating of the propellant was involved, was known colloquially as a 'gas jet'.

The design concept of the OSO-1 spacecraft incorporated both dual-spin stabili-sation and cold-gas thrusters. Its base section or service module was spin-stabilised, while a de-spun payload module allowed its instruments to remain pointing towards the Sun. Three arms, complete with spherical glass-fibre propellant tanks containing pressurised nitrogen gas, were deployed from the base section once the satellite was in orbit and the nitrogen-powered thrusters imparted a spin rate of 30 rpm [54].

Eventually, as with any other technical development, it was felt necessary to improve the efficiency of propulsion systems by substituting higher energy propellants and by heating those propellants. The most commonly used thruster in the 1960s and 1970s used the chemical compound hydrazine (N_2H_4) as a propellant. It was stored as a liquid and usually fed to the thruster by a pressurant such as gaseous nitrogen or helium. It was then decomposed to ammonia, nitrogen and hydrogen with the aid

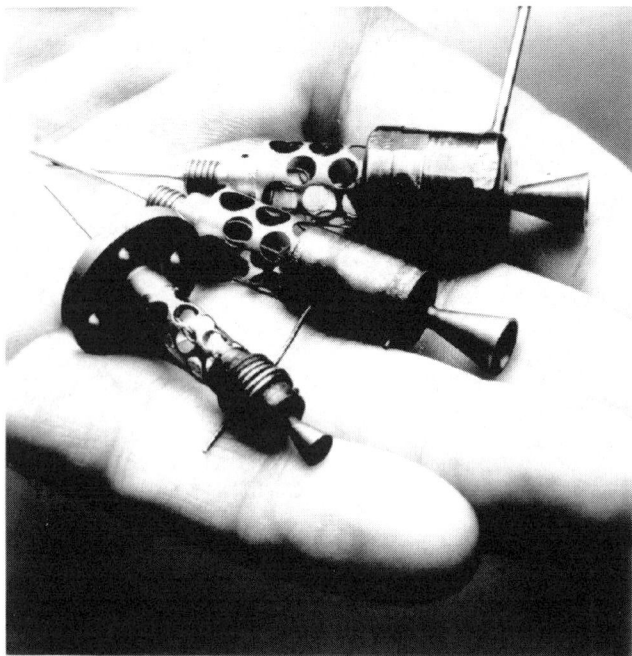

Figure 3.11   A handful of hydrazine thrusters [EADS, formerly British Aerospace, Space and Communications Division]

of a catalyst in a combustion chamber and the resultant hot gases were ejected from a conical nozzle (Figure 3.11). Later variations utilised electrical heaters to increase the exhaust velocity and performance of the thruster.

Of course, the most important part of OSO's attitude control design was the part that kept its payload – comprising X-ray, UV and IR instruments – pointing towards the Sun. Since it was designed to face the Sun, the sun-pointing side of the fan-shaped payload section was the obvious place to mount the array of solar cells, which provided about 38 W of power for OSO-1 (and some 110 W for OSO-8) [55]. It is interesting to note, for comparison purposes, that 110 W is little more than the power consumed by an average light bulb here on Earth, an indication of the power constraints on spacecraft designers and the supreme efficiency of space-based equipment.

The spacecraft's spin axis was kept perpendicular to the solar vector, for the benefit of the payload and the solar array, by magneto-torquer coils in the base section. Magneto-torquers operate on the principle of physics which states that a freely-suspended, current-carrying coil of wire will align itself with a local magnetic field: in this case, the device acts to realign itself with the Earth's field, thereby damping out any small disturbances in a satellite's attitude; larger disturbances are corrected by thrusters.

In addition, the OSO spacecraft carried a gyroscope in its upper section which acted as a memory to ensure that it acquired the Sun quickly after emerging from the Earth's shadow [56]. A spinning gyroscope has this 'memory' because its spin axis tends to remain fixed in inertial space – fixed relative to the stars – which means that it can be used to detect rotation in both spacecraft and launch vehicles (its application to the inertial platform was discussed in Chapter 2).

In common with most other spacecraft of the time, the OSO spacecraft were injected into low Earth orbits (around 500–600 km in altitude), which meant that they orbited into the Earth's shadow on every revolution. Although solar observations were, of course, impossible during these eclipse periods, the temperature of the spacecraft and the operation of its subsystem equipment had to be maintained. For this reason, the observatories carried rechargeable nickel–cadmium batteries to provide power during eclipse [54].

OSO-1, stationed in a near-circular orbit inclined at 33° to the equator, operated for nearly 2 years, during which time it transmitted data on more than 140 solar flares [56]. The data from its 13 experiments were tape-recorded on board the satellite and transmitted to Earth in a 5-minute period on each orbit. Later spacecraft in the series carried more complex and more capable payloads and subsystems. Notable among their achievements was the provision, by OSO 4 (launched in October 1967), of the first complete UV map of the Sun [57] and the discovery that solar flares have temperatures of over 30 million degree Celsius and can release as much energy as the Earth uses in 100,000 years [56].

Although the OSO satellites undoubtedly qualified as space science spacecraft, their discoveries were not limited to pure science. High energy protons and other particles emitted by the Sun are not only dangerous to astronauts, but also act to expand the Earth's atmosphere and increase its drag on orbiting satellites. This was realised as early as the late 1950s when data on the orbit of Sputnik 3 were correlated with solar activity: it was found that the drag on the satellite increased every 27 days or so, when the Sun's rotation brought a large solar flare into line with the Earth [58]. Of course, the Sun's activity also affects the Earth's magnetic field, causing aurorae, and in recent years has been identified as the culprit in several satellite failures since, as semiconductor chips are made smaller, they become more prone to the charges deposited by solar particles.

With only one of the nine OSO missions failing in an unsuccessful launch attempt, the OSO series was an extremely successful introduction to the new regime of spacecraft standardisation and improved reliability.

Orbiting Astronomical Observatory

Unfortunately, the series of astronomical spacecraft which followed the first two OSOs was initially less of a success. The first Orbiting Astronomical Observatory (OAO-1), launched on 8 April 1966, suffered a battery failure on its second day in orbit [59] and was declared a 'write-off' after an overheating problem caused failures in other equipment.

It was more than two years before its replacement, OAO-2, was launched on 7 December 1968, the same month that the Apollo 8 astronauts orbited the Moon. Once again, space astronomy had been upstaged. Nevertheless, the OAO was a sophisticated satellite carrying no less than eleven telescopes and two scanning spectrometers enabling it to observe stars in the infra-red, ultra-violet, X-ray and gamma-ray parts of the spectrum [60]. It was easily 'the Hubble Space Telescope of its time'.

Having realised that a spin-stabilised platform – like that of OSO – could not provide the stability required for accurate astronomical observations, NASA's Ames Research Center designed and built a full-sized prototype of a three-axis stabilised platform: instead of spinning a large part of the spacecraft to provide stability, its attitude would be controlled by a number of rotating wheels, known generally as inertia wheels, mounted inside the spacecraft and aligned with each of its three axes. Rotating these wheels by means of electric motors would, by the law of conservation of angular momentum, rotate the body of the satellite in the opposing direction. For a sun-pointing satellite, the instructions or inputs for the wheels' operation would normally be derived from sun sensors or trackers, while for an Earth-pointing satellite an infrared Earth sensor would be used.

The use of 'flywheels' for attitude control was proposed by Ernst Stuhlinger in 1955 in a paper in *Jet Propulsion*, the Journal of the American Rocket Society; in this case, however, the suggested inputs for Earth-pointing correction signals came from Geiger counters arranged to detect the shadowing effect of the Earth on cosmic rays [61].

Three-axis stabilisation is now standard for the majority of manned and unmanned spacecraft, which use one of two types of wheel whose names reflect how the individual systems work. In the 'zero momentum' system, for example, a set of *reaction wheels* – one for each of the orthogonal spacecraft axes – can be rotated in either direction to provide full three-axis attitude control. It is common to install a fourth wheel, mounted at an angle to all three axes, as a spare, so that if one of the first three wheels fails, the fourth can be used in conjunction with the remaining wheels.

In the 'momentum bias' system, a constantly spinning *momentum wheel*, aligned with the spacecraft's north-south (pitch) axis, acts like a gyroscope to provide a resistance to perturbing forces in roll and yaw while allowing the spacecraft body to rotate about the pitch axis. An alternative system uses two wheels set at a slight angle to each other but symmetrically on either side of the north-south axis: run together the wheels will act as a double-sized single wheel, but changing the speed of one will alter the balance to provide a degree of attitude control [62].

Having completed the basic design, Ames commissioned Grumman Aircraft Engineering Corporation, the company that later built the Apollo lunar module, to manufacture the initial flight models of the OAO (Figure 3.12) [63]. The satellite itself was an octagonal box about 3 m high and 2 m wide with a mass of about 1.6 tonnes, a large satellite for its day. Telescope mirrors up to about a metre in diameter could be mounted within a 1.2 m-diameter cylindrical core section running the length of the satellite [64]. In fact, OAO-2 carried four 31 cm telescopes and four photometers, while OAO-3 (otherwise known as 'Copernicus'), which followed OAO-2 in 1972, carried a 82 cm telescope [65].

Figure 3.12   OAO under construction at Grumman [Northrop Grumman Corp., formerly Grumman Aircraft Engineering Corp.]

The body also contained the usual stabilisation, power, telemetry and command subsystems while two square solar panels were mounted, like wings, on the sides of the spacecraft (Figure 3.13). Once again, the power capability of the satellites, at up to about 1600 W depending on the version, was impressive for its day [65].

As spacecraft and their payloads became larger, their power requirements also grew and they carried increasing numbers of solar cells placed on any and every available surface to maximise the potential on-board power. When the surface area of the spacecraft proved a limiting factor to these 'conformal arrays', it became common practice to mount the arrays on deployable panels that were rotated into position once the spacecraft reached orbit. This was usually engineered by firing a pyrotechnic device to cut a cable, or cables, which tied the array to the satellite body (thus allowing it to fit within the rocket's payload fairing).

The arrays on the OAO satellites were of the basic deployable type, but they were not the first deployable solar arrays; they belonged to Explorer 6, which was launched on 7 August 1959. It was a roughly spherical satellite with four square panels which hinged out from the body at various angles to enable at least one of them to 'catch the sun' at any given time; its appearance earned it the nickname of 'paddle-wheel

Figure 3.13   OAO 2 complete with solar arrays [Northrop Grumman Corp., formerly Grumman Aircraft Engineering Corp., and NASA]

satellite'[66]. A similar paddle-wheel design was used by the Pioneer 5 deep space probe, launched in March 1960 (see Figure 6.3).

Although the solar array was but one part of many in a typical spacecraft, it was an expensive part. In the early 1960s, the efficiency of the silicon cell was between 8 and 14 per cent, depending on how much the purchaser wanted to pay. Since the cost of a single cell was between £2 and £3, according to one contemporary author, this could amount to up to half a million pounds for a complete satellite array and as much again for the protective coverslips [67]. According to another source, the development cost of the OAO was about $200 million [68], which puts the solar array costs in perspective.

The most important challenge in designing an astronomical satellite, however, is maintaining its pointing accuracy. In the same way that terrestrial telescopes rely upon the stability of their mounting, a space telescope – however good its optics – relies on the stability of the spacecraft platform. Moreover, a crucial aspect of this stability is the inherent stiffness of the platform itself. This meant that, from the outset, a satellite's structure came to be recognised as a subsystem in its own right.

In most cases, this led to a structural design based on the cylinder, a form which exhibits a natural strength along its main axis and, importantly, helps to protect a satellite from the forces, or structural loads, experienced during launch. As a result, the primary structure of a typical 1960s spacecraft consisted of a central cylinder surrounded by a number of triangular segments formed by platforms and panels made from light-weight aluminium honeycomb. The segments would typically contain key subsystem equipment, while the cylinder itself would often house propellant tanks or, depending on the design, a solid propellant motor used to boost the spacecraft to a higher orbit. The remaining structural elements, such as support brackets for propellant tanks and mountings for telescopes and other payload equipment, would form the secondary structure, thus completing the stable platform required for accurate pointing.

The angular resolution of terrestrial telescopes is limited, by turbulence in the atmosphere, to about 1 arcsec (or 0.2 arcsec for short periods under the most favourable conditions) [69]. In orbit, angular resolution translates, in effect, to angular stability because a space-based platform is not naturally stable (being pulled by gravitational effects and pushed by solar radiation pressure or the solar wind). So, in the development of space-based observatories, once platform stability problems could be overcome, the angular resolution could exceed that of ground based telescopes.

The OSO-1 platform had a pointing accuracy of about 1 arcmin (some 60 times worse than the arc-second precision of ground-based telescopes), but by the time OSO-8 was launched, technology had equalled the effective angular resolution of terrestrial telescopes [70]. For the more sophisticated OAO, this was improved by a factor of ten (to about 0.1 arcsec [71]), reaching 0.03 arcsec for OAO-3 [72]. Naturally enough, technology did not stop improving in the early 1970s: the Hubble Space Telescope has a pointing stability of 0.007 arcsec and ESA's Hipparcos satellite (an astrometry satellite designed to compile an accurate star-position catalogue) had an incredible 0.001 arcsec stability (better than one billionth of a complete circle).

Later developments aside, the achievement – with the OAO – of exceeding the angular resolution of terrestrial telescopes after only 11 years of the Space Age provided yet another demonstration that space was a good site for an astronomical observatory. Indeed, the OAO was the first satellite to provide 'versatile three-axis pointing' [73]. Its exemplary stability was produced in two stages, the first using star trackers and the second using the telescope itself.

The satellite's attitude was controlled using six gimballed star trackers (four around the body and one at each end), which fed error signals to inertia wheels mounted on each of the three axes [74]. The heritage of the star tracker dates back to MIT's work on its application to long-range bombers in 1945 and they were first used operationally in space on the Mariner 4 Mars probe in 1964; their use on OAO-1 marked their first application to Earth-orbiting satellites [75].

Since the pointing direction of each star tracker relative to the telescope axis (its gimbal angle) could be telemetered to the ground controllers, several trackers could be aligned with a given guide star and the telescope rotated to point towards the object of study. A potential difficulty occurred when an individual tracker reached the end of its gimbal movement, at which point it stopped tracking the star and slewed back to its centre position to select another guide star. To ensure that spacecraft orientation would not be lost, only one tracker was permitted to change guide stars at a given time [76]. Although this process could produce a baseline pointing accuracy of about half a minute of arc, this was still relatively crude compared to the arc-second precision of ground-based telescopes [77].

The figure of 0.1 arcsec was produced by using the telescope itself for fine guidance. Light from the main telescope mirror system was passed through an entrance slit to one of the payloads, known as the Goddard spectrometer. A small mirror, called a pick-off mirror, was mounted in front of the main spectrometer mirror to reflect light into the small optical system of the fine guidance telescope, which focused it onto a light modulator. The light modulator comprised two tuning fork scanners vibrating at slightly different frequencies (356 Hz and 455 Hz) at right angles to each other, one coincident with the pitch axis and one with the yaw axis of the observatory. Photomultiplier tubes detected the two modulated beams producing an electrical signal-input to the OAO's guidance and control system [78].

John Naugle, another early proponent of space astronomy and later NASA's chief scientist, illustrated OAO's 0.1 arcsec accuracy by pointing out that 'A dime 12 miles away subtends an angle of about 0.1 [arcsec]'. 'When you think about pointing something that weighs as much as an ordinary car at a dime 12 miles away', he went on, 'you can begin to see why we said the OAO was the most complex and difficult spacecraft in the space program' [79].

Apart from telemetry data from the star trackers and inertia wheels, overall verification of the observatory's pointing direction was derived from a small TV camera (with an accuracy of about 1 arcmin). In telescope terminology, this was akin to the wide-angle 'finder' telescope attached to most terrestrial instruments.

In addition to the TV pictures, the telemetry system was capable of transmitting digital data from both the payload and the subsystems. The OAO was particularly innovative in replacing the usual tape recorder with magnetic core memories, in its case having a storage capacity of 204,800 bits (equivalent to 8192 data words of 25 bits each) [80]. The memory would store data for downlink transmission when the satellite was in range of an earth station and, in the uplink direction, the system would receive pointing commands which could be stored for execution up to two hours later [81].

During its first month in orbit, OAO-2 gathered a total of 65 hours of astronomical data, including 20 times more information about the ultraviolet characteristics of stars than had been obtained during 15 years of sounding rocket flights [82]. According to one contemporary writer, 'Within 50 more days, it returned so much diverse information that scientists were bewildered' [83]. Designed for a one-year mission, the Observatory operated for four years, among other things making the first observations of UV emissions from the planet Uranus and providing new insights into the structure and composition of the Earth's upper atmosphere [72].

Once again the facility of an orbiting platform had proved itself. The OAO had carried a variety of instruments which could be repointed at will, and offered the possibility of relatively long-term, repeatable observations free from the distortion of the Earth's atmosphere and the vagaries of its weather. Lyman Spitzer's vision had been well and truly realised.

Since the OAOs were launched, there have been literally hundreds of space astronomy missions, culminating, as far as the general public and a large number of professional astronomers are concerned, in the well-publicised Hubble Space Telescope mission. The success of this and other space telescopes indicate that remote sensing of the Universe from Earth orbit continues to be an important application of space technology.

Conclusions

Although it is true that the Earth's radiation belts and pear-shaped figure would have been discovered eventually, by one space power or another, it is significant that they were discovered within the first few months of the Space Age by the simplest of satellites. Whether this indicates that simple scientific experiments are always worth doing, or only in the early days of a given branch of science, is a moot point. What is not at issue is the fact that actually placing scientific payloads in orbit is of benefit to the understanding of both terrestrial and extraterrestrial phenomena.

To enable these discoveries, it was necessary to develop spacecraft technology in a fast-track manner. Spacecraft electronics, particularly, has developed apace since the late 1950s, both stimulating and benefiting from parallel developments in terrestrial electronics. Improvements in the capacity and reliability of electronics subsystems have allowed an increase in spacecraft complexity and lifetime, while continual reductions in the size of components has meant that payloads of greater complexity can be shoe-horned into ever smaller volumes.

A direct result of this increase in packaging density is the more recent development of smaller, cheaper, but extremely capable satellites called microsatellites, which by definition weigh between 10 kg and 100 kg. Although this is small by commercial standards, so-called nanosatellites weighing less than 10 kg are already being developed. The continual development of spacecraft electronics is crucial to the success of these ventures, and may yet herald the return of the 'grapefruit' satellite ... at which point history will have turned full circle.

References

1 Williamson, Mark: 'The Cambridge Dictionary of Space Technology' (Cambridge University Press, Cambridge, 2001) p. 353
2 ibid., p. 176
3 Fraser, Ronald: 'Once Around The Sun: the Story of the IGY 1957–58' (Hodder & Stoughton, London, 1956) p. 97
4 Whipple, Fred L.: 'Recollections of Pre-Sputnik Days' in Ordway, Frederick, I. and Liebermann, Randy (Eds): 'Blueprint for Space' (Smithsonian Institution, Washington, 1992) p. 130
5 Fraser, Ronald: 'Once Around The Sun: The Story of the IGY 1957–58' (Hodder & Stoughton, London, 1956) p. 144
6 Siddiqi, Asif, A.: 'Challenge to Apollo: The Soviet Union and the Space Race, 1945–1974' (NASA SP-2000-4408, Washington, 2000) p. 168
7 Smolders, Peter, L.: 'Soviets in Space' (Lutterworth Press, Guildford, 1973) p. 73

8   ibid., p. 74

9   ibid., p. 75

10   Osman, Tony: 'Space History' (Michael Joseph, London, 1983) p. 38

11   Clark, Philip: 'The Soviet Manned Space Programme' (Salamander Books Ltd, London, 1988) p. 10

12   Stoiko, Michael: 'Soviet Rocketry: The First Decade of Achievement' (David & Charles, Newton Abbot, 1970) p. 82

13   Shternfeld Ari: 'Soviet Space Science' (Basic Books Inc, New York, 1959) p. 152

14   Williamson, Mark: 'The Communications Satellite' (Adam Hilger, Bristol, 1990) p. 32

15   Allward, Maurice (Ed.): 'The Encyclopedia of Space' (The Hamlyn Publishing Group Ltd, Feltham, 1968–69) p. 290

16   Hudson, Wayne R., and Gartrell, Charles F: 'Spacecraft Technology Trends: A View from the Past' (IAF-87-07): 38th IAF Congress, Brighton, UK, 1987

17   Ministry of Defence: 'An Introduction to Space' (Her Majesty's Stationery Office, London, 1968) pp. 49–50

18   Siddiqi, Asif A.: 'Challenge to Apollo: The Soviet Union and the Space Race, 1945–1974' (NASA SP-2000-4408, Washington, 2000) p. 174

19   Shternfeld Ari: 'Soviet Space Science' (Basic Books Inc, New York, 1959) p. 216

20   ibid., p. 153

21   Martin, Charles-Noel: 'Satellites Into Orbit' (George G Harrap & Co Ltd, London, 1967) p. 72

22   Siddiqi, Asif A.: 'Challenge to Apollo: The Soviet Union and the Space Race, 1945–1974' (NASA SP-2000-4408, Washington, 2000) p. 172

23   ibid., p. 174

24   Furniss, Tim: 'The History of Space Vehicles' (Grange Books, Rochester, 2001) p. 31

25   Emme, Eugene M.: 'Aeronautics and Astronautics: An American Chronology of Science and Technology in the Exploration of Space 1915–60' (NASA, Washington, 1961) p. 140

26   Thompson, Tina D. (Ed.): 'TRW Space Log 1996' (TRW Space & Electronics Group, Redondo Beach, Volume 32, 1997) p. 65

27   Shelton, William: 'Soviet Space Exploration: The First Decade' (Arthur Barker Ltd, London, 1969) p. 53

28   Martin, Charles-Noel: 'Satellites Into Orbit' (George G Harrap & Co Ltd, London, 1967) p. 79

29   Williamson, Mark: 'The Cambridge Dictionary of Space Technology' (Cambridge University Press, Cambridge, 2001) pp. 342–3

30   Martin, Charles-Noel: 'Satellites Into Orbit' (George G Harrap & Co Ltd, London, 1967) p. 80

31   Williamson, Mark: 'The Cambridge Dictionary of Space Technology' (Cambridge University Press, Cambridge, 2001) p. 88

32   Williamson, Mark: 'Extreme Contact', *IEE Communications Engineer*, 2004, 2(5), pp. 12–15

33 Smolders, Peter L.: 'Soviets in Space' (Lutterworth Press, Guildford, 1973) p. 79

34 Ley, Willy: 'Epilogue' in Shternfeld Ari: 'Soviet Space Science' (Basic Books Inc, New York, 1959) p. 348

35 Martin, Charles-Noel: 'Satellites Into Orbit' (George G Harrap & Co Ltd, London, 1967) p. 73

36 ibid., p. 75

37 ibid., p. 77

38 ibid., pp. 76–7

39 Thompson, Tina D. (Ed.): 'TRW Space Log 1996' (TRW Space & Electronics Group, Redondo Beach, Volume 32, 1997) pp. 65–9

40 Allward, Maurice (Ed.): 'The Encyclopedia of Space' (The Hamlyn Publishing Group Ltd, Feltham, 1968–9) p. 206

41 ibid., p. 207

42 Williamson, Mark: 'The Cambridge Dictionary of Space Technology' (Cambridge University Press, Cambridge, 2001) p. 343

43 Makins, Marian (Man. Ed.): 'Collins Concise English Dictionary' 3rd ed. (HarperCollins Publishers, Glasgow, 1992) p. 415

44 Pardoe, G. K. C.: 'The Challenge of Space' (Chatto & Windus, London, 1964) pp. 56–7

45 Allward, Maurice (Ed.): 'The Encyclopedia of Space' (The Hamlyn Publishing Group Ltd, Feltham, 1968–69) p. 274

46 Clarke, Arthur C: 'Extra-Terrestrial Relays', *Wireless World*, October 1945, pp. 305–7

47 Corliss, William R.: 'Scientific Satellites' (NASA SP-133, Washington, 1967) p. 59

48 Heppenheimer, T. A.: 'Countdown: A History of Space Flight' (John Wiley & Sons, New York, 1997) p. 173

49 Bester, Alfred: 'The Life and Death of a Satellite' (The Scientific Book Club, London, 1967) pp. 73–4

50 Capart, J. J. and O'Sullivan, D. M.: 'Uniform Power Distribution Interfaces for Future Spacecraft', ESA Bulletin, (42), May 1985, pp. 64–9

51 Roman, Nancy Grace: 'Exploring the Universe: Space-Based Astronomy and Astrophysics' in Logsdon, John, M.: 'Exploring the Unknown, Volume 5: Exploring the Cosmos' (NASA SP-2001-4407, Washington, 2001) p. 503

52 'NASA Renames SIRTF for Princeton Astrophysicist', Space Log, *Space News*, **15** (1), 5 January, 2004, p. 10

53 Singer, S. F.: 'A Minimum Orbital Instrumented Satellite – Now', *Journal of the British Interplanetary Society*, **13**, 1954, p. 74

54 Gatland, Kenneth W.: 'Astronautics in the Sixties' (Iliffe Books Ltd, London, 1962) p. 47

55 Caprara, Giovanni: 'The Complete Encyclopedia of Space Satellites' (Portland House, New York, 1986) pp. 160–1

56 Turnill, Reginald (Ed.): 'Jane's Spaceflight Directory, 1987' (Jane's Publishing Co Ltd, London, 1987) p. 90

57 Stubbs, P.: 'Research Satellites' in Fishlock, David (Ed.): 'A Guide to Earth Satellites' (Macdonald, London, 1971) p. 109

58 Gatland, Kenneth W.: 'Astronautics in the Sixties' (Iliffe Books Ltd, London, 1962) p. 39

59 Thompson, Tina D. (Ed.): 'TRW Space Log 1996' (TRW Space & Electronics Group, Redondo Beach, Volume 32, 1997) p. 98

60 Johnson, Nicholas: 'Space Science' in Gatland, Kenneth (Ed.): 'The Illustrated Encyclopedia of Space Technology' (Salamander Books, London, 1981) pp. 120–1

61 Stuhlinger, Ernst: 'Control and Power Supply Problems of Instrumented Satellites', *Jet Propulsion*, **26** (5 part 1) May 1956, pp. 364–8

62 Williamson, Mark: 'The Cambridge Dictionary of Space Technology' (Cambridge University Press, Cambridge, 2001) pp. 306–7

63 Gatland, Kenneth W.: 'Astronautics in the Sixties' (Iliffe Books Ltd, London, 1962) p. 49

64 Naugle, John E.: 'Unmanned Space Flight' (Holt, Rinehart and Winston, New York, 1965) p. 146

65 Caprara, Giovanni: 'The Complete Encyclopedia of Space Satellites' (Portland House, New York, 1986) p. 162

66 Allward, Maurice (Ed.): 'The Encyclopedia of Space' (The Hamlyn Publishing Group Ltd, Feltham, 1968–9) p. 188

67 Pardoe, G. K. C.: 'The Challenge of Space' (Chatto & Windus, London, 1964) p. 54

68 Naugle, John E.: 'Unmanned Space Flight' (Holt, Rinehart and Winston, New York, 1965) p. 144

69 Lemaire, Philippe: 'Science in Space: Solar Physics' in Rycroft, Michael (Ed.): 'The Cambridge Encyclopedia of Space' (Cambridge University Press, Cambridge, 1990) p. 217

70 ibid., p. 216

71 Allward, Maurice (Ed.): 'The Encyclopedia of Space' (The Hamlyn Publishing Group Ltd, Feltham, 1968–9) p. 244

72 Turnill, Reginald (Ed.): 'Jane's Spaceflight Directory, 1987' (Jane's Publishing Co Ltd, London, 1987) p. 88

73 Roman, Nancy Grace: 'Exploring the Universe: Space-Based Astronomy and Astrophysics' in Logsdon, John, M. (Ed.): 'Exploring the Unknown, Volume 5: Exploring the Cosmos' (NASA SP-2001-4407, Washington, 2001) p. 505

74 Naugle, John E.: 'Unmanned Space Flight' (Holt, Rinehart and Winston, New York, 1965) p. 148

75 Corliss, William R.: 'Scientific Satellites' (NASA SP-133, Washington, 1967) pp. 57–8

76 Naugle, John E.: 'Unmanned Space Flight' (Holt, Rinehart and Winston, New York, 1965) p. 149

77 Stubbs, P.: 'Research Satellites' in Fishlock, David (Ed.): 'A Guide to Earth Satellites' (Macdonald, London, 1971) p. 110

78　Naugle, John E.: 'Unmanned Space Flight' (Holt, Rinehart and Winston, New York, 1965) p. 154

79　ibid., pp. 154–5

80　Corliss, William R.: 'Scientific Satellites' (NASA SP-133, Washington, 1967) p. 382

81　Gatland, Kenneth W.: 'Astronautics in the Sixties' (Iliffe Books Ltd, London, 1962) p. 50

82　Von Braun, Wernher, Ordway, Frederick, I., and Lange, Harry H.-K.: 'Science of the Seventies: History of Rocketry and Space Travel' (Encyclopaedia Britannica Educational Corp, 1969) p. 188

83　Sharpe, Mitchell R.: 'Satellites and Probes' (Aldus Books, London, 1970) p. 122

Chapter 4

Looking at Earth – the development of the Earth observation satellite

Tell the people of Tierra del Fuego to put on their raincoats. Looks like a storm out there

William Anders, Apollo 8

Although the concept of using newly developed satellite technology to study space itself was fascinating to scientists of the late 1950s, it was evident that an orbiting satellite also offered a unique perspective of the Earth's surface and its atmosphere. As a result, satellite payloads were designed to take advantage of this new viewpoint and the sciences of Earth remote sensing and satellite meteorology were born.

Mankind's interest in the Earth as a planet dates back to the Ancients, whose scientists and philosophers strove passionately to understand its many mysteries. As far as observing their environment was concerned, few can have failed to note the advantage of elevated terrain in surveying a scene and placing it in context with the surrounding area. Indeed, the advantage of 'taking the high ground' has long been known to military leaders, hunters and other aggressors. However, in order to view the Earth as a planet, it was necessary to develop the technology to lift our viewpoint high above its surface.

The most obvious first step was the development of the passenger-carrying balloon, which is credited to the Montgolfier brothers, who made the first successful ascent in a hot-air balloon over Paris on 21 November 1783. Not surprisingly, only a few years later, during the French Revolution of 1789–99, balloons were deployed as military observation platforms. Moreover, it was inevitable that, once photography had been invented, the two technologies should be combined, as they were by Gaspard Félix Tournachon, who took the first aerial photograph while ballooning over Paris in the late 1850s [1]. He could hardly have imagined how the technology of Earth remote sensing, Earth observation, or Earth imaging as it is now commonly called, might develop in the twentieth century (Table 4.1).

Table 4.1   Selected Earth observation satellites

Launch date	Name	Nation	Comments
28 February 1959	Discoverer 1	USA	First satellite in polar orbit (Corona spy satellite disguised as biomedical science programme)
13 April 1959	Discoverer 2	USA	First recoverable capsule: ejected on orbit 17, but lost in Arctic (Corona)
7 August 1959	Explorer 6	USA	First photograph of Earth from space
13 October 1959	Explorer 7	USA	First satellite radiometer
1 April 1960	Tiros 1	USA	First dedicated weather satellite
10 August 1960	Discoverer 13	USA	First capsule recovery (Corona)
12 July 1961	Tiros 3	USA	First hurricane (Anna) detected by a satellite
26 April 1962	Cosmos 4	USSR	First announced Cosmos recovery
21 December 1963	Tiros 8	USA	First use of Automatic Picture Transmission (APT)
28 August 1964	Nimbus 1	USA	Earth observation technology demonstrator; first Advanced Vidicon Camera System (AVCS)
28 August 1964	Cosmos 44	USSR	Experimental weather satellite
22 January 1965	Tiros 9	USA	First photomosaic of entire Earth
3 February 1966	ESSA 1	USA	First in Tiros Operational System (TOS); operated by Environmental Science Services Administration
25 June 1966	Cosmos 122	USSR	First dedicated Soviet weather satellite
7 December 1966	ATS 1	USA	First meteorological payload in geostationary orbit (GEO): Spin-Scan Cloud-Cover Camera (SSCC)
5 November 1967	ATS 3	USA	Colour version of Spin-Scan Cloud-Cover camera
26 March 1969	Meteor 1-1	USSR	First Soviet operational weather satellite
14 April 1969	Nimbus 3	USA	First Satellite Infra-Red Spectrometer (SIRS)
23 January 1970	ITOS 1	USA	First Improved TOS
11 December 1970	NOAA 1	USA	First National Oceanic and Atmospheric Administration satellite
23 July 1972	Landsat 1/ERTS 1	USA	Return Beam Vidicon (RBV) camera and Multi-Spectral Scanner (MSS)
11 December 1972	Nimbus 5	USA	First Electrically Scanning Microwave Radiometer (ESMR)
17 May 1974	SMS 1	USA	First US dedicated GEO metsat: Synchronous Meteorological Satellite; Visible and Infrared Spin-Scan Radiometer (VISSR)
11 July 1975	Meteor II-01	USSR	First second-generation Soviet metsat
16 October 1975	GOES 1	USA	First Geostationary Operational Environmental Satellite (GOES)
14 July 1977	GMS 1	Japan	First Japanese metsat (GEO)
23 November 1977	Meteosat 1	ESA	First European metsat (GEO)
24 October 1978	Nimbus 7	USA	Total Ozone Mapping Spectrometer (TOMS)

In addition to our interest in the Earth's surface, the weather that occurs above it has always been a preoccupation. In ancient times it was used to indicate the mood of the gods or predict the abundance of a crop, and the ability to predict the weather was an important skill. In more enlightened times, when it became necessary to disseminate the results of a forecast to a widespread population, communications was the limiting factor. In the days of Thomas Jefferson and Benjamin Franklin, for example, the exchange of meteorological information was limited by the speed of the postal service, which meant that anything beyond short-term, regional forecasting was pointless. Communications improved with the development of telegraphy by Morse and wireless telegraphy by Marconi, but observation techniques were limited by what could be achieved from the ground or from kites and balloons.

Not surprisingly, the development of modern weather observation and forecasting techniques has been intimately related to the ability to observe large-scale meteorological phenomena from above. Thus, in turn, the aeroplane, weather balloon and meteorological rocket have each enhanced the efficacy of weather observation. The chief advantage of the rocket, in its ability to reach higher altitudes than the other technologies, was exemplified by the use of V-2s liberated from Germany by the US Army following the Second World War. A camera carried by a rocket in 1947 produced images of large-scale cloud formations that could not have been observed by any other means [2]. However, the V-2 and the rockets that followed could provide neither a stable nor a permanent observation platform. They travelled on a ballistic trajectory which brought them back to Earth after only a few minutes of observing time. A satellite in a permanent orbit was the only practical solution.

The first photograph of the Earth from space was taken by the Explorer 6 satellite, launched on 7 August 1959 [3]. Compared with later efforts, the image was extremely crude, but it illustrated well the potential of the satellite platform (Figure 4.1).

Weather satellites

The fact that the early satellite images were of cloud formations, weather fronts and other similar phenomena made meteorology one of the first practical application of the Earth observation satellite. Indeed, many meteorologists were convinced that satellites would revolutionise their subject, even before the first satellite was launched.

Among them were W. G. Stroud and W. Nordberg, who contributed a chapter on 'Meteorological Measurements from a Satellite Vehicle' to a book entitled *Scientific Uses of Earth Satellites*, edited by none other than James Van Allen [4]. 'No other technique could possibly do what will ultimately be possible from a satellite', they wrote, adding that they expected the satellite to be able to track weather systems by means of cloud movement. They also predicted 'that "bad weather" systems, such as hurricanes, [would] undoubtedly be observable'.

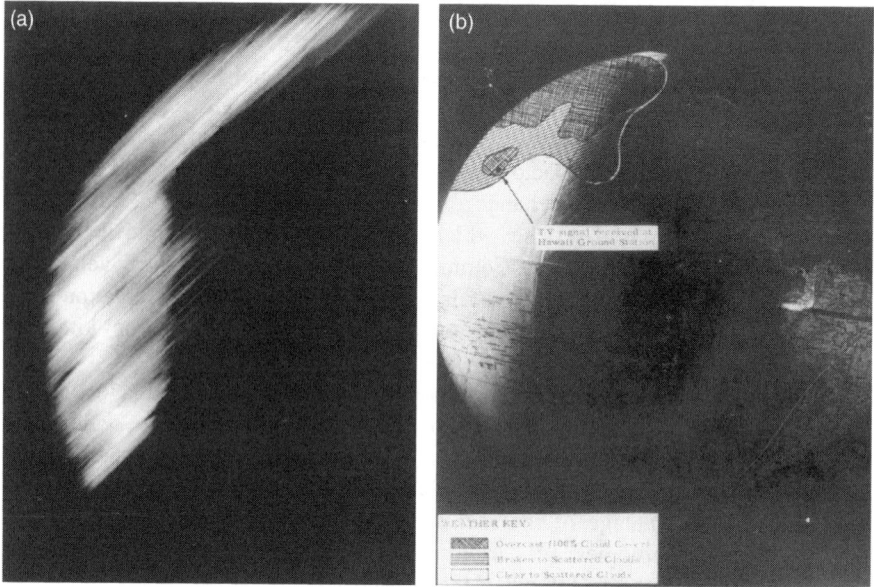

Figure 4.1   *(a) First image of Earth (from Explorer 6 on 14 August 1959) showing a sun-lit area of the central Pacific Ocean and its cloud cover, (b) by way of explanation, lined areas at left represent a cloud-cover map superimposed on a globe [NASA]*

Although such comments seem rather self-evident today, it should be noted that Stroud and Nordberg were writing at least two years prior to the launch of Sputnik 1 in 1957, and five years before the first weather satellite. At that time, as they lamented, less than five per cent of the Earth's surface was covered by observations from ground-based stations and specialised weather balloons, or radiosondes. Although weather reports were also available from ships at sea, oil platforms and permanently anchored vessels such as lightships, information about weather over the oceans was miserably inadequate – especially considering that most weather systems originate over the sea. It was the meteorologists' frustration at this and the realisation that space-based observation platforms could provide superior coverage that led to the development of a series of satellites dedicated to meteorology.

In the early days of spacecraft development, the predominant view was outwards, towards the conquest of the new frontier: in the two years following Sputnik 1, spacecraft were dispatched not only to map the Earth's radiation belts and photograph the previously unknown farside of the Moon, but also to carry a living creature into Earth orbit. It is interesting to note, however, how quickly the view was returned to the home planet: only 25 successful space launches had occurred before the launch of the first dedicated meteorological satellite, Tiros 1. Thus the weather satellite was an earlier innovation of the Space Age than is generally realised, even within the space community.

Tiros

The Tiros programme itself owes its origins to a 1951 paper by two research engineers with The Rand Corporation,[1] W. W. Kellog and S. M. Greenfield, which proposed taking photographs from a satellite and transmitting them to ground stations using a television signal [2].

It was not until May 1958, however, that a study of the Tiros system was begun by the Advanced Research Projects Agency (ARPA) of the US Department of Defense. In October 1958, the study was taken over by the newly formed National Aeronautics and Space Administration (NASA) and Tiros became a space project [2]. The satellites themselves were engineered for NASA by the Radio Corporation of America (RCA), which evolved into one of the country's leading satellite builders.

Tiros 1 was launched on 1 April 1960. It was the model for a series of 19 satellites orbited by America in the 1960s: the first ten were experimental in nature, in recognition of the immaturity of the new technology, and the remainder were part of an 'operational' system designed to guarantee the provision of weather data on a day-to-day basis. Together, they laid the groundwork for the revolution in meteorology that was to follow.

Tiros 1 was drum-shaped – 0.42 m high and 1.07 m in diameter – and spin-stabilised about the major axis of the cylinder, the spin being imparted on separation from the launch vehicle. To enable the satellite to retain the alignment of its spin axis relative to the Earth's magnetic field, it carried a magneto-torquer coil. The sides and top surface of the drum were covered with some 9200 solar cells to provide power for the electronic equipment that was mounted mainly on its base (Figures 4.2 and 4.3). When the satellite was eclipsed by the Earth, it drew power from a rechargeable nickel–cadmium battery [6].

The heart of Tiros 1 was its sensor payload, a pair of half-inch vidicon[2] TV cameras capable of producing images in both the visible and infrared parts of the spectrum. It was this capability which led to the satellite's name, an acronym for Television and Infra-Red Observation Satellite. The cameras, which were mounted 180° apart, consisted of an optical system, a vidicon tube and a focal plane shutter. One camera was equipped with a 104° wide-angle lens providing coverage of some 2000 km[2], the other with a 12° narrow-angle lens covering 190 km[2] [6] . When the shutter was opened, the camera's optical system formed an electric charge pattern, corresponding to the image, on the photoconductive surface of the tube. The image was then converted into a TV signal by scanning the pattern with an electron beam.

Each camera could take 16 pictures per orbit at 128 s intervals, though the interval could be reduced to 32 s when required [7]. Images could be either transmitted directly to a ground station, if one was in range, or up to 32 images per camera could be

---

[1] The Rand Corporation was a development of 'Project RAND', which was established in 1945 by General H. H. Arnold, commander of the Army Air Forces, to perform operations research and provide advice; it became a non-profit consulting group in 1948. The acronym RAND is thought by some to stand for research and development and by others 'Research for America's National Defense' [5].

[2] The generic term for such a device is an iconoscope; vidicon is a contraction of video iconoscope.

Figure 4.2   Tiros 1 [NASA]

recorded on magnetic tape for later transmission. This meant that even the most remote regions of the Earth, which would probably never have a receiving station, could be imaged.

The magnitude of the technological development involved in placing a satellite such as Tiros in orbit cannot be overstressed. Only two years and two months had separated the launch of Tiros from that of America's first satellite, Explorer 1. Whereas Explorer was a unsophisticated cylindrical canister weighing only 5 kg, Tiros was a fully functional, spin-stabilised Earth observation satellite which, depending on the instruments carried, weighed between about 120 kg and 135 kg [6,7].

It seems likely that even the meteorologists who had called for the introduction of satellites would have been surprised by the advances in miniaturisation and capability. In 1956, Stroud and Nordberg had stated that they did not expect the early satellites to carry 'photographic devices, television or observers'[4]. In terms of space history, this mention of observers is interesting, since it reflects the assumptions of the time that

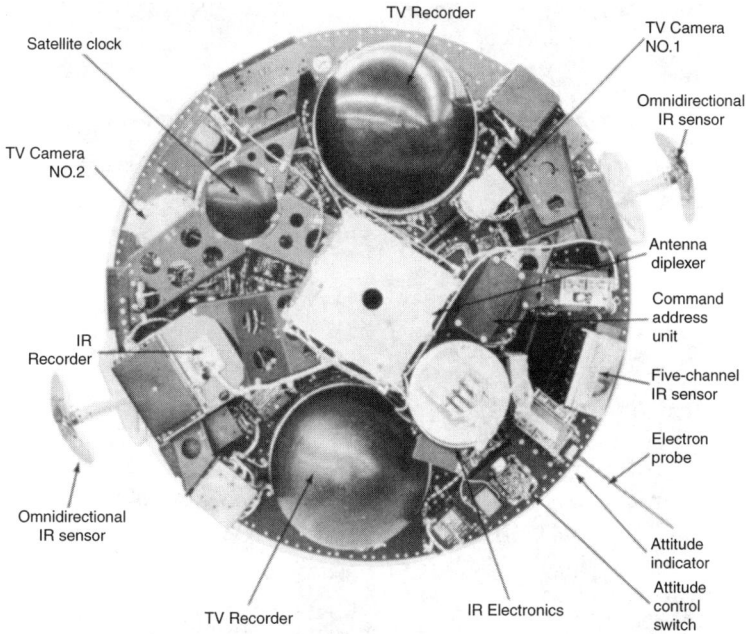

Figure 4.3   Tiros top view showing component arrangement [NOAA]

instruments would be unlikely to match the observational capabilities of a man, and that machines, especially spacecraft orbiting high above the Earth, would never be sufficiently autonomous (or reliable) to conduct such complex and demanding tasks.

The meteorologists did, however, expect the satellite to carry a photodetector of some description to measure the Earth's albedo (the ratio of reflected radiation to that received from the Sun), and this was the basis of their contribution. Although their expectations for the payload were somewhat conservative, those for other aspects of the satellite's design reflected the outcome more closely. They described a cylindrical spacecraft, spinning 'at a high rate' about an axis fixed in space, in an orbit measuring 320 km × 1280 km and inclined at 30° to the equator. Although the actual numbers were different, this was the concept followed for Tiros 1, as explained later.

In common with other satellite programmes, as the Tiros series progressed, the payload was gradually improved and augmented. Tiros 3, for example, carried two wide-angle cameras instead of one wide- and one narrow-angle camera as on Tiros 1 and 2. A more important innovation, included for the first time on Tiros 8 in 1963, was Automatic Picture Transmission (APT), which allowed the immediate transmission of images to inexpensive ground stations installed anywhere beneath the satellite's ground track. The pictures were not of the same quality as those transmitted to the main stations on command, but they were sufficient for local uses. The eventual 871 known stations in 123 countries did much to improve the dissemination and application of satellite images [6].

Figure 4.4   Tiros (showing downward pointing camera lenses) and components
           [NOAA]

In addition to the cameras, numbers 2, 3, 4 and 7 of the Tiros series also carried a pair of infrared (IR) radiometers to measure the radiation balance of the Earth. This ratio between incoming and outgoing radiation is extremely important to meteorologists because of its strong dependence on latitude. It is the greater solar input at the equator, compared with higher latitudes, that drives the atmospheric circulation and thus the planet's weather systems.

The efficacy of the radiometer had been shown by the scientific spacecraft Explorer 7, launched on 13 October 1959. It carried the first radiometer, and thus the first meteorological payload, to be flown in space. Designed and built by Professor Verner Suomi of the University of Wisconsin, it consisted of two hemispheres, one painted black to absorb all wavelengths and the other white to reflect solar energy and absorb only terrestrial infrared radiation. Thus the radiometer would measure the imbalance between incoming solar energy and outgoing terrestrial radiation [8]. By means of another experiment, Explorer 7 also established a link between solar activity and the Van Allen radiation belts (discovered by Explorer 1) by showing that solar flares energise the particles in the belts to produce magnetic storms and aurorae [9].

The Tiros radiometers confirmed and extended the results of the early instruments and formed the basis of a programme of satellite radiometer development that continues to this day.

Results

Two hours after it was launched, Tiros 1 transmitted its first TV photographs. Although crude by today's standards, the images revealed that cloud systems were more highly organised than had formerly been supposed. According to science writer Nigel Calder, the 'pictures of the Earth's cloud cover ... astounded even the meteorologists'. They expected chaos; instead they saw what one called 'nature's weather map, drawn for us'[10].

Some pictures contained huge spiral-shaped formations whose existence was previously unknown; others showed fronts and storm centres which were easily recognisable from space, enabling their positions and subsequent development to be accurately plotted. Intended only as an experimental satellite, Tiros 1 operated for 78 days (until its batteries expired), and transmitted 22,952 photographs of the Earth [11].

Following the launch of Tiros 1, in a paper called 'Tiros program results' [12], the chief of NASA's meteorological programmes considered the three basic questions that the Tiros experiments were designed to answer:

- First, could a satellite system with TV cameras and IR radiation detectors be developed to transmit 'with satisfactory fidelity' the measurements of those instruments? The answer was 'yes', although it was noted that Tiros had suffered some reliability problems.
- Second, would the measurements contain useful information? The answer was an unqualified 'yes'.
- And third, could the information in the data be extracted and transmitted to the weather services in time to be useful? Again the answer was 'yes', although the IR data had proved more difficult to reduce for display and interpretation.

Despite the inevitable problems, some photographs, showing icebergs and ice floes, were the first satellite images to indicate that sea-ice boundaries could be discerned and monitored on a continual basis. This was particularly important for the sea-lanes, like the Gulf of St Lawrence on Canada's east coast, because the availability of regular updates would allow them to be navigated more safely. In addition to the use of individual photographs, successive images could be used, to an extent, to forecast the rate of ice-melting.

Stroud and Nordberg's confidence in the potential of the weather satellite was justified by Tiros 3, launched in July 1961, when it captured an image of Hurricane Anna, the first hurricane to be detected by a satellite. When the satellite repeated the achievement in September 1963, by detecting and tracking Hurricane Carla, it made possible the largest mass evacuation at the time ever to take place in the United States. More than 350,000 fled to safety and relatively few deaths were attributed to the devastating hurricane [13]. Thus very early in the American satellite programme, the claim could legitimately be made that expenditure on at least this part of the space programme was worth it, not only in terms of the reduction in damage to property and infrastructure, but also in lives.

Initially, the images received from Tiros showed only natural phenomena and landforms, providing no clue that the Earth was an inhabited planet (Figure 4.5). Then one day a photograph of the Arctic was received with what appeared to be an unnaturally straight line extending hundreds of miles across the ice. Later analysis revealed it as the shadow of the contrail of a transpolar jet aircraft cast by the sun [14].

More importantly, Tiros showed that the use of satellite imagery was not confined to meteorology: photographs of deforestation due to logging and the evolution of forest fires and locust swarms indicated the range of possibilities for the observation and monitoring of land-based activities and phenomena. These and other applications were later to be addressed by dedicated land observation satellites.

Orbits and orientation

Whereas the orbital parameters and orientation of the early satellites had often not been crucial to the success of the mission, a satellite designed to produce images of a given subject was, rather obviously, required to point towards that subject. Here on Earth, or even in the air, a vehicle's ability to change course within reasonable limits is taken for granted, as is the ability to point an imaging device or camera. In space, matters are not quite so straightforward.

For example, a satellite's initial orbit is directly related to the capabilities of the launch vehicle, in that it requires more energy to lift a heavy satellite into a high altitude orbit than it does to lift a lighter satellite into a lower orbit. In addition, the orbital altitude determines the satellite's lifetime, since satellites in low orbits are dragged back to Earth, by friction with the upper reaches of the atmosphere, sooner than those in higher orbits (as discussed in Chapter 2). The early satellites may have been equipped with rocket thrusters to maintain their orbital altitude, or change the orbit's inclination to the Earth's equator, but their orbits were governed largely by the launch vehicles.

As such, the orbit is an integral part of the mission plan for the satellite's lifetime: for example, a satellite in a high orbit will have a longer orbital period than one in a low orbit, and a satellite in an equatorial orbit will not be able to observe the polar regions. Tiros 1 was placed in a slightly elliptical orbit, with an altitude varying between 677 km and 722 km, an inclination of 48.4° and a period of 98.8 minutes [15]. Its altitude governed the field of view of its cameras, and its inclination – itself constrained

(a)

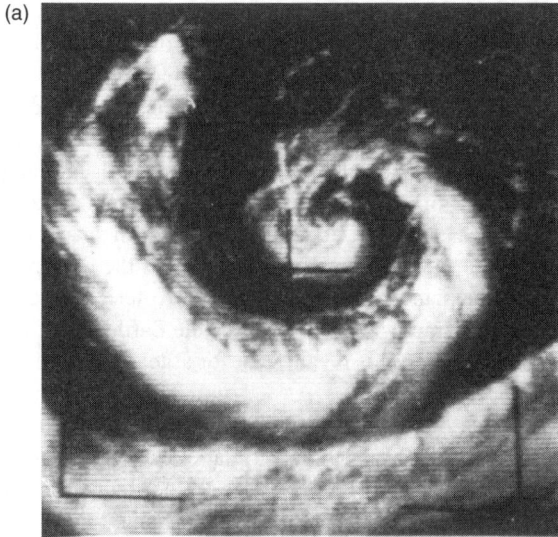

a. Large Spiraling Cloud System,
TIROS VI

(b)

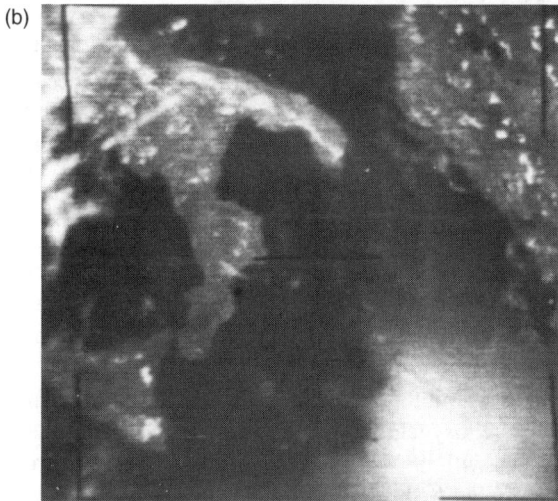

ITALY AND SICILY
TIROS VII ORBIT 4569 R/O 4568
DATE 4-23-64 AT 101830 Z
MODE: TAPE

Figure 4.5   Tiros results: (a) swirling storm clouds from Tiros 6; (b) southern Italy
from Tiros 7 [NOAA]

by the latitude of the launch site and the capabilities of the launch vehicle – governed its view to the north and south of the equator. An inclination of 48.4° means that the satellite's ground track across the Earth's surface extends from 48.4°N to 48.4°S.

Once all these parameters are set, and the satellite is in orbit, additional hardware systems, propellant and power are required to change them. In general, because the early satellites were relatively small and light, they could not carry the resources necessary to change their orbital parameters. As a result, the Tiros 1 design was a compromise. Since it was spinning about an axis fixed in space, the satellite could only take pictures of the Earth when its base, in which the cameras were mounted, was pointing towards the planet. And since it was an American satellite, the spin axis was arranged so that the cameras pointed towards the Earth when the satellite orbited over the northern hemisphere. This meant that, for at least half the time, the cameras pointed into deep space.

Later Tiros satellites (Tiros 9 and 10) were also spin-stabilised but with their spin axes oriented perpendicular to the orbit plane, causing them to 'roll along the orbit' with their side-walls facing the Earth. As a consequence the cameras were moved from the base to the side of the drum, which enabled them to image the Earth successively and increase the number of images that could be taken.

Throughout the first half of the 1960s, developments in all aspects of space technology were moving apace and it was inevitable that meteorological satellites would be a part of this development. Largely as a result of launch vehicle capabilities, the early Tiros spacecraft were injected into orbits that were not ideal for observing the whole Earth, even if the cameras had been pointing towards it, because the inclination of the orbit plane was too low to enable coverage of the higher latitudes.

The ideal orbit for the Earth observation satellite is a high-inclination (near-polar) orbit, which enables coverage of the entire planet: effectively, the satellite's orbit forms a ring inside which the Earth rotates, exposing successive lines of longitude to the satellite's imagers. The first satellite launched to a polar orbit was Discoverer 1, the initial example of a series of US military reconnaissance satellites designed to return a capsule containing film to Earth. Although the Discoverer series was ostensibly a biomedical science programme, it was in fact a cover for a 'spy satellite' series known to the military as Corona.

More specifically useful for Earth observation is a type of polar orbit known as a sun-synchronous or heliosynchronous orbit (Figure 4.6). Since the satellite's orbital plane is synchronised with the Sun, its ground track remains approximately fixed at the same local time on Earth, thus providing the same lighting conditions on each pass over a given area. This type of orbit has proved extremely useful for many generations of spacecraft, from the later Tiros satellites to those of the present day.

The launch of Tiros 9, on 22 January 1965, was the first attempt to place a Tiros satellite in this type of orbit. The orbit attained was of the correct inclination, but an unplanned additional boost from the launch vehicle's second stage extended it into an ellipse of 700 km × 2580 km rather than a circle of diameter 644 km. In the event, however, the high point of the orbit gave the satellite a greater coverage of the Earth and, on 13 February, Tiros 9 completed the first photomosaic of the Earth's entire cloud cover in 450 pictures (Figure 4.7) [11].

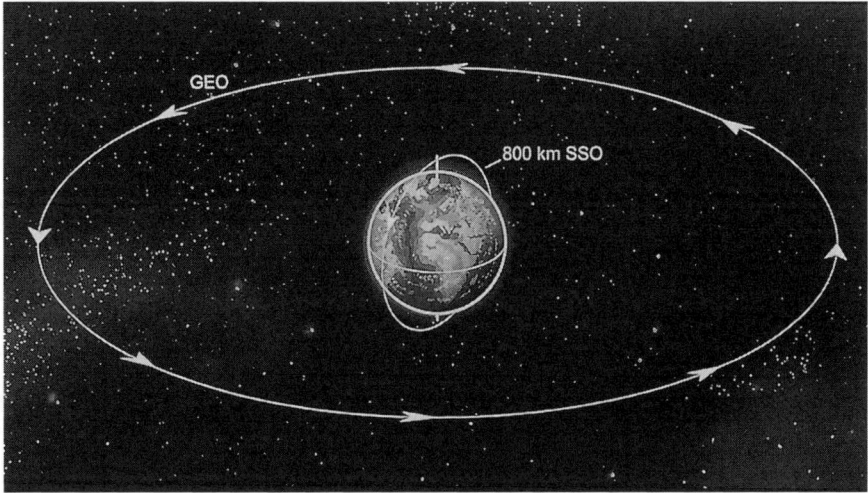

Figure 4.6   Sun-synchronous orbit (SSO) compared – approximately to scale – with geostationary orbit (GEO) [Mark Williamson]

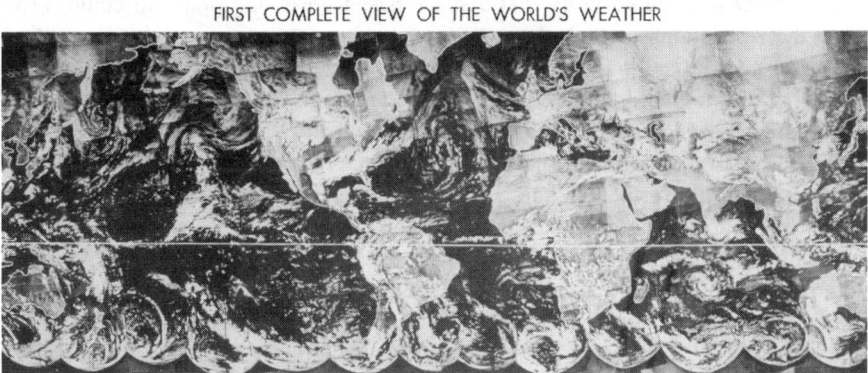

Figure 4.7   First complete view of the world's weather assembled from 450 individual images from Tiros 9 [NOAA]

Military aspects

Although the Tiros programme is regarded as the precursor to today's weather satellites, it also established itself as a capable military reconnaissance satellite. This was to be expected, since the initial studies had been undertaken by the US Department of Defense (DoD).

Figure 4.8   U-2 spyplane [Lockheed Martin]

The DoD realised early on that a space-based observing platform could over-come the restrictions of high-flying reconnaissance aircraft, since the concept of transgressing national airspace did not extend to orbital altitudes. Indeed, some of the early satellite photographs of the Soviet Union and China were so detailed that aircraft runways and missile launch sites could be readily identified [11].

The proof of the concept of space-based reconnaissance, innocently undertaken by the Tiros satellites, further encouraged America to develop a comprehensive system of what came to be known as spy satellites, the very existence of which was denied by the authorities until 1992 [16]. The satellites largely replaced the U-2 spyplanes (Figure 4.8) which, apart from being exceptionally provocative towards the Soviet Union, by the late 1950s were within the range of Soviet ground-to-air missiles – even when cruising at 70,000 ft (21 km). This was shown publicly, to great embarrassment in the United States, by the shooting down of a U-2 and subsequent imprisonment of its pilot Francis Gary Powers in May 1960 [17] (he was exchanged for a US-held Soviet agent two years later). The argument that satellites were orbiting safely out of range of the aggressor's missiles was hard to refute and led, beginning in February 1959, to the launch of a series of spy satellites under the Corona programme [18].

Apart from the infamous spy satellites, both the American and Soviet military fielded constellations of weather satellites to provide secure meteorological data to their respective armed forces. One such was the Defense Meteorological Satellite Program (DMSP), which was operated by the US Air Force under a number of differ-ent project names and designations. Civilian meteorologists received an unexpected bonus in 1972, when real-time DMSP data were declassified and archived images dating back to 1965 were made available.

Manned Earth observation

In parallel with the development of Tiros, NASA was also pursuing its manned space programme in competition with the Soviet Union, and even before the one-man Mercury capsule was entrusted with a human cargo, several unmanned versions had carried cameras on a suborbital (ballistic) trajectory. In September 1961, for instance, Mercury MA-4 took photographs of Hurricane Debbie and other meteorological phenomena.

Like the Tiros images, these photographs showed the potential of the space-based platform for Earth observation. When the first astronauts reached orbit they reported being able to see ships' wakes, roads and even trains. At first considered the result of overactive imaginations engendered by the excitement of spaceflight, the clarity of the view from low Earth orbit was soon capitalised upon by manned and unmanned spacecraft alike.

By the time the two-man Gemini capsules were launched in the mid-1960s, astronauts were equipped with top-of-the-range Hasselblad cameras as a matter of course. As their pictures found their way onto magazine covers, knowledge of the usefulness of space for observing the Earth extended from the scientific and military realm to the public domain.

Nimbus technology demonstrator

Also in parallel with Tiros, NASA was engaged in the development of Nimbus, a satellite intended primarily as a technology demonstrator. Given the early success of Tiros, it was inevitable that it should develop from an experimental to an operational system. The problem with operational systems, then as now, is that users expect them to be reliable and balk at the inclusion of new and unproven technology. This can lead to a stagnation of technological development, which was something America could not afford to countenance in the early years of the Space Age.

The solution then, as now, was to build a satellite to demonstrate and prove (or 'space qualify') the new technology that would be incorporated into the next generation of operational satellites. The first of the new series, called Nimbus, was launched a few months before Tiros 9, on 28 August 1964.

The Nimbus series was originally intended to develop instruments and techniques for meteorological observations, as its name (from the Latin for 'raincloud' or 'storm-cloud') implies. However, in a run of seven satellites launched between August 1964 and October 1978, its role widened until it became a testbed for a wide range of Earth science and remote sensing technology. In fact, the Nimbus programme begat ERTS, the Earth Resources Technology Satellite, which was later renamed Landsat, arguably the most important satellite remote sensing system produced by the United States.

Built for NASA by General Electric, the Nimbus satellites were placed in sun-synchronous orbit to provide full Earth coverage. In the same way that astronomy satellites migrated from the spin-stabilised platform to three-axis stabilisation (to improve their pointing stability and power generation capability), Nimbus was

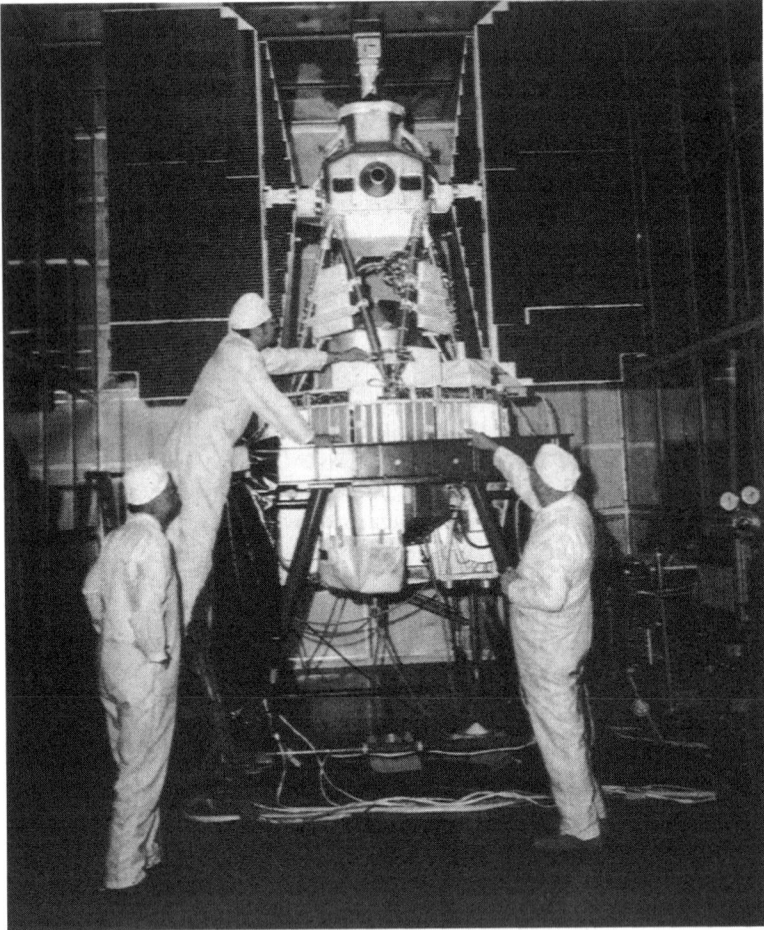

Figure 4.9   Nimbus 7 [NASA]

designed around a three-axis platform. A system which sensed the spacecraft's attitude in relation to the Earth's horizon allowed it to point towards the planet at all times, the necessary adjustments being made by a combination of reaction wheels and compressed nitrogen gas-jets [19].

Nimbus 1 was about 3 m high and 2 m wide, weighed 380 kg and produced some 550 W of power, 140 W of which was consumed by the payload, the remainder by supporting subsystems. A modular compartment, designed to house the majority of the satellite's payload and subsystem equipment, contained as many as six TV cameras as well as IR radiometers, tape recorders and telemetry equipment (Figure 4.9).

Mounted above this platform were two unusually shaped solar array panels, supported by a skeletal truss-structure, which gave the spacecraft a distinctly butterfly-like appearance. Considering how often photographs of the design appeared in books and magazines of the time, it is hardly surprising that solar panels mounted in this way came to be known as array 'wings'.

Nimbus carried an improved vidicon camera system called, somewhat predictably, the Advanced Vidicon Camera System (AVCS). It was capable of producing images with wider coverage and better resolution than those on the experimental Tiros satellites as a result of the new 1-inch vidicons developed by RCA, which had a resolution of about 1 km from an altitude of around 900 km [20]. This increase in spatial resolution, otherwise known as geometric resolution, is a key measure of the capability of an Earth observation payload, since it defines the ability of a system to 'resolve' or distinguish between separate parts of an image. A related quantity in imaging technology is the signal-to-noise ratio of the sensor, a measure of how good it is at discerning the wanted signal above the unwanted noise. The noise may originate from the equipment itself or from terrestrial or astronomical sources and its reduction or suppression is an important challenge in the development of all electronic equipment.

The Nimbus AVCS comprised three separate cameras, one mounted so that it pointed straight down, the others angled outwards by 35° in a plane perpendicular to the direction of flight. This produced a fan-shaped coverage which swept along the satellite's ground track. The angular size of each camera's image was 37° square and a 2° overlap between the field of adjacent cameras gave a composite field of 37° by 107°. Thus from an orbital height of about 1100 km, attained by Nimbus 2 and subsequent spacecraft, the AVCS was able to image a 3600 km wide strip around the Earth in 1250 km segments. And since it was in a polar orbit, the satellite could cover the whole Earth twice a day, once in daylight when images were provided by the AVCS, and once at night using infrared imagers [21]. As intended, once the AVCS technology had been tested successfully on Nimbus it was used in the operational Tiros satellites (see Tiros derivatives below).

Like the later, experimental Tiros satellites, Nimbus was also equipped with Automatic Picture Transmission (Figure 4.10). The camera associated with the APT system was a wide angle device with lower resolution than the AVCS (4.4 km at the image centre) which produced images covering 2320 km on a side [22].

As far as results were concerned, Nimbus 1 provided the first 'high resolution' optical and IR cloud and surface images from a satellite. Of course, the term high resolution is entirely subjective and date-related, in that while a 1000 m optical resolution might have been considered good in the 1960s, by the end of the 1990s it would need to be around 1 m to be classed as high resolution. Nevertheless, the Nimbus imagery was sufficiently precise to enable cartographers to correct the location of Mount Siple, a 3000 m peak in Antarctica used by pilots as a navigational aid. Previous measurements from aircraft had produced an error in the mountain's position of 72 km [23], and maps had shown two mountain groups where only one existed.

Unfortunately, the spacecraft's solar array drive mechanism failed after a month in orbit and operations ceased due to lack of power. Nimbus 2, however, operated for almost three years from its launch in May 1966, well beyond its 6-month design lifetime [24].

Nimbus 3, launched in April 1969, carried a new type of spacecraft instrument called a Satellite Infra-Red Spectrometer (SIRS). It was used for vertical sounding,

Figure 4.10   APT equipment for use with Nimbus [NASA]

a term borrowed from nautical practice, to determine the temperature variation of the atmosphere with altitude. SIRS made it possible to collect temperature data over the entire Earth to an accuracy, or radiometric resolution, of about 1°C above 6100 m and to within about 2° below that altitude [25].

This was acclaimed as one of the most significant developments in the history of meteorology, because little previous data from above the oceans had been recorded. According to meteorologist Dr B. J. Mason, the operation of SIRS exceeded expectations, since the determination of the vertical distribution of temperature and humidity through the atmosphere provided 'the first ... quantitative global data of the kind required for numerical weather prediction' [26].

Although, by the time Nimbus 3 was operational, there were estimated to be some 8000 ground stations and 4000 merchant ships making regular weather observations, only a few hundred stations were launching radiosondes to collect data at altitude [26]. According to one author writing in the early 1970s, radiosondes were being launched from about 700 'upper air stations' on land and sea, and dozens of aircraft were engaged in meteorological observations, but this was an extremely inefficient method of providing only partial coverage. Apparently, some half a million balloons were 'literally thrown away every year ... and [incidentally] often mistaken for flying saucers' [10].

Although it had been possible for some time to measure the temperature of the cloud tops and therefore gauge their approximate height, SIRS' ability to combine observations in selective frequency bands made it possible to map areas of high and low humidity, small differences in ground surface temperature, and so on [27]. Following developments in spatial and radiometric resolution, this added the concept of spectral resolution to the Earth observation toolkit, leading first to multispectral sensors capable of producing images in several spectral bands and later to hyperspectral sensors that could resolve hundreds of discrete frequencies.

SIRS used a fixed grating to select seven narrow spectral intervals 5/cm-wide in the 15 μm carbon dioxide band and one interval in the 11 μm (far infrared) window. By means of a 'chopper' system, an alternating signal corresponding to the temperature of the atmosphere and an on-board reference source, in the form of a 'black body' at liquid nitrogen temperature, was produced. The signal was detected by a set of thermistor bolometers (resistance thermometers used to detect radiant energy at IR wavelengths).

The spacecraft also carried a Michelson infrared interferometer spectrometer (IRIS) to accomplish the same task for a much broader section of the spectrum from 6 to 20 μm [28]. For comparison, the visible band extends from 0.4 to 0.7 μm, and the near infrared from 0.7 to 0.9 μm.

Microwave imagers

Extremely useful though these optical devices were, they suffered one major drawback: the inability to image through cloud. All clouds except cirrus are opaque to visible and IR radiation, so satellite sensors operating at these wavelengths show only the cloud tops and are unable to distinguish between clouds producing heavy precipitation and those that are not.

For this reason, engineers developed a microwave device (operating at a wavelength of 1.55 cm) which could penetrate most clouds apart from those containing large water droplets, snow or hail. The device developed for Nimbus was known as an Electrically Scanning Microwave Radiometer (ESMR). It marked an evolution from passive sensors, which simply received the natural radiations reflected from the Earth, to active sensors, which generated their own radiation, transmitted it towards the Earth and received the reflections. This was only possible with satellites with greater power generation capabilities, since it required relatively high levels of DC power to generate the microwave energy.

The ESMR's antenna differed from the common dish-shaped paraboloid in that it consisted of a number of separate radiating elements arranged in a flat (or planar) array. This is known as a phased-array antenna, since signals can be sent to the elements in phase (or in any phase relationship, depending on requirements, which may include the formation of multiple beams or electronic steering of a given beam).[3] The ESMR formed a beam 1.5° wide, which was capable of scanning up to 50° on each side of the flightpath. Its spatial resolution at nadir, directly below the satellite, was 30 km.

---

[3] In the 1980s and 1990s, the phased array antenna became increasingly popular among communications satellite engineers because of its flexibility.

The scanning nature of this device was an important development for Earth observation spacecraft, especially the three-axis stabilised types like Nimbus. Removing the spin meant that onboard sensors were no longer scanned across the face of the Earth by the natural motion of the satellite; the solution, of course, was to spin or scan the sensor (or, more accurately, to spin or scan a mirror to reflect the image into the sensor).

However, since the scanning motion was essentially one-dimensional, this would provide an image of only a single line, unless the mirror was also stepped in the perpendicular axis. Luckily, since the satellite was moving in its orbit track around the Earth, its orbital motion provided that second dimension. Imagers which operated in this manner became known as whiskbroom scanners, after the rapid sweeping motion of a household broom. Initially, the imaging occurred on a cell-by-cell basis in the cross-track direction, while later imagers used a number of cells in the along-track direction in parallel to make the process more efficient.[4]

The ESMR microwave radiometer, carried for the first time on Nimbus 5 in December 1972, extended the comparative abilities of the weather satellite. Now, by imaging in the visible, IR and microwave parts of the spectrum, which have different degrees of cloud penetration, it was possible to investigate in greater detail the energy and momentum transfer in large storm systems such as cyclones [29]. And since the band of sensitivity of the instrument was outside the visible, as with infrared detectors, it was also possible to produce images when the spacecraft was orbiting over the dark side of the Earth.

Among other things, the ESMR proved particularly useful for the mapping of ice boundaries. Land areas on radiometer images generally appear dark due to their relatively high emissivity; water by contrast appears light. When mapping sea ice it was soon discovered that new ice appears dark, like land, because it contains pockets of brine frozen into it and has a high emissivity. Glaciers, on the other hand, are much more closely packed and have a low emissivity [30]. Thus in the microwave band an image of a landmass such as Greenland provides an unusual picture: the permanent icecap appears light, new ice surrounding it appears dark and open water beyond that is light, an image which at first glance appears more like a coral atoll than an icecap.

The microwave radiometer also proved useful for the observation of precipitation. Although precipitation appears dark over land, which makes it difficult to see, its dark signature makes it easy to distinguish over the ocean, where the satellite really comes into its own due to the lack of observing stations. Indeed, the development of microwave sensors led to the launch, in 1978, of Seasat, America's first radar satellite dedicated to the study of the sea surface.

---

[4]   Later imaging payloads used pushbroom scanners, which gather data a line at a time in the cross-track direction, again using the orbital motion of the satellite to build up an image. This method was contingent on the development of the linear CCD (charge-coupled device), a technology invented at AT&T's Bell Labs in 1969. The first spaceborne pushbroom imager (MSU-E) was flown on the Russian Meteor-Priroda 5 (Meteor I-30) spacecraft launched on 18 June 1980.

Later spacecraft in the Nimbus series carried colour scanners to map the meanders of the Gulf Stream by registering the high chlorophyll content associated with plankton, as well as instruments to measure the Earth's radiation budget (the balance of incoming and outgoing radiation). Nimbus 7, launched in 1978, was the first satellite to measure air pollution and, after nearly 15 years, was still returning data on ozone levels in the upper atmosphere; unfortunately its Total Ozone Mapping Spectrometer (TOMS) suffered a malfunction in May 1993.

Apart from its scientific applications, the Nimbus series came to be regarded as indispensable to US Navy shipping operations in the Arctic and Antarctic, since satellite images allowed ships to navigate through the ice fields for several additional months of the year [31]. But perhaps the most unusual application of the Nimbus series was elk tracking. An electronic collar worn by an elk in the National Elk Refuge of Wyoming was interrogated twice daily by a transceiver payload on Nimbus 3 in a study of migration habits [25]. Whether wearing a 10.4 kg collar affected the habits of the wild elk is not recorded.

Similarly, Nimbus 4 carried a payload designed to track weather balloons and ocean buoys around the world, but as a spin off it also tracked the aviator Sheila Scott on her solo flight over the North Pole [25]. Thankfully, it was not necessary for Ms. Scott to wear a 10.4 kg collar!

Tiros derivatives

The operation of the Nimbus series as a technology demonstrator was extremely successful and many of the instruments proven by Nimbus were later incorporated in operational satellites, including derivatives of the original Tiros series.

The satellites of the Tiros Operational System (TOS) were called ESSA, after the Environmental Science Services Administration which managed the programme. Nine ESSA satellites were launched into sun-synchronous orbits between February 1966 and February 1969. Based on its experience with Tiros, RCA produced satellites which were very similar in appearance to the earlier Tiros satellites and ESSA 1 carried a similar payload. Subsequent odd-numbered satellites, however, were equipped with the advanced vidicon camera system (AVCS) first demonstrated on the Nimbus series, while even-numbered satellites carried automatic picture transmission (APT) cameras (Figure 4.11) [32]. The ESSA satellites were capable of producing between 144 and 156 images per day, while also monitoring solar flares, solar particle densities and magnetic storms in the Earth's ionosphere.

In 1969, a picture from ESSA 7 made history by helping to prevent a potential disaster. It revealed that the snow cover in Minnesota and the Dakotas, in America's Midwest, was three times thicker than normal and equivalent to 15–25 cm of water covering thousands of square kilometres. A disaster alert was called before the snow melted and, by the time the inevitable floods arrived, much had been done to alleviate the extent of the damage [11]. By this time, with justification, weather satellites were being widely publicised as property and life savers, and the goal

(a)

(b)

AVCS
Camera

IR
Sensors

IR
Sensors

Sensor-data
transmitting
antenna

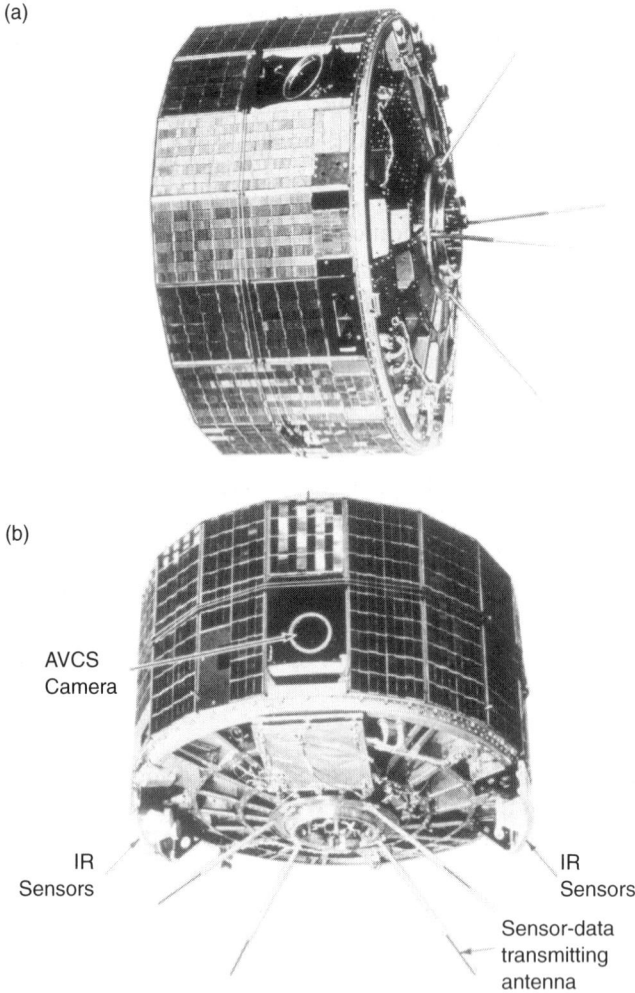

Figure 4.11   (a) ESSA 1, (b) ESSA 3 [NOAA]

in the United States was to develop a national operational meteorological satellite system.

In 1970, a second generation of meteorological satellites, designated ITOS for Improved TOS, was introduced (Figure 4.12). They were box-shaped (measuring 1 m × 1 m × 1.24 m) with three solar panels, which were deployed from the sides of the satellite in orbit. The ITOS satellites were three-axis stabilised using a spinning momentum wheel to provide gyroscopic stability, and weighed in at 309 kg – over twice the mass of the original Tiros spacecraft.

The ITOS series capitalised on the technology of the experimental Tiros satellites, the Nimbus technology demonstrators and the operational ESSA series, carrying both

Figure 4.12   ITOS satellite [NOAA]

AVCS and APT, but many devices and techniques were improved. Indeed, ITOS-1 was able to photograph the entire world's cloud cover every 12 hours, instead of every 24 hours as with ESSA.

As it turned out, ITOS-1, launched on 23 January 1970, was the only satellite to carry the new designation. That same year ESSA was taken over by the National Oceanic and Atmospheric Administration (NOAA) and subsequent satellites were renamed after this organisation.

In addition to the now familiar vidicon system, the ITOS/NOAA satellites carried the first operational two-channel (visible and IR) scanning radiometer (SR). Some also carried a very high resolution radiometer (VHRR), a device that is now standard equipment aboard Earth observation spacecraft. The early VHRRs had a resolution of about 1km, which was sufficient to identify all types of cloud system including local systems and fog. Radiometer data were both transmitted in real time and stored for later transmission to other stations.

The Nimbus series had, to a large extent, set the standards for polar-orbiting weather satellites, the descendents of which continue to provide a service today. But perhaps the most significant achievement of the Nimbus series was its demonstration of technology for imaging the planet's land surfaces.

Landsat

Nimbus had shown, among other things, that the sun-synchronous orbit was ideal for regular observation of the Earth's surface. Once large-scale, high-resolution images of the Earth became available, it was evident to many in the field that the potential of satellites for Earth remote sensing was almost limitless. This impression led to the development of Landsat 1, which was based heavily on Nimbus and, to the uninitiated, appeared to be its virtual twin. Formerly called the Earth Resources Technology Satellite (ERTS), the change in name was, then, a rare example of the American space community's adoption of a descriptive name for a satellite at the expense of a cherished acronym (Figure 4.13).

The Landsat satellites have been known for many years as providers of crop-health surveys. In fact, Robert N. Colwell – a professor of forestry at the University of California, Berkeley – demonstrated in the early 1950s that infrared imaging could be used to distinguish diseased plants from healthy ones. He showed that wheat infected with a fungus called black stem rust appeared darker in IR photographs [33].

Although little attention was paid to his work until the 1960s, when airborne and satellite imagers were being developed, the ability to colour code images according to their temperature signatures – for crop-health surveys or other applications – is now central to remote sensing technology. When Colwell was interviewed for National Geographic in 1969, these techniques had still to be tested from spacecraft, and by no means did everyone agree that the instruments would prove effective. Colwell, however, was convinced that they would work, and has subsequently been shown not only to have been correct but, in common with many observers of the time, to have underestimated the capabilities of the satellite.

Another widely publicised aspect of satellite sensing was the ability to identify geological structures that were likely to harbour oil reserves and other mineral resources. This was part of what we might now call a resource observation programme,

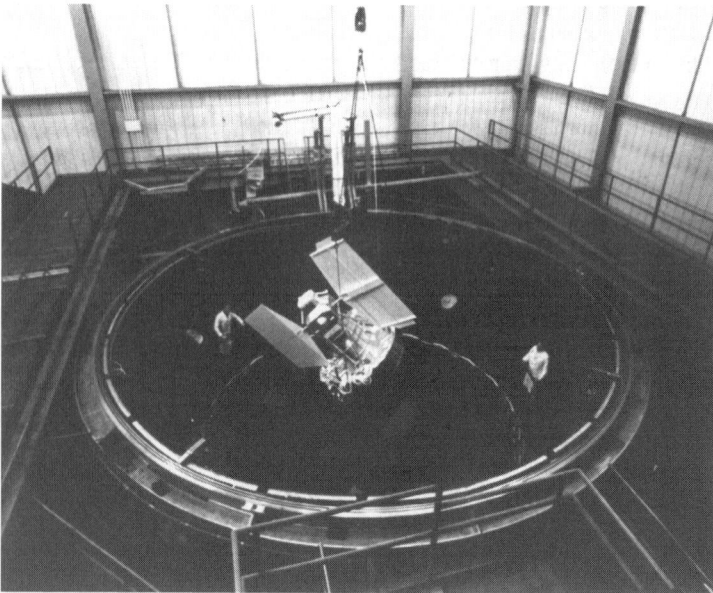

Figure 4.13   ERTS mock-up in a vacuum chamber test at General Electric Space Division [NASA]

and explains why the Landsats and others were called Earth resources satellites. As the role of satellites expanded to cover a wider variety of aspects, including the mapping of deserts and oceans and the investigation of the ice caps and the atmosphere, the more generic term remote sensing came to be used – a term subsequently superseded, as the population became more globally aware, by Earth observation.

One of the people who understood the potential of the satellite in the Earth resources realm was William Pecora, Director of the US Geological Survey (USGS), who lobbied NASA to develop such a satellite for several years before the Agency finally developed ERTS. It is believed that the USGS's announcement of its intention to request its own budget to design and launch an Earth resources satellite stimulated NASA into action.

Although the key characteristic of any Earth observation spacecraft was the resolution of its principal sensor, it was not until the launch of Landsat 1, on 23 July 1972, that the all-important factor of spatial resolution broke the 100 m barrier. Until then, any feature smaller than this could not be readily differentiated from the background. This improvement in resolution was crucial to the development of commercial satellite remote sensing, since fields, rivers and estuaries are typically three or four orders of magnitude smaller than the average weather system.

Landsats 1, 2 and 3 had two main payloads: a Return Beam Vidicon (RBV) camera and a Multi-Spectral Scanner (MSS) (Figure 4.14). The RBV system included on the first two satellites was developed by RCA from its earlier vidicon cameras. The return beam device added a photomultiplier tube to a two-inch vidicon, giving the system exceptional sensitivity and a high signal-to-noise ratio in addition to its high

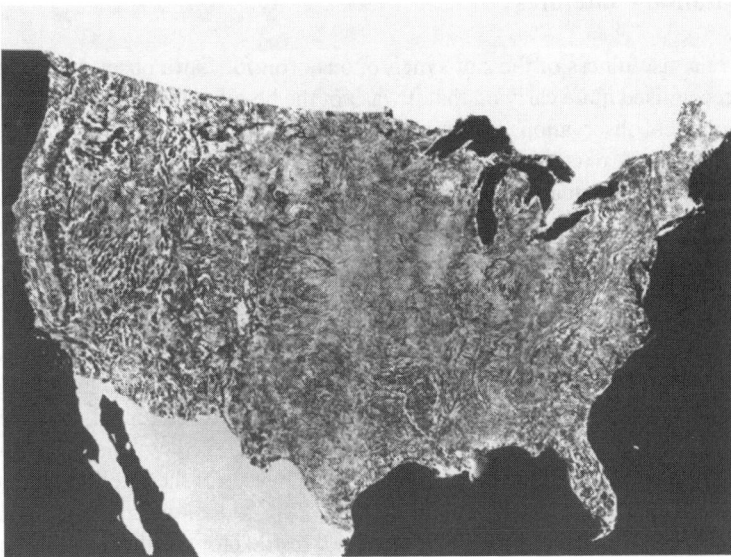

Figure 4.14   First cloud-free map of continental USA (made from 595 separate ERTS-1 MSS images) [NASA]

resolution [34]. The system produced images made up of 4500 lines (which compares extremely well with the several hundred lines of domestic TV systems), outputting a 185 km-square image with a resolution of about 80 m.

To produce hard-copy images on the ground, RCA also developed a helium–neon laser scanner. The laser beam was scanned from side-to-side, in the same way as the electron beam in the RBV itself, as the unexposed film was driven forward to build up the image. The intensity of the beam was modulated, through a ferro-electric crystal, by the video signals received from the satellite [34].

Unfortunately, the RBV system on Landsat 1 operated for less than a month due to tape recorder problems and Landsat 2's was operated only for engineering evaluation purposes. Thus the satellites' multispectral scanners became the principal source of data [35].

Landsat's Multi-Spectral Scanner (MSS) was developed by Hughes Aircraft Company, another major player in the nascent satellite manufacturing industry. It used photomultiplier tubes for three spectral bands between 0.5 and 0.8 μm (visible to near IR); silicon photodiodes between 0.8 and 1.1 μm (near IR); and an intrinsic crystal detector for 10.4–12.6 μm (thermal IR). The detectors were cooled to reduce thermal noise.

The MSS was the first space-based scanner to utilise the whiskbroom concept, developed as part of the Nimbus technology programme. In common with other scanning imaging systems, the image was produced by scanning the instantaneous field of view of the sensor at right angles to the direction of travel of the satellite, the satellite's orbital motion providing the second dimension.

Geostationary satellites

Despite the usefulness of the sun-synchronous orbit for Earth observation satellites, it was recognised quite early on that it was not the be-all and end-all, particularly for meteorological observation. The early weather satellites, placed in low-altitude orbits, circled the planet once every 90 minutes or so, providing fleeting glimpses of the changing weather systems below them. By contrast, a stable platform which remained stationary with respect to a given region would allow uninterrupted observation of events below.

As with many aspects of space technology, this was dependent on the development of launch vehicles with sufficient capability to place a satellite in geostationary orbit, a circular orbit in the same plane as and 35,786 km above the Earth's equator. Since a satellite in this orbit has an orbital period matched with the Earth's rotation, it appears stationary with respect to the Earth and allows continual, 24-hour monitoring of a selected portion (approximately one-third) of the planet. The fact that, as one commentator pointed out, 'the weather moves, not the satellite' allowed the geostationary satellite to make a serious study of the global weather circulation.

In the same way that Nimbus had acted as a testbed for weather satellites in polar orbit, the instrumentation for future American geostationary satellites was tested by NASA's Applications Technology Satellite (ATS) series (Figure 4.15).

Figure 4.15   ATS-1 [Boeing, formerly Hughes Aircraft Corp.]

ATS-1, launched on 7 December 1966, carried the Spin-Scan Cloud-cover Camera (SSCC) – a 13cm-diameter telescope and photomultiplier tube – which scanned the Earth as the satellite spun at 100 rpm. Building up a picture line by line, it produced a 4 km-resolution image of a wide strip of the Earth, between 52°N and 52°S, every 20 minutes. When linked with subsequent images to produce a time-lapse film, it allowed the motion and development of weather systems to be plotted. The SSCC operated successfully until 16 October 1972.

ATS-3, the second satellite of the series intended for geostationary orbit, was launched on 5 November 1967 and stationed over Brazil. It carried a multicolour version of the SSCC, a telescopic photometer with red, green and blue filters which allowed the first full colour images of the Earth's disk to be transmitted every 24 minutes. Successive images were once again processed into time-lapse films to compress a day's weather into a few seconds [10].

Useful though the system was, meteorologists realised its limitations: as B. J. Mason reported, 'at present the remoteness [of geostationary orbit] restricts observations to the visible spectrum and therefore to daylight hours' [36]. He also added, however, that 'improvements in the design and operation of infrared detectors would eventually remove this limitation' and, indeed, within a few years IR radiometers were also being carried on geostationary satellites.

Moreover, the utility of being able to image simultaneously in separate bands was already appreciated: by using visible and IR images together, snow and ice could be distinguished from cloud, for instance. And since it was possible to scan simultaneously in several bands, it was considered desirable to combine the two bands in a single radiometer rather than carry a separate vidicon TV camera which was only capable of imaging in the visible [27].

This view led to the development of a multi-frequency radiometer known as VISSR (Visible and Infrared Spin Scan Radiometer), which was flown on the first American dedicated weather satellites to operate from geostationary orbit – those of the SMS (Synchronous Meteorological Satellite) system. The VISSR payload contained eight identical photomultipliers operating simultaneously in the visible band and a redundant pair of cooled mercury–cadmium–telluride (Hg–Cd–Te) detectors for the infrared band. Coupled with the 100 rpm spin of the satellite, a scanning mirror took just 18 minutes to build up an image of the complete Earth, scanning 14,568 lines to give a resolution of about 0.9 km in the visible band and 1821 lines for a 7 km resolution in the infrared [37].

The two SMS satellites, launched in May 1974 and February 1975 respectively, were the forerunners of the current Geostationary Operational Environmental Satellite (GOES) system (Figure 4.16). GOES 1, launched into geostationary orbit on 16 October 1975, provided a complement to the NOAA satellites in sun-synchronous orbit and initiated the American operational weather satellite system that exists today.

Global meteorology

Although the early development of Earth observation satellites was dominated by the United States, by the mid-1970s other nations had begun to develop their own weather satellites and the scene was set for a truly global meteorological research programme.

Like America, the USSR had begun experiments with weather satellites in the early 1960s, certainly by the time the spacecraft designated Cosmos (or Kosmos) 44 was launched on 28 August 1964, and possibly as early as Cosmos 23 (launched on 13 December 1963) [38]. These early technology demonstration satellites carried infrared sensors to obtain low resolution photographs of cloud cover. The first official Soviet weather satellite to be equipped with a TV camera was probably Cosmos 122, launched on 25 June 1966 [39], which would place the Soviets some six years behind the technology of the American Tiros. Cosmos, incidentally, was a cover-all designation used by the Soviets to disguise the payload carried by individual satellites and, by the early 1970s, at least 50 per cent are known to have carried military 'spy-satellite' payloads [40].[5]

By the time the second in the series of dedicated weather satellites, Cosmos 144, was launched on 28 February 1967, the Soviet metsat had become of standardised

---

[5] In fact, by 2005, the Cosmos series had surpassed 2400 [41].

Figure 4.16   The Earth from GOES [NOAA]

cylindrical design, about 5 m long, 1.5 m in diameter and weighing around 2 tonnes. It was three-axis stabilised, using inertia wheels, and had two deployable solar array panels which were electrically driven to face the Sun. Its payload included two TV cameras for observation in the visible part of the spectrum and an infrared camera. The former imaged a 1000 km strip or 'swath' at a resolution of 1.25 km, while the latter, operating from 8 to 12 μm, had a resolution of 15 km [42].

The Soviet Union's first operational weather satellite, Meteor 1-1, was launched on 26 March 1969, and the first generation series continued until the late 1970s, by which time it was being replaced by the Meteor 2 series. Also, beginning in 1974, Meteor satellites began to include Earth resources instruments and were designated Meteor-Priroda (Priroda being the Russian word for nature) [43]. Like the American Nimbus satellites, the Meteor spacecraft were placed in near-polar orbits and carried a variety of cameras and other sensors for both day and night observation (Figure 4.17).

Despite the similarities between the satellite systems of the two superpowers, any thoughts of coordination in the early 1960s were hampered by the all-pervading cold-war mentality and the difficulty of obtaining information on Soviet intentions.

Figure 4.17   Soviet Meteor satellite display at Le Bourget Air and Space Museum, Paris [Mark Williamson]

Moreover, any potential cooperation that might have ensued between western nations was obstructed by America's decision to develop the Tiros system independently.

However, by August 1966, the back-door route of scientific cooperation had surmounted political differences and the United States and USSR had begun exchanging weather satellite data: America was sending the USSR between six and ten images a day from ESSA 1 and receiving 5–15 Cosmos photos per day in return [24]. Following the operational deployment of the Meteor system, the World Meteorological Organisation (WMO), an organ of the United Nations, was able to take a more proactive role in integrating the two national systems into its World Weather Watch programme [44].

Although, as in other aspects of space technology, Europe was some years behind the superpowers, it began to develop its own meteorological satellites in the mid-1970s, culminating in the launch of Meteosat 1 in November 1977. In fact by 1992, following the launch of five Meteosats, Europe was in a position to loan one of its older spacecraft to the United States which, as a result of satellite and programmatic failures, had been left with only a single operational spacecraft in orbit. Likewise, in the field of Earth observation, France's SPOT satellite series challenged Landsat's supremacy, offering both higher resolution and superior data continuity.

The fundamentally global nature of meteorology has led to a number of cooperative research programmes, for which satellite systems are an essential tool. One of the best known and most wide ranging of these programmes was the Global Atmospheric Research Programme (GARP), itself the research component of the World Weather Watch programme. Launched under the auspices of the WMO and the International Council of Scientific Unions (ICSU) in 1978, GARP was the biggest meteorological research programme ever undertaken. It involved the coordination of a hundred ground stations, ships, platforms and aircraft, and ten satellites, six from the USA, two from the USSR, one from Europe and one from Japan.

Over a period of 11 months, data were distributed to 22 processing centres around the world. The result was a significant increase in knowledge and understanding of meteorological processes and evolution which made it possible to extend the forecasting period to seven days for farming, transport, tourism and other general applications [6]. The global nature of meteorology, and its sister subject climate, has been a subject of international interest and cooperation ever since.

Conclusions

Earth observation satellites have been used, in one form or another, since the early years of the Space Age and have been instrumental, not only in weather forecasting, but in providing a better understanding of the planet as a whole. Indeed, it was the ability of meteorological satellites to image the Earth and show details never before seen that prompted the development of higher resolution sensors for monitoring the planet's resources and mapping.

Add to this the observations made over the years from manned spacecraft, like the American Gemini, Apollo, Skylab and Space Shuttle, and the Russian Vostok, Soyuz, Salyut and Mir, and the importance attached to observing and imaging the Earth from space becomes clear.

Since the first satellite image of the Earth was received in 1959, the technology for producing these images and transmitting them to Earth has seen great advances. By developing sensors that not only allow imaging in the visible waveband, but also in the infrared and microwave regions, the scope of remote sensing from space has expanded considerably beyond its early beginnings. Images of Earth are now widely used for land use surveys, geological prospecting, pollution monitoring and a wide variety of other applications.

Moreover, the repeatability of the observations, as a natural consequence of the satellite's orbit, means that long-term observation and monitoring has become a reality. Since 1966, for example, the entire Earth has been photographed at least once a day and, since then, no tropical storm has escaped detection and daily tracking. As Nigel Calder wrote in his book *The Weather Machine*, 'the modern satellites open up the possibility of a new level of weather service – not forecasting, but "now-casting"' [10].

Moreover, storms are usually detected by satellite some time before a weather observation aircraft could detect them, because of an aircraft's limited range. And high quality satellite photographs can also be used to show and track small scale weather conditions, such as snow storms, summer hailstorms and tornadoes, which cannot be predicted by computer models but may prove devastating for communities in their path.

The wealth of images from weather satellites has not only been used in daily operations, but archived as part of a global database for the investigation of climate development. The existence of the ozone hole was confirmed and is continually monitored by satellites, which, by tracking hurricanes and other tropical storms, have also helped to save lives.

We are now so used to seeing pictures of the Earth from space as part of our TV weather forecast, that we take them for granted. Despite this, Earth observation satellites of all types have earned their rightful place in the history of technology.

References

1 Weaver, Kenneth F.: 'Remote Sensing: New Eyes to See the World', *National Geographic*, **135** (1), January 1969, pp. 46–73

2 Allward, Maurice (Ed.): 'The Encyclopedia of Space' (The Hamlyn Publishing Group Ltd, Feltham, 1968–69) p. 227

3 Thompson, Tina D. (Ed.): 'TRW Space Log 1996' (TRW Space and Electronics Group, Redondo Beach, Volume 32, 1997) p. 67

4 Van Allen, James A. (Ed.): 'Scientific Uses of Earth Satellites' (Chapman and Hall, London, 1956) Chapter 13

5 Cargill Hall, R.: 'The Origins of US Space Policy: Eisenhower, Open Skies, and Freedom of Space' (IAA-92-0184), 43rd IAF Congress, Washington DC, USA, 1992, p. 1

6 Caprara, Giovanni: 'The Complete Encyclopedia of Space Satellites' (Portland House, New York, 1986) p. 23

7 Turnill, Reginald, (Ed.): 'Jane's Spaceflight Directory, 1987' (Jane's Publishing Co Ltd, London, 1987) p. 85

8 Villevieille, Adelin: 'Satellite Applications: Meteorology' in Rycroft, Michael, (Ed.): 'The Cambridge Encyclopedia of Space' (Cambridge University Press, Cambridge, 1990) p. 230

9 Lewis, Richard S.: 'The Earth' in Lewis, Richard S. (Ed.): 'The Illustrated Encyclopedia of Space Exploration' (Salamander Books, London, 1983) p. 103

10 Calder, Nigel: 'The Weather Machine' (BBC, London, 1974) pp. 57–9

11 Turnill, Reginald (Ed.): 'Jane's Spaceflight Directory, 1987' (Jane's Publishing Co Ltd, London, 1987) p. 86

12 Tepper, Morris: 'Tiros program results', *Astronautics*, **6** (5), May 1961, pp. 28–9, 63–8

13 Blow, Michael, and Taylor, L. B.: 'Satellites' (Nelson Doubleday Inc, Garden City, 1966 for US Science Service Science Program) pp. 21–2

14 Bono, Philip, and Gatland, Kenneth: 'Frontiers of Space' (Blandford Press, London, 1969) p. 97

15 Thompson, Tina D. (Ed.): 'TRW Space Log 1996' (TRW Space and Electronics Group, Redondo Beach, Volume 32, 1997) p. 68

16 Kiernan, Vincent: 'DoD Admits Secret Space Office', *Space News*, **3** (35), September 21–27, 1992, p. 1

17 Peebles, Curtis: 'Guardians: Strategic Reconnaissance Satellites' (Ian Allan Ltd, London, 1987), pp. 38–9

18 Day, Dwayne A., Logsdon, John M., and Latell, Brian: 'Eye in the Sky: The Story of the Corona Spy Satellites' (Smithsonian Institution Press, Washington, 1998), p. 6

19 Gatland, Kenneth W.: 'Astronautics in the Sixties' (Iliffe Books Ltd, London, 1962) p. 56

20 Sharpe, Mitchell R.: 'Satellites and Probes' (Aldus Books, London, 1970) p. 131

21 Porter, Richard W.: 'The Versatile Satellite' (Oxford University Press, Oxford, 1977) p. 59

22 ibid., p. 60

23 Lewis, Richard S.: 'The Earth' in Lewis, Richard S. (Ed.): 'The Illustrated Encyclopedia of Space Exploration' (Salamander Books, London, 1983) p. 113

24 Von Braun, Wernher, Ordway, Frederick I., and Lange, Harry H.-K.: 'Science of the Seventies: History of Rocketry and Space Travel' (Encyclopaedia Britannica Educational Corp, 1969) p. 184

25 Turnill, Reginald (Ed.): 'Jane's Spaceflight Directory, 1987' (Jane's Publishing Co Ltd, London, 1987) p. 84

26 Mason, B. J.: 'Meteorological Satellites' in Fishlock, David (Ed.): 'A Guide to Earth Satellites' (Macdonald, London, 1971) p. 46

27 Porter, Richard W.: 'The Versatile Satellite' (Oxford University Press, Oxford,1977) pp. 62–3

28 Mason, B. J.: 'Meteorological Satellites' in Fishlock, David (Ed.): 'A Guide to Earth Satellites' (Macdonald, London, 1971) p. 57

29 Campbell, W. J.: 'Satellites: A New Look at Nature' in Fuchs, Vivian (Ed.): 'Forces of Nature' (Thames and Hudson, London, 1977) p. 269

30 Porter, Richard W.: 'The Versatile Satellite' (Oxford University Press, Oxford, 1977) pp. 71–2

31 Turnill, Reginald (Ed.): 'Jane's Spaceflight Directory, 1987' (Jane's Publishing Co Ltd, London, 1987) p. 83

32 Caprara, Giovanni: 'The Complete Encyclopedia of Space Satellites' (Portland House, New York, 1986) p. 25

33 Weaver, Kenneth F.: 'Remote Sensing: New Eyes to See the World', *National Geographic*, **135** (1), January 1969, pp. 62–4

34 Laing, Watson: 'Earth Resources Satellites' in Fishlock, David (Ed): 'A Guide to Earth Satellites' (Macdonald, London, 1971) p. 82

35 Earth Observation Sciences Ltd: 'A Guide to Earth Observing Satellites', Issue 1.0, (National Remote Sensing Centre Ltd, Farnborough, 1990) p. 2.A.2

36 Mason, B. J.: 'Meteorological Satellites' in Fishlock, David (Ed.): 'A Guide to Earth Satellites' (Macdonald, London, 1971) p. 53

37 ibid., p. 65, 69

38 Sharpe, Mitchell R.: 'Satellites and Probes' (Aldus Books, London, 1970) p. 132

39 Smolders, Peter L.: 'Soviets in Space' (Lutterworth Press, Guildford, 1973) p. 82

40 Jung, Philippe (personal communication, 9 May 2005)

41 Clark, Phillip S.: 'Satellite Digest -387', Spaceflight, **47** (1), January 2005, p. 10

42   Caprara, Giovanni: 'The Complete Encyclopedia of Space Satellites' (Portland House, New York, 1986) pp. 25–6

43   Harvey, Brian: 'The New Russian Space Programme' (John Wiley and Sons, Chichester, 1996) p. 192

44   Sharpe, Mitchell R.: 'Satellites and Probes' (Aldus Books, London, 1970) p. 28

Chapter 5

Keeping in touch – the development of the communications satellite

A transmission received from any point on the hemisphere could be broadcast to the whole of the visible face of the globe

Arthur C. Clarke in Wireless World, October 1945

Now that satellite communications is an accepted commercial application of space technology, embedded within the global communications infrastructure, it is difficult to appreciate that, in the early 1960s, it was largely an experimental technology. In common with space science and Earth observation, communications was just another space application with its own dedicated proponents striving to earn its wings.

However, once the pictures of large areas of the planet began to be beamed down from weather satellites, it must have become clear to even the most sceptical that placing a communications transponder on a spaceborne platform had undeniable advantages for global communications.

In common with remote sensing, a key advantage was height above the Earth's surface. Since radio waves travel in straight lines, the coverage of a TV transmission tower or microwave telecommunications mast is directly dependent on its height: the higher the mast, the greater the coverage. In essence, the orbiting satellite was seen as an extremely high transmission tower with, potentially, extremely wide coverage. The radio signals transmitted by Sputnik 1 in October 1957 had, after all, been 'heard' right around the world.

Although Sputnik 1 could not be considered a communications satellite, it had shown that broadcasting radio signals from low Earth orbit was a practical endeavour. A similar attempt to show the potential of radio transmissions from space was made by the US Advanced Research Projects Agency (ARPA) in December 1958. A 68-kg communications package called SCORE (for Signal Communication by Orbiting Relay Experiment) was launched into a low altitude orbit by a converted Atlas ICBM. From an onboard tape recorder, it broadcast President Dwight D. Eisenhower's

Christmas message to the American nation for 13 days, before re-entering the Earth's atmosphere 34 days later.

As discussed earlier, one of the problems with American rockets in the early years of the Space Age was the small amount of payload they could deliver to orbit – in fact, given the poor reliability of the early launch vehicles, engineers were happy to get anything at all into orbit. It was for this reason that project SCORE, using

Table 5.1　Selected communications satellites

Launch date	Name	Nation	Comments
18 December 1958	SCORE	USA	ARPA Signal Communication by Orbiting Relay Experiment (rocket-borne transmitter)
12 August 1960	Echo 1	USA	First passive communications satellite (reflective balloon)
4 October 1960	Courier 1B	USA	First active communications satellite (radio repeater)
21 October 1961	Midas 4	USA	Project West Ford: copper dipole ejection failed (Midas: Missile Defense Alarm System)
10 July 1962	Telstar 1	USA	First privately owned satellite (LEO)
13 December 1962	Relay 1	USA	Experimental communications satellite
14 February 1963	Syncom 1	USA	First intended geosynchronous comsat (exploded following apogee motor firing)
9 May 1963	Midas 6	USA	Project West Ford: copper dipole ejected to form passive reflector
26 July 1963	Syncom 2	USA	First geosynchronous communications satellite
25 January 1964	Echo 2	USA	Passive satellite (balloon)
19 August 1964	Syncom 3	USA	First geostationary communications satellite
22 August 1964	Cosmos 41	USSR	Technology demonstrator for Molniya
11 February 1965	LES 1	USA	First Lincoln Experimental Satellite for military communications (LEO)
6 April 1965	Early Bird/ Intelsat I	USA	First commercial comsat; technology demonstrator for Intelsat (GEO)
23 April 1965	Molniya 1-1	USSR	First Soviet comsat (Molniya orbit)
16 July 1965	Cosmos 71-75	USSR	First Soviet milcomsats (5 launched together)
14 October 1965	Molniya 1-2	USSR	USSR-France communications link
16 June 1966	IDSCS 1-7	USA	Initial Defense Satellite Communications System (7 launched together)
7 December 1966	ATS-1	USA	First Applications Technology Satellite (GEO): comms and meteorology demonstrator
11 January 1967	Intelsat II-F2	USA	First second-generation Intelsat to reach orbit

(Continued)

Table 5.1 Continued

Launch date	Name	Nation	Comments
1 July 1967	IDSCS 16-19 LES 5 DODGE 1	USA	IDSCS 19 demonstrated electronically de-spun antenna; DODGE gravity gradient stabilisation experiment (7 satellites launched together)
5 November 1967	ATS-3	USA	Geostationary comms demonstration satellite
26 September 1968	LES 6	USA	First military communications satellite in GEO
18 December 1968	Intelsat III-F2	USA	First third-generation Intelsat to reach orbit
9 February 1969	Tacsat	USA	VHF/SHF Milcomsat
21 November 1969	Skynet 1A	UK	Second military satellite in GEO/first from United Kingdom
20 March 1970	NATO 1A		First NATO milcomsat
26 January 1971	Intelsat IV-F2	USA	First fourth-generation Intelsat to reach orbit
3 November 1971	DSCS II	USA	First operational satellites in Defense Satellite Communications System (2 launched together)
10 November 1972	Anik F1	Canada	First operational Canadian comsat
19 January 1974	Skynet 2A	UK	First comsat built outside the United States and USSR (placed in incorrect orbit)
30 May 1974	ATS-6	USA	Demonstrated large (9 m) deployable antenna; first TV direct broadcasting satellite
23 November 1974	Skynet 2B	UK	First second-generation Skynet in GEO
19 December 1974	Symphonie 1	France/ Germany	First three-axis stabilised geostationary communications satellite

LEO = Low Earth orbit, GEO = Geostationary orbit

a stripped-down Atlas booster to deliver a tape recorder to low Earth orbit for little more than a month, was deemed such a success.

This relatively inauspicious beginning to space-based communications held few clues to the future development of communications satellite technology, but the genesis of a communications revolution was at hand.

Active and passive satellites

Although satellite communications, as a satellite application, was waiting in the wings behind space science and Earth remote sensing, its basic techniques were being used on almost every space mission. Spacecraft telemetry, after all, was enabled by the transmission of radio signals from the satellite to the ground, while instructions or 'commands' to the satellite and its equipment were transmitted in the opposite direction. This led to the development of an integrated communications system known as a telemetry and command system, which as a by-product of the signal emitted by the satellite allowed it to be tracked by a ground station. For this reason, the system became known as a telemetry, tracking and command (TT&C) or telemetry, command and ranging (TC&R) subsystem.

To qualify as a practical communications satellite, however, a spacecraft had to be able to do more than simply receive commands on the uplink path and return separate data to a ground station on the downlink: it had to be able to intercept a signal transmitted from the Earth, amplify and filter that signal, and pass it back to Earth, to one or more receiving stations in widespread locations. Clearly, the further apart the locations, the greater the advantage of using a satellite, since it avoided the need for chains of land-based radio masts or long, undersea cables.

This reception, amplification and retransmission is an 'active' rather than 'passive' process, since it requires onboard equipment to actually *do* something. In the early days of space communications, however, experiments with passive reflectors were an obvious starting point, and perhaps the most obvious satellite to use was the Earth's own natural satellite, the Moon.

In the spring of 1946, the US Army Signal Corps successfully demonstrated that a short-wave radar beam could be reflected from the surface of the Moon. There was no attempt to incorporate a message into the transmission, but it showed that long-range communications was possible with standard and comparatively low-powered radar equipment [1]. Strangely, considering this early start, it seems that there were no serious experiments in 'lunar communications' until the late 1950s. The 75m-diameter Jodrell Bank radio telescope, which had become operational just in time to track Sputnik 1 in 1957, played a large part in these pioneering experiments, which included the transmission of intelligible messages, including speech and music, across the Atlantic between the United Kingdom and the United States [2]. Teleprinter communications at about 50 bits/s was also demonstrated using much smaller, 3.5m-diameter, earth stations with 2 kW transmitters [3].

A link between Maryland and Hawaii, operated by the US Navy, was in regular use for teleprinter communications in 1960 [4]. One of the more visual experiments featured the transmission, from Hawaii to the US mainland, of an aerial photograph of the aircraft carrier *Hancock* with the words 'MOON RELAY' stamped across its deck in aircraft-sized letters [5].

Experience showed that the optimum frequency range for the successful reception of these Moon-bounced signals was between 3 and 5 GHz, in the lower part of the electromagnetic spectrum known as the microwave band.[1] The experiments

---

[1] The microwave band is defined as encompassing the frequencies $1 \times 10^9 - 3 \times 10^{11}$ Hz or 1–300 GHz (wavelengths between 0.001 and 0.3 m) [6]; under an alternative definition the microwave band covers 1–30 GHz and 30–300 GHz is termed the millimetre-wave band.

used a device developed in 1953 by the American physicist Charles Hard Townes. Known as a maser – an acronym for microwave amplification by stimulated emission of radiation – it was designed to amplify signal frequencies in the microwave band, which includes the high-frequency end of the familiar UHF band, plus SHF and EHF (ultra, super and extra high frequency, respectively).

In 1958, incidentally, Townes and his brother-in-law, Arthur Schawlow, went on to publish a paper on the possibility of an 'optical maser' for the amplification of infrared or visible light instead of microwaves. The first practical laser – meaning *light* amplification by stimulated emission of radiation – was constructed by Theodore Harold Maiman, another American physicist, in 1960. It is now, of course, a key component in the optical fibre links of terrestrial communications systems and in domestic consumer devices such as CD and DVD equipment.

Balloons and needles

Active and passive satellites were developed in parallel in the early 1960s. The best known passive satellites were NASA's Echo 1 and Echo 2, placed in low Earth orbit on 12 August 1960 and 25 January 1964, respectively. They were spherical, aluminium-coated Mylar balloons (30 m and 41 m in diameter), used simply as radio reflectors (Figure 5.1).

During its first day in orbit, Echo 1 was used to reflect another of President Eisenhower's tape-recorded messages from a transmitting station in Goldstone, California (part of NASA's Deep Space Network) to a receiving station in Holmdel, New Jersey. The following day, it relayed a telephone conversation between engineers

Figure 5.1   Echo balloon [NASA]

in the respective earth stations. After a year or so of successful operation in orbit, the ultra-thin material of the balloon had been degraded by cosmic rays, UV radiation and micrometeorites and its surface had become wrinkled. However, it was still able to reflect radio signals carrying telephone conversations and even pictures, albeit less efficiently [7]. Despite the obvious limitations of this technique, feasibility studies had suggested that a chain of some 20 such spheres around the Earth would be sufficient to provide global coverage [4].

Other experiments in passive space communications ranged from using existing 'space junk' as a peripatetic orbiting reflector to sending up even more garbage! An exponent of the latter practice was a US Air Force communications programme known as Project West Ford. Its intention was to girdle the Earth with a passive reflector formed from 18 mm copper dipoles (lengths of 'hair-like' copper wire), towards which digitised speech signals could be transmitted at a frequency of 8 GHz. The experiment was first attempted on 21 October 1961 when the Midas 4 satellite, placed in a polar orbit, ejected a container from which the 350 million dipoles, or 'needles', were intended to disperse. The needles were embedded in naphthalene, which was supposed to evaporate, slowly releasing them to form an orbital belt some 8 km wide and 40 km deep. Ground radar subsequently located the container but failed to detect the dipoles and it is assumed that they did not disperse [8].

In May 1963, however, a 23 kg payload of copper dipoles was ejected from the Midas 6 satellite at an altitude of about 3600 km. Although the experiment was deemed a success at the time, the long-term consequences of a malfunction in the needle dispensing system has cast a shadow on the endeavour. While individual needles, correctly dispensed, were expected to re-enter the atmosphere relatively quickly as a result of solar radiation pressure, the malfunction is believed to have formed clumps of needles which remain in orbit, as a potentially destructive force, to this day [9].

Courier

In parallel with NASA's experiments in passive satellite communications, the US Army – which had earlier launched America's first satellite Explorer 1 – was developing the world's first active communications satellite, or 'comsat' as they came to be called. Named Courier, it was spherical in shape, about 1.5 m across, and weighed about 230 kg. Apart from a section round its midriff, it was covered in solar cells. Unfortunately, during the first launch attempt – of Courier 1A on 18 August 1960 – the launch vehicle developed a fault about two-and-a-half minutes after lift-off and had to be destroyed by range safety officers [10].

The first satellite to carry an active radio repeater was therefore Courier 1B, launched on 4 October 1960, three years to the day after Sputnik 1; sadly it operated for only 17 days, four less than the Sputnik [11]. It was, however, a true communications satellite, insofar as it carried a communications payload incorporating four receivers, four transmitters and five tape recorders [10]. The inclusion of the latter identifies Courier as a 'delayed repeater' satellite (now termed 'store-and-forward'), since uplinked signals were recorded on tape while passing over a transmitting ground station, then replayed later when in range of a receiving station. This method was prescribed mainly by the satellite's relatively low (968 km × 1215 km [11]) orbit, which meant that two ground stations were less likely to be in view at the same time. For an instantaneous, active satellite communications link, the world would have to wait until 1962.

Telstar

On 10 July 1962, America placed Telstar 1 in orbit (Figure 5.2). As a result of the publicity surrounding it, the satellite immediately became a household name and, as an icon of the new technological age, it even had an electronically synthesised pop-tune named after it. More importantly, Telstar 1 was the first satellite to receive and transmit simultaneously, using an uplink frequency of around 6 GHz and a downlink frequency of about 4 GHz (known as C-band – see Box). This straightforward reception and onward transmission was known by those working in the field as 'bent-pipe communications', because they saw the satellite repeater, or transponder,[2] as a conduit for a signal received from one direction and transmitted to another. Later, more complex satellites provided a degree of onboard signal processing to which the term was not applicable.

Telstar 1 was the first satellite to carry TV signals across the Atlantic. In fact, according to a paper by F. J. D. Taylor, within a few days of launch the satellite had been used to transmit black and white TV and sound, colour test slides, multi-channel telephony, telegraphy and facsimile between Europe and the United States [13]. Taylor was the head of operations at the Goonhilly Downs satellite station and, later, the first director of technology for the Communications Satellite Corporation (see later). His paper was published as part of the proceedings of an International Conference on Satellite Communications held in November 1962 at the Institution of Electrical Engineers (IEE) headquarters in London. Apart from being one of the earliest conferences on the subject, Taylor's paper was extremely topical, being written only seven weeks after Telstar 1 was placed in orbit. The same ground-breaking conference featured at least three papers on Telstar, one on Moon Bounce communications and others on a variety of satellite communications topics [3,14–22].

Telstar 1 was also the first privately owned satellite, thus marking the beginning of an evolution of the comsat from scientific curiosity to revenue generator. The satellite was owned and operated by the American telecommunications giant, AT&T (American Telephone and Telegraph Company), whose satellites bore the name Telstar for many years thereafter. The spacecraft itself was designed by John Pierce, an engineer at AT&T's Bell Telephone Laboratories who became a leading light in

---

[2] By definition, if a spacecraft's communications payload comprises a single transponder chain, the term 'transponder' (a contraction of transmitter and responder) includes all the devices of that payload with the exception of the receive and transmit antennas. Usually, however, a communications satellite has many transponder chains in its payload which, when grouped together (without the antennas), are known as a repeater [12].

Satellite communications frequency bands

The radio spectrum has been divided into a number of bands by the International Telecommunication Union (ITU). Communications satellites use frequencies in the SHF band (Super High Frequency: 3–30 GHz), but may also use UHF (Ultra High Frequency: 300–3000 MHz) and EHF (Extra High Frequency: 30–300 MHz).

For many communications applications the sub-bands are given letter-designations, for example:

P-band: 0.23–1 GHz
L-band : 1–2 GHz
S-band : 2–4 GHz
C-band : 4–8 GHz
X-band : 8–12 GHz (in USA, 8–12.5)
Ku-band: 12–18 GHz (in USA, 12.5–18)
K-band : 18–27 GHz (in USA, 18–26.5)
Ka-band: 27–40 GHz (in USA, 26.5–40)
O-band: 40–50 GHz
V-band: 50–75 GHz

These designations are intended as a guide to common usage in the satellite communications industry and should not be taken as definitive for all applications. The above compilation is broadly based on the IEEE (Institute of Electrical and Electronic Engineers) Radar Standard 521, which also defines 40–100 GHz as the millimetre-waveband (mm-wave) and 0.3–1 GHz as UHF (at variance with the ITU definition). Readers may also come across, among others, R-band (1.7–2.6 GHz), H-band (3.95–5.85 GHz), and designations which differ in that Ku-band is shown as J-band and Ka-band as Q-band (although the range is 33–50 GHz). The designations Ku-band and Ka-band are derived from their position 'under' and 'above' K-band, respectively. Still other designations refer to U-band as 40–60 GHz and W-band as 75–110 GHz, so band designations are far from standardised throughout the world [6].

the field of satellite communications (Figure 5.3). Like Courier it was roughly spherical, but much smaller and lighter: about 0.9 m in diameter and 77 kg. It drew its power from 3600 surface-mounted solar cells, which provided just 15 W with which to operate the satellite and recharge its 19 nickel cadmium battery cells. The solar cells were mounted, in flat panels, on most of the 72 facets of the satellite's outer skin, which was made of aluminium; three of the facets held mirrors designed to reflect tiny flashes of sunlight to optical tracking telescopes on the ground [23].

Although solar arrays had, by then, become an accepted part of satellite technology, their design and construction was far from trivial. The silicon cells on Telstar 1, for example, were bonded to a ceramic base within a platinum frame to provide

Figure 5.2   Telstar [Used with permission of Lucent Technologies Inc/Bell Labs]

Figure 5.3   John Robinson Pierce, former director of research at AT&T Bell Telephone Laboratories [NASA]

a sufficiently rugged mounting structure for the fragile cells. And to protect them from the Sun's high energy radiation and particle flux, which degrade the cells' efficiency with time, transparent coverslips of man-made sapphire were bonded to their top surfaces [23]. The engineering challenge, of course, was to ensure that all the different materials would expand and contract, in response to the substantial temperature changes experienced in space (typically several hundred degrees Celsius), without cracking, debonding or severing their electrical connections. Such was the complexity of this and other design challenges that engineers began increasingly to specialise in various aspects of spacecraft design, such as power, materials and thermal engineering.

Moreover, as satellite users began to expect their spacecraft to last more than a few weeks in orbit, the longevity of all onboard equipment became an increasingly important issue. The lifetime of Sputnik 1, for example, was constrained by its non-rechargeable battery, but once satellites began to carry solar cells their potential lifetimes increased substantially, as shown by the 'inexhaustible' Vanguard 1 which transmitted for six years (see Chapter 3).

As a result of this increasing longevity, it became necessary to ensure that all spacecraft equipment was designed to the same standards, in order to meet the design lifetime of the spacecraft. This was simply an example of the 'weak link in the chain philosophy': if one key item failed, it could lead to the failure of the entire mission. So, as equipment reliability became more of a concern, it too evolved into a spacecraft engineering discipline in its own right [24].

Although the foundations of reliability, such as life tests, failure statistics and materials fatigue, have a longer history, the discipline of reliability engineering is not much older than the space industry itself. It is traditional to set its birth date in 1952 when Robert Lusser, an engineer at the Redstone Arsenal in Huntsville, Alabama, presented a formal definition of reliability at a San Diego symposium: 'The reliability of an object is the probability that it will perform correctly for an assigned period of time and under specific conditions' [25]. Of course, this begs the questions of what is 'correct performance' and what are the 'specific conditions', but these are factors that can be defined before a reliability assessment is made. Certainly, as far as space development was concerned, it was clear by the time Telstar was being designed, that the conditions mentioned by Lusser were very special indeed.

Part of the reliability assurance process was to take the utmost care in everything from selection of materials to the manufacturing of components; another was the rigorous testing at every stage of the process, including component selection. Testing a batch of components in a process known as 'burn in' allowed those that were prone to early failure (so-called 'infant mortality') to be weeded out. This automatically selected, as flight hardware, those that might be expected to operate for the satellite's lifetime. In addition, once satellite lifetimes began to increase, it became necessary to consider the wear-out mechanisms of components as well, a much less important factor when lifetimes were short.

Of course, once the components had been selected, it was important to ensure that they would survive the ride into orbit. In a concerted attempt to protect Telstar's delicate electronic equipment from shocks and vibrations, it was set in polyurethane foam and then hermetically sealed in a 0.5 m cylindrical aluminium canister, which itself was suspended inside the satellite's spherical outer shell by nylon cords [23]. Telstar contained 2528 semiconductor devices – 1064 transistors and 1464 diodes – and one travelling wave tube, the satellite's high-power amplifier [26].

Relay

On 13 December 1962, Telstar was joined in orbit by Relay 1, an experimental communications satellite developed by NASA's Goddard Space Flight Center and the Radio Corporation of America (RCA), which had also built the Tiros weather satellites (as described in Chapter 4). As an indication of the growing interest in satellite design and manufacturing, its specifications had been issued to no fewer than 41 US companies at a briefing held at Goddard in January 1961. RCA's design proposal was chosen from those of seven finalists, rewarding it with a contract for more than $3 million (Figure 5.4) [27].

In the interests of reliability, the satellite contained two communications transponders, each with its own travelling wave tube amplifier. Each was capable of handling television, telephony or other forms of data transmission, but if one should fail the other could operate independently. This was an early example of a reliability assurance technique known as redundancy.

One of Relay's claims to fame was its transmission of the first 'live' colour TV pictures across the Pacific on 22 November 1963. Japanese audiences were

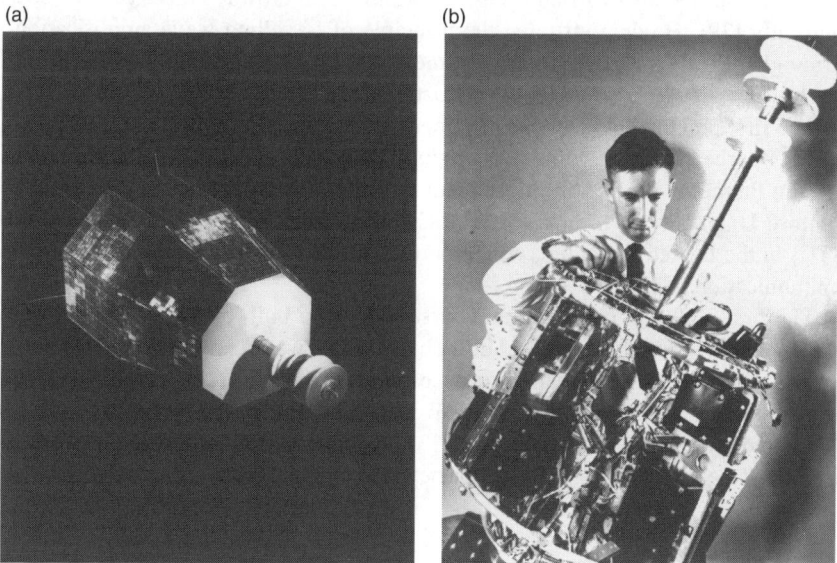

Figure 5.4   *(a) Artist's impression of Relay in space, (b) Relay 1 assembly at RCA Astro-Electronics Division [NASA]*

treated to a discussion between NASA Administrator James Webb and the Japanese Ambassador to Washington, Ryuji Takeuchi. They were to have seen a taped greeting from President John F. Kennedy, but instead were informed of his assassination in Dallas a few hours earlier [28].

In addition to its communications payload, Relay carried experiments to continue the work on space radiation, particularly with regard to its effects on semiconductors. These included silicon diodes and samples of the two available types of solar cell (p/n and n/p semiconductors). It was experiments such as this which led to the conclusion that the n/p cell was the more resistant to radiation. As part of a scientific package, the satellite also carried radiation monitors to measure electron and proton energy spectra and the anisotropy (directional dependence) of the radiation [19].

The travelling wave tube amplifier

Although the typical communications satellite contained a number of crucial elements, the heart of the beast was the travelling wave tube (TWT), which is part of a family of microwave tubes – including klystrons, crossed-field amplifiers, magnetrons and gyrotrons – used for high power generation or amplification in fields as varied as military radar and particle physics research. When a TWT is combined with its power supply unit, or electronic power conditioner (EPC), it is known as a travelling wave tube amplifier (TWTA).

The TWT was invented in 1943, during the latter part of Second World War, by Rudolf Kompfner, an Austrian refugee working on microwave tubes for the British Admiralty [29]. He demonstrated the principle of travelling wave amplification at Birmingham University later that year and published the first results of his work in the November 1946 issue of *Wireless World* magazine [30].

The first practical device was developed at Bell Telephone Labs (BTL) in 1945 by J. R. Pierce and L. M. Field and a detailed theory of its operation was published by John Pierce in 1947 [31]. Subsequent development work was done at BTL and Stanford University in the United States, and by Standard Telephones and Cables (STC) in the United Kingdom, with a particular eye to potential applications in the communications field [29].

The first TWTs to enter operational service were built by STC for a television relay link between Manchester and Edinburgh, which was operated by the Post Office and used by the British Broadcasting Corporation (BBC) from 1952. These early tubes produced an output of about 2 W across a band of frequencies centred on 4 GHz and had a gain of about 25dB[3] [32]. Meanwhile, the military services had discovered the potential for TWTs in the newly expanded fields of radar and electronic

---

[3] The dB (decibel) is used in communications technology to denote the gain (amplification factor) of an antenna, amplifier, etc. It represents a logarithmic ratio between two signal power levels, such that $dB = 10 \log_{10} P_1 / P_2$, where P_1 and P_2 are the output and input powers, respectively. For example, a gain of 3 dB is equivalent to a doubling in signal strength; 60 dB represents a factor of 1,000,000.

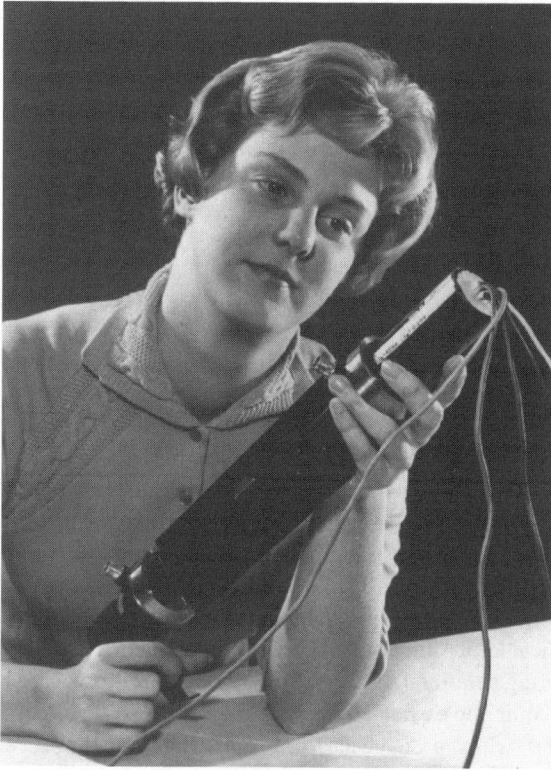

Figure 5.5   *Relay travelling wave tube (TWT): weight 1.6 kg; output RF power 11 W;*
             gain 33 dB [NASA]

counter-measures (ECM), for which the high gain and wide signal bandwidth of the
TWT were ideally suited.

Another type of electron tube, the klystron, was developed at Stanford University
in the late 1930s by two brothers, R. H. and S. P. Varian. The klystron power amplifier
(KPA) provides a useful alternative to the TWTA in radar systems and satellite earth
stations. However, its narrower operational bandwidth and the fact that it cannot be
easily re-tuned has made the klystron relatively unattractive for space applications.
By contrast, the technical attributes of the TWT made it a key technology for satellite
communications and it was eagerly adapted to suit this new role.[4]

---

[4] From the mid-1980s, the TWT began to be replaced by the solid state power amplifier (SSPA),
typically based on gallium arsenide field effect transistors (GaAsFETs). Although the performance and
reliability of SSPAs has been improved since their introduction, they are limited to relatively low-power
applications at a given frequency since the solid-state medium is a much poorer conductor of heat than the
materials used in a TWT.

Much of the early development work on 'space tubes' was done by Hughes Aircraft Company in the late 1950s, eventually leading to the formation of its Electron Dynamics Division and the inclusion of its devices in many communications satellites and other spacecraft [33].

Indeed, one could argue that the development of the TWTA was as important to the development of space technology as any other device, simply because of the large distances involved in space communications. All transmission systems which rely on radiated RF energy have to combat what is known as spreading loss, the decrease in power flux density – or signal strength – resulting from the divergence of a beam of radiation with distance from the transmitter. With terrestrial microwave links the distance between towers is measured in tens of kilometres, so the spreading loss is relatively small, but for satellites hundreds or thousands of kilometres away the loss becomes a crucial factor in the link design. The governing factor is the immutable inverse-square law, which rules that signal intensity decreases with the square of the distance.

For example, a typical earth station antenna with a beamwidth of $0.25°$ delivers all of its usable 'amplification power' within that narrow beam, which by the time it reaches a satellite can be tens to hundreds of kilometres across. The fact that a spacecraft antenna can intercept typically one part in 10^{10} of the energy in that beam makes spreading loss by far the most important factor in a spacecraft communications link budget (in technical terms amounting to a loss of more than 200 dB in signal strength).

The engineering solution in the early days, when electronic circuits were in their infancy, was to equip the earth stations with high power transmitters and very large-diameter antennas (which produce narrow beams). The high power amplifiers (HPAs) employed in the stations were either travelling wave tubes or klystrons, both of which were capable of producing output powers from several hundred to several thousand watts (in terms of the link budget, a thousand-watt HPA contributes 30 dB to the positive balance). This was the solution embodied in NASA's Deep Space Network (DSN), which was discussed in relation to science spacecraft in Chapter 3.

By contrast with the ground-based amplifiers, the TWTs installed on both the Courier and Telstar satellites had output powers of only 3 W, while Relay's were 11 W devices (Figure 5.5); and to make matters worse, their antennas were non-directional. The receive and transmit antennas on Telstar, for example, comprised two rings of simple dipoles housed in cavities around the equatorial section of the satellite which were sized to match the wavelength of the signals, the smaller ones for the 6 GHz uplink and the larger ones for the 4 GHz downlink. The telemetry and command antenna was a simple, helical wire antenna mounted on top of the satellite (as shown in Figure 5.2).

Because of the difficulty in arranging for an antenna to remain pointing towards the Earth, especially on a spin-stabilised satellite, the early spacecraft all had non-directional or 'omnidirectional' antennas, which meant that only a fraction of the already miniscule amount of RF power available from the amplifiers was radiated towards the Earth. This placed the onus very much on the ground station, not only to receive the satellites' tiny downlink signals but to radiate a sufficiently strong uplink signal for the spacecraft to receive.

The ground segment

In the early days of the Space Age, the severe limitations placed on what became known as the 'space segment' had a direct impact on the development of the 'ground segment' or 'earth segment' – the stations designed to transmit and receive either communications traffic or command and telemetry data. First, the mass of the spacecraft that could be placed in orbit was limited by the payload capability of the launch vehicles. Second, and as a direct result, the size of the spacecraft placed a limitation on the area of solar array, which in turn limited the power available to the payload. Together, the mass and power limitations of spacecraft made the development of the ground segment equally as important for this new era of telecommunications as the development of the satellites themselves.

From the outset, the great majority of earth station antennas took the form of a parabolic 'dish' with a transmit/receive feedhorn at its focus. However, two historic earth stations, built in the United States for the early satellite experiments, were based on a design that has since been relegated to history. They took the form of a horn, shaped much like an ice cream cornet with a hand cupped over the open end to reflect signals into it. The first was built by Bell Telephone Labs at Holmdel, New Jersey, for experiments with Echo 1 and the second at Andover, Maine, for Telstar (Figure 5.6).

The Andover horn was the larger of the two and a true giant: it was 29 m high, 54 m long and mounted on a track with a diameter of 55 m. The whole antenna was rotated to follow the satellite in its path across the sky, its moving parts weighing some 380 tonnes [34]. The inflatable radome that housed the antenna measured 49 m wide and 64 m high and was made of 1.6mm-thick Dacron and synthetic rubber, containing some three acres of material [35].

The main signal amplifier at each of the two stations was a ruby maser. The horn antennas were designed to focus a satellite's signals onto a detector, which was cooled in a bath of liquid helium to little more than two degrees above absolute zero (which is $-273.15°C$) [36]. This was done to reduce the detector's thermal noise – a result of the intrinsic random motion of its constituent electrons – which would otherwise obliterate the satellite's diminutive signals.

Interestingly, it was this ability to detect extremely weak radiation which reserved a place for the Holmdel station, not only in the annals of satellite communications, but also in the history of radio astronomy. While working with the horn in 1965, two BTL physicists, Arno Penzias and Robert Wilson, discovered a weak but persistent level of microwave radiation, at an effective temperature of 3 K, emanating from the sky. The radiation was the same in all directions and invariant with the time of day. At first, the astronomers had suspected the constant level of background noise to be due to what they called 'a white dielectric material' (pigeon droppings), which had been deposited in the horn. However, it was later deduced to be the radiation remnant of the 'Big Bang' which brought the Universe into existence, and is now referred to as the 'microwave background' or '3 degree-K background' radiation.

In preparation for experiments with transatlantic satellite links, a duplicate of the larger Andover horn was built at Pleumeur-Bodou in Brittany, France, but in England it was decided to install what is now seen as the more conventional, parabolic antenna.

(a)

(b)

Figure 5.6   *(a) Horn antenna at Andover, built for experiments with Telstar [Used with permission of Lucent Technologies Inc/Bell Labs]. (b) Horn antenna at Holmdel station, New Jersey. In 1990 it was dedicated to the National Park Service as a National Historic Landmark [NASA]*

Figure 5.7 Goonhilly Aerial 1 [Mark Williamson]

The decision, made in early 1961, to build a 26m-diameter earth station at Goonhilly Downs was based on experience with the Jodrell Bank Mark I radio telescope, itself a fully steerable parabolic antenna. Perhaps equally as important, the parabolic solution was also much cheaper than the French horn [37].

The site of the Goonhilly station, on Cornwall's Lizard peninsula at the southern-most tip of mainland England, was chosen to give the best possible coverage of satellites orbiting the Earth's equatorial and middle latitude regions, and also because the area was 'electronically quiet' in terms of RF interference. The design, manufacture, construction and testing of 'Goonhilly Aerial 1' was completed in less than a year and the station was fully operational in time for the launch of Telstar (Figure 5.7) [17].

Once the satellite was in orbit, the records came thick and fast: at 7.28 pm US east-coast time on 10 July 1962, Telstar 1 handled the first telephone call through an active communications satellite; at 7.31 pm it carried the first TV signal via satellite; and at 7.47 pm the first *transatlantic* TV signal, transmitted from the United States and received in France. The first TV picture to grace a satellite channel showed the US flag, flapping in the breeze, against the backdrop of the inflatable radome of the Andover station; the picture was accompanied, perhaps predictably, by the stirring melody of 'The Star Spangled Banner'. Then, on the final test of its sixth orbit, Telstar transmitted a Polaroid photo of itself, held in front of the TV camera by two steady hands; as a fitting tribute, it was selected by the Press Association as its 'news shot of the day' [38].

Although most of the glory was rightly due to the US, the competitive spirit between the British and the French to be first to receive the transatlantic signals was rife. The target date for both countries' experiments with Telstar was originally 20 July, but shortly before launch it was decided to bring this forward to launch day, 10 July [39]. The British proudly declared their readiness to receive Telstar's signals several hours before the French, but, when it came to the test, the French station at Pleumeur-Bodou was the first to receive the TV transmission. Goonhilly was initially unable to receive the signal because its antenna was set to receive a signal of the opposite polarisation.

It appears that it was a telephone call between British and American engineers on the definition of circular polarisation that led to a misunderstanding of which polarisation would be used [37]. According to F. J. D. Taylor, the Goonhilly engineers initially had the polariser the right way round, but changed it after talking to the engineers at Andover; it was difficult for the Goonhilly team to change it back at short notice because the polariser was '90 ft above ground' and for safety reasons they had to wait for daylight [40]. The problem apparently arose due to a fundamental difference in the way circular polarisation was defined on either side of the Atlantic: the CCIR[5] recommendation (used in Europe) defined polarisation 'looking away from the transmitter', whereas the IEEE[6] definition, used in the United States, defined it 'looking towards the transmitter' [41].

Chalking it up to experience, the British offered their warmest congratulations to the French for receiving the first transatlantic TV signal. But congratulations turned to indignation the next day when Pleumeur-Bodou made the first east–west telecast to the United States. It was declared that the French had broken an agreement among the 16 European nations of the Eurovision network that the first programme televised from Europe would be a joint effort. The French replied that their telecast, introduced by their Minister of Communications and featuring songs from a number of French entertainers, was not 'a programme' but 'a test'. Whatever it was, the transmission was not immune to 'finger trouble': with a full two minutes of transmission time left, the horn at Pleumeur-Bodou lost its 'lock' on the satellite when someone pressed the wrong button [42].

The fact that the French telecast had been pre-recorded on tape allowed the British to redress the balance on Telstar's next orbit. Their nine minute telecast from Goonhilly was, for the most part, 'live'. It showed engineers and technicians in the Goonhilly control room and gave Britain the distinction of being the first to transmit live TV across the Atlantic [43]. F. J. D. Taylor remembered this, in 1991, as follows:

> It featured Capt. Charles Booth (Deputy Engineer-in-Chief), John Bray (Head of the Radio Branch) and myself (the three 'Downhilly Goons'). We accepted the invitation from Andover that we should 'go live'. Our intention was merely to take the opportunity to thank our many BTL friends. In actual practice, they were so pleased with the quality of reception that they offered it immediately to CBS so it was seen throughout the States [44].

---

[5] French abbreviation for International Radio Consultative Committee, a body established within the International Telecommunication Union (ITU) in 1927 to consider and issue recommendations on technical, operational and tariff matters relating to radio frequency use.

[6] Institute of Electrical and Electronics Engineers.

Today, we are so used to seeing satellite earth stations dotted around our towns and countryside that it is easy to forget that throughout the 1960s they were very few and far between. In 1965, three years after Goonhilly Aerial 1 was commissioned, there were only four earth stations in operation worldwide, and even by 1970 there were still only 22 in operation, another 20 under construction and some 20 more in the planning stages [45]. Since then, of course, the number has become practically uncountable.

Geostationary orbit

Prior to 1963, all satellites were placed in what is now referred to as low Earth orbit (LEO). Far from being a single, clearly defined orbit, the term encompasses a multitude of different orbital paths from several hundred to over a thousand kilometres above the planet, including both circular and elliptical orbits with inclinations from equatorial to polar.

Anything placed in orbit is a slave to the laws of orbital dynamics. As examples, the Tiros 1 weather satellite (discussed in the previous chapter) had an almost circular orbit – apogee 722 km and perigee 677 km – giving it an orbital period of 98.8 min [46], while Telstar 1 was boosted into a somewhat higher and more elliptical orbit (5653 km × 936 km) producing a 157.8 min period [47] (their orbital inclinations were similar, at 48.4° and 44.8°, respectively). This meant that all satellites in LEO spent only a limited time in sight of a given earth station and had to be tracked continuously to maintain communications, as indeed they do today. It was obvious that a single satellite in LEO could not form the basis of a permanent communications service.

The solution to providing a permanent communications link was to place the satellite in geostationary orbit, where the majority of today's communications satellites reside. Geostationary orbit (commonly abbreviated to GEO) is a circular orbit in the same plane as the Earth's equator with an altitude of 35,786 km (Figure 5.8). A satellite orbiting at this altitude in the same direction as the Earth's rotation has an orbital period exactly the same as the Earth's sidereal period of rotation (23 h 56 min 4 s), which means that it appears stationary from the point of view of an earth station. This offers a considerable reduction in the cost and complexity of the earth segment, and, perhaps more than any other single factor, has been responsible for the success of satellite communications.

The usefulness of this unique orbit in the development of a global satellite communications network was first brought to public attention by Arthur C. Clarke in the October 1945 issue of *Wireless World* magazine. Among other things, he suggested that near-global coverage could be provided with just three geostationary satellites. As a sign of the times, Clarke's article, entitled 'Extra-Terrestrial Relays', prompted several indignant letters from readers who 'disapproved of "jokes" of this kind' [48]. Typically, it was the editor who was responsible for this title, having changed Clarke's prosaic 'The Future of World Communications' to something a little more lively. He did, however, retain the subtitle 'Can Rocket Stations Give World-Wide Radio Coverage?' [49,50]. Whatever the title, it is to the credit of the editor that he chose

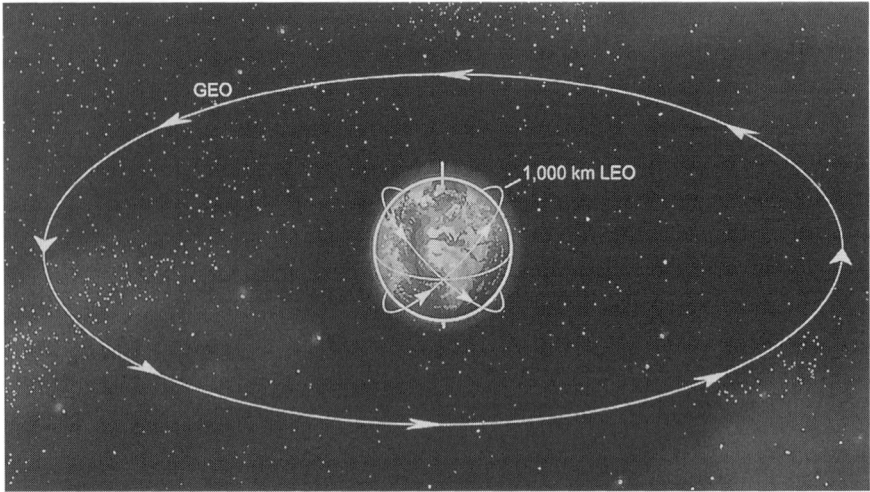

Figure 5.8   Low Earth Orbit (LEO) compared – approximately to scale – with geostationary orbit (GEO) [Mark Williamson]

to accept the paper, from a budding science fiction writer, some 12 years before the launch of Sputnik 1.

Clarke's article gives a good summary of the system aspects – orbital positions, coverage, signal strength and so on – but is less accurate regarding the technology required to realise the system (hardly surprising, since satellite technology was non-existent!). For example, he predicted that 'all space-rockets will be launched from very high country' to minimise air resistance, completely missing the fact that placing launch sites near to the equator would be far more beneficial in terms of payload capability. The same launch vehicle could launch at least 10 per cent more payload from the equator than from Kennedy Space Center, for instance [51].

As for on-board power, Clarke foresaw the use of 'solar engines' employing 'mirrors to concentrate sunlight on the boiler of a low-pressure steam engine', though he did concede that 'photoelectric developments may make it possible to utilise the solar energy more directly' (a nod to the solar cell). Finally, as a result of receiving the article proofs a few days after the *Enola Gay* had dropped its atom bomb on Hiroshima [52], he added an epilogue on 'atomic power', which he felt had 'brought space travel half a century nearer' [50].

Although, according to Clarke's biographer, he can recall no response, positive or negative, to the appearance of his article, stating that it was received with 'monumental indifference', Clarke later learned that the US Navy, which had begun a study of rockets and spaceflight in 1944, had noted the article [53] and no doubt incorporated its ideas into the study.

In fact, Clarke had proposed 'the idea of geostationary relays' in a letter to *Wireless World* of February 1945 and his original proposals for a geostationary satellite system were presented a few months later in the form of a memorandum to the Council of the British Interplanetary Society, which supported the wider dissemination of the ideas. The document, dated 25 May 1945, is now in the archives of the Smithsonian Institution in Washington, DC [54[7]]. Coincidentally, that same year, Pierce and Field had developed the first practical TWT, the amplification device that formed the heart of the communications satellite in decades to come. Clarke, himself, writes glowingly of Pierce in his book 'The Promise of Space' as 'Bell Laboratories' energetic director of communications research' who, he says, 'had dragged [AT&T] singlehanded into the space-communications field' [55].

Clarke's *Wireless World* article remained the only one on the subject for some ten years, until Pierce published a paper titled 'Orbital Radio Relays' in the April 1955 issue of *Jet Propulsion*, the journal of the American Rocket Society. The paper proposed a method of transoceanic communication that involved relaying signals 'by means of a satellite revolving about the earth above the atmosphere' [56]. This almost pedantic reference to the position of the satellite 'above the atmosphere' (as if it could be anywhere else) is an indication of the unfamiliar nature of the idea. Even then, two-and-a-half years pre-Sputnik, there were quite a few people who ridiculed the very idea of the artificial satellite. Pierce was not among them, though his paper is clear to point out that, in 1955, there was 'no unclassified information in regard to how long it will be before a satellite could be put up or what it might cost to do so'.

Not surprisingly, Pierce wrote of the possibility of including a TWT on the satellite, citing an output (RF) power of just 30 mW from an input (DC) power of 'five watts or less'. He also mentioned the usefulness of 'a 24-hr orbit (at an altitude of around 22,000 miles)', but made no reference to Clarke's paper. One American author puts this down to *Wireless World* being 'a little-read British publication' of which Pierce did not learn 'until he was well along in his own studies'. He also reports that although 'Clarke reiterated his ideas in his 1952 best seller, *The Exploration of Space*, ... Pierce missed that one, too'[57].

Pierce did, however, state in his final sentence: 'It is interesting to note that the problem of radio links to the planets has been treated by Richey as early as 1938 (7) and in considerable detail by J. J. Coupling in 1952 (8)', the relevant references being:

'7 "Are you There, Mars," by J.R.Richey, *The Sky*, September 1938, pp. 20,30'

and

'8 "Don't Write: Telegraph!" by J.J.Coupling, *Astounding Science Fiction*, vol.49, March 1952, pp. 82–96.' [56].

What Pierce omitted to say was that his *Jet Propulsion* paper was not the first public airing of his ideas on space-based communications: J. J. Coupling was, in fact, the nom de plume of J. R. Pierce [58].[8]

---

[7] This reference also includes Clarke's May 1945 Memorandum and his *Wireless World* paper as appendices.

[8] It seems likely that Pierce derived his pen name from his knowledge of physics, since 'J' is a vector quantity relating the electron spin and orbit vectors in the field of spin-orbit interaction, or coupling.

Signal delay

Towards the end of the 1950s, when geostationary satellites were first being contemplated in serious engineering terms, there were still many obstacles to overcome. For a start, the restrictive payload fractions of the early launch vehicles made placing a satellite in a high altitude orbit impossible and, second, there was significant disagreement as to whether geostationary orbit was, in fact, the 'ideal orbit' for the communications satellite.

Although the altitude of GEO brought about the advantage of the 'stationary satellite', it also introduced a previously unheard of signal delay. Satellites in LEO are close enough to the Earth for transmission delays to be comparable with transcontinental terrestrial links, but a signal transmitted from the Earth takes at least 0.12 s to reach a satellite in GEO. Taking into account additional delays in the terrestrial network means that a signal transmitted via GEO from one side of the Atlantic to the other takes just over a quarter of a second, with another quarter of a second delay for the response. In the early years of the Space Age, it was thought that this delay would render normal speech impossible, and even Arthur Clarke suggested that it might be necessary to terminate each side of the conversation with the word 'over' [59].

The effects of the delay were exaggerated by terrestrial operators around the world, who spread the disinformation that satellite communication was unacceptable to telephone users and would remain unsuitable for data transmission. The British Post Office, for example, keen as ever to protect its telecommunications monopoly, was one of the main proponents of the dogma that the delay would make satellite communications unworkable. According to a 1962 paper, based on joint studies by the Post Office engineering department and the Ministry of Aviation space department at the Royal Aircraft Establishment (RAE), Farnborough, 'tests by the Post Office indicate, even in the absence of echoes, that with 600 ms one-way delay, some 16 per cent of conversations are unsatisfactory' [14]. It is difficult to say whether the authors were being deliberately misleading, but it appears that their 'one-way delay' did, in fact, include the two passes through the satellite required for a reply; an example, perhaps, of being economical with the truth. It is also faintly amusing that the Post Office quoted a percentage of satellite conversations that were classed 'unsatisfactory', when it seems likely that a similar percentage of local, terrestrial calls could have attracted equal criticism.

Those with a more independent viewpoint were more positive. A report on a satellite link test, made by IBM to the US Federal Communications Commission (FCC) in 1974, suggested that the delay would increase politeness. One speaker at the IEE Conference on Satellite Communications in 1962 outlined a theoretical 'worst case scenario': if both parties began to talk at the same time, he suggested, after 250–300 ms they would hear the other and stop, and after another 300ms all would be silent. He observed, no doubt with tongue-in-cheek, that this 'may go on until the exhaustion of one of the [speakers]' [16].

In practice, it is often the case that a single satellite link is used with a terrestrial link in the opposite direction to reduce the total delay. One author reports an interesting effect that used to occur when a satellite path and a transocean cable were used together

[60]: caller A can hear caller B distinctly over the satellite link, as if he is in the next room; B, however, can hear A only poorly, so he shouts. A reacts to this by talking more softly, which only makes B shout louder!

Arthur Clarke's summary of the debate on delay provides a useful, if not politically correct, conclusion: 'in practice it was found that few people even noticed it, though it may occasionally cause trouble to excitable, interruption-prone Celts and Latins' [61].

LEO versus GEO

The overall result of the 'delay debate' and the poor performance of contemporary launch vehicles was a host of proposals for satellite communications systems based on a number of satellites in low to medium altitude orbits. Placing several satellites in a single orbit would ensure that one was always above the horizon and available for communications.

A proposal for a worldwide satellite communications system was included in another paper at the 1962 IEE conference, in which the author, John Bray, proposed a system of 12 attitude-controlled satellites in a 14,000 km equatorial orbit with an eight hour period [14]. The Earth was divided into seven overlapping zones, each about 4.5 hours wide. An individual satellite could be used for about an hour as it traversed the zone, all earth stations communicating via that satellite in a single hop; the satellite would then switch to the next zone. Communication links between zones were to be made by terrestrial means.

Yet another paper at this forward-looking conference proposed a system of nine satellites in a similar orbit [15]. They would have a lifetime of at least four years, which would be expected to improve, as hardware reliability improved, throughout the programme. As an inevitable sign of the times, especially since the author, Geoffrey Pardoe, was an avid supporter of Britain's Blue Streak launcher, the proposal was based on a launch by Europa, the European vehicle for which Blue Streak was the first stage. Pardoe estimated the cost of an individual 'launch attempt' at about £1.4 million (for the record, inflation has since increased this estimate by up to two orders of magnitude, when insurance and other costs are taken into account).

Now that satellites are equipped with sophisticated attitude and orbital control systems designed to maintain accurate antenna pointing and confine spacecraft to designated geostationary orbital positions, it is instructive to recall that these capabilities were not always taken for granted. In the early 1960s, engineers considered the development of 'random', as opposed to 'station-keeping' satellites, which would be randomly distributed in a number of orbits instead of stationed in GEO [14]. The arguments for 'random satellites' included their simplicity, the lack of an attitude and orbital control system making them potentially more reliable, and the fact that they could be launched with existing launch vehicles. There would, however, need to be between 50 and 100 of this type of satellite to provide full coverage.

The trade-offs which led to the eventual design solutions were much the same as they are today and were well understood. As Clarke had pointed out, for a geostationary system, only three satellites are required to give worldwide coverage between about 60° north and south of the equator. Station keeping allows the use of directional satellite antennas which reduce power requirements and equipment mass, and thereby launch costs. Moreover, the satellite's signals are not Doppler-shifted by excessive motion relative to the earth station, as they are with low orbiting satellites.

Interestingly, the concept of a constellation of many small, relatively cheap satellites returned nearly 30 years later. In 1990, Motorola proposed its Iridium system for communications with hand-held, mobile receivers which originally featured 77 active satellites in seven different polar orbits. In addition, a system similar to that proposed by Bray and Pardoe in 1962 – in that it featured a mixture of satellite and terrestrial links – was established by Globalstar Corp. Unfortunately, although they were successful from an engineering point of view, Iridium and Globalstar were financial failures and survived only by shedding their considerable debts through bankruptcy proceedings.

In the early 1960s, however, the concept of communications with mobile receivers with which these latter-day systems were concerned, was confined to the science fiction of Dan Dare's wristwatch radio. All practical satellite receivers were unquestionably 'fixed', bolted down and set in massive concrete foundations. The development of satellite applications was therefore tied to satellites in an orbit that simplified the operation of the earth station – geostationary orbit.

The use of the geostationary satellite removed the requirement for the smaller-diameter earth stations to track the satellite and eventually reduced the cost of individual satellite receivers to the point where they have become consumer items. It was this orbital advantage that led to the widespread use of satellites for TV broadcasting direct to the home and business communications using very small-aperture terminals (VSATs).

In fact, this was realised at the time of the IEE conference, as shown by a paper by Donald Bond, who considered GEO as the only feasible solution for direct TV broadcasting [21]. Fellow speaker, Geoffrey Pardoe, was of the opinion that the medium altitude orbit might be best for commercial telephony because of the delay problem with GEO, although he thought geostationary satellites would be used, in parallel, for TV, telegraph and 'certain telephone circuits for special use'. He believed that 'experience in the latter [would be] of great value in determining customer reaction to the time delay question' and that, if this could be resolved, it was 'inevitable' that future communications satellites would reside in GEO [15]. On that conclusion, he was certainly correct.

Geostationary comsats

As early as 1959, a team of engineers at Hughes Aircraft Company in Los Angeles, led by Harold Rosen, had begun to realise the advantages of geostationary orbit and set about the development of a satellite that would operate from this orbit [62]. Since the satellites of the time were too heavy to be carried to such a high-altitude orbit by contemporary launch vehicles, the team designed a light-weight, spin-stabilised satellite – the first in a long line of 'Hughes spinners' (Figure 5.9).

Figure 5.9   Harold Rosen (right) and Thomas Hudspeth with model of 'Hughes spinner' on the Eiffel Tower [Boeing, formerly Hughes Aircraft Corp.]

In view of the mass limitations, there was no possibility of the satellite carrying the much-desired directional antenna, so a degree of directivity was achieved through the design of a slot array antenna which produced a flat, 'pancake beam' [63] or 'squinted beam'. With the satellite's spin axis parallel to the Earth's, this would at least confine the radiated energy to a plane, rather than, as F. J. D. Taylor put it, 'warming the Universe' [44]. As a further weight-saving measure, the command receiver and telemetry transmitter were integrated into the design of the communications repeater, which also provided the tracking beacon [63].

Since the basic stability of the spacecraft was provided by its spin, its attitude and orbital control subsystem was also simplified, comprising only two pulsed 'gas jets' (cold-gas thrusters). One was aligned perpendicular to the spin axis (with its line of action passing through the satellite's centre of mass) and was used to adjust the orbital altitude. The other was aligned parallel to the spin axis to make adjustments of the orbital plane when used in a sustained manner and, when pulsed, to readjust the spin axis [63].

Syncom

By August 1961, Hughes' internal development programme had been recognised by NASA with a contract for three spacecraft called Syncom (an abbreviation for

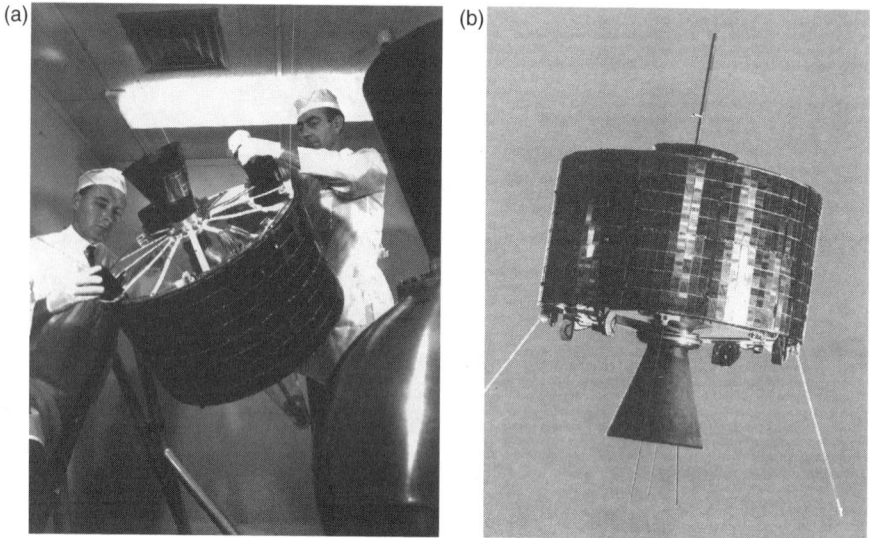

*Figure 5.10   (a) Syncom in integration (note integral apogee motor at top),
(b) artists' impression of Syncom in space [NASA]*

synchronous communications). Compared with Courier at 230 kg and Telstar at
77 kg, Syncom was indeed a lightweight: complete with its integral apogee motor –
designed to fire at the apogee of the elliptical geosynchronous transfer orbit to
circularise the orbit – the satellite weighed 68 kg; without the 41 kg motor it was only
27 kg [64].

It was drum-shaped, 64 cm high and 71 cm in diameter, the periphery of the
cylinder being covered with around 4000 solar cells producing a power of 29 W
[64] (Figure 5.10). Its operating voltage was 28 V, which became the standard for
communications satellites, and many other spacecraft, for several years thereafter. It
carried a single transponder, powered by the increasingly ubiquitous travelling wave
tube amplifier and capable of handling a single transmission circuit [65].

In common with all geostationary satellites, Syncom would enter the Earth's
shadow when its orbit plane coincided with the ecliptic, the plane in which the
Earth orbits the Sun. The eclipses occur during two annual periods, spanning
44 days around the equinoxes, when satellites are in shadow for up to 72 minutes
per day. To enable uninterrupted service, Syncom was equipped with rechargeable
nickel-cadmium batteries, thus providing a 'role-model' for the future *commercial* communications satellite, which would need to do exactly this if it was to be
accepted.

Syncom 1 was launched on 14 February 1963 and an onboard timer successfully
ignited its apogee motor, but twenty seconds later it was ripped apart by an explosion
[66]. Syncom 2 was launched to geostationary altitude on 26 July 1963, but occupied
a non-equatorial, 33° inclination orbit, in which the satellite made large excursions

Figure 5.11   USNS Kingsport [NASA]

north and south of the equator. Thus it was the first satellite to be launched to what is known as a geosynchronous rather than geostationary orbit. Syncom 3, however, was placed in a nearly equatorial orbit following its launch on 19 August 1964, thus becoming the first communications satellite to reach a reasonable approximation to geostationary orbit.

As ever, there were a number of high points on the transmission agenda. On 13 September 1963, Syncom 2 and Relay 1 were used together to link Nigeria, the USA and Brazil in a three-continent conversation. Signals were transmitted from the USNS Kingsport in Lagos harbour via Syncom 2 to New Jersey, and from there via Relay 1 to Rio de Janeiro (Figures 5.11 and 5.12) [67]. The first 'famous' telephone call via Syncom was made in 1963 between US President Kennedy and Nigerian prime minister Abubaker Balewa and, among other communications experiments, Syncom 3 was used to transmit TV coverage of the 1964 Tokyo Olympic Games.

By this time it was already apparent that communications satellites had commercial potential, which is why, on 31 August 1962, the US Congress had voted in the Communications Satellite Act, which created the Communications Satellite Corporation (universally known as Comsat) to exploit satellite communications in the United States. Comsat thus became the first company to be founded by the US Government since an act in 1862 established the Union Pacific Railroad [68]. The Act was, in effect, a device to end the monopoly in US satellite communications that AT&T appeared to be developing with Telstar, which apparently included 'a specific proposal that would feature fifty Telstars, to ensure that at least one would always be in view' [69].

Figure 5.12   *9 m parabolic antenna developed by International Telephone and Tele-graph Federal Laboratories as a Project Relay space communications terminal [NASA]*

Early Bird

Having begun operations in February 1963, Comsat was sufficiently impressed by the performance of Syncom 2 and 3 to commission Hughes to design and build the first *commercial* communications satellite [70], which was identified in the contract as 'HS [Hughes Satellite] 303' but later became widely known as Early Bird. It was built for eventual use by the International Telecommunications Satellite organisation (Intelsat) – itself formed in August 1964 – which gave the satellite its less well known and formal name, Intelsat I. It thus became the first of a continuing series of ever more complex satellites which evolved through ten successive generations and beyond (Figure 5.13).

Early Bird was capable of handling 240 two-way transatlantic telephone circuits, which represented an impressive increase in capability over the one-circuit Syncom despite being similar in mass. This was partly due to its larger solar array, but also a function of the steadily increasing experience of the satellite designers and their

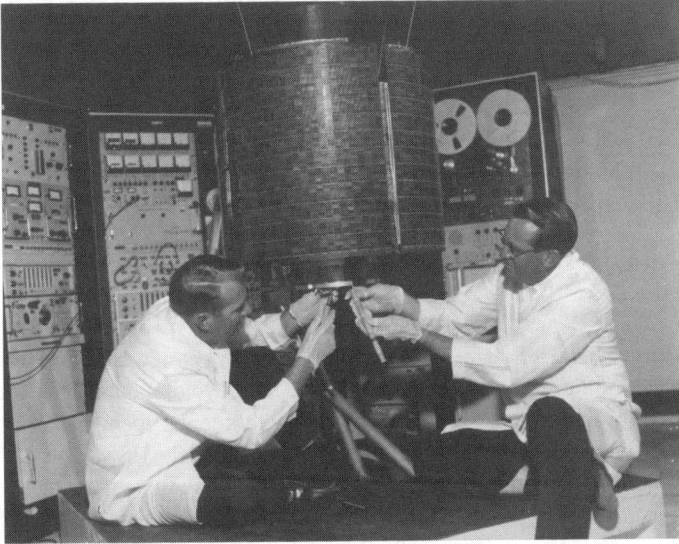

Figure 5.13   Early Bird (Intelsat I) showing some of the 6000 solar cells with which it was covered [NASA]

growing skills in manipulating the available technologies. As an alternative to the telephone circuits, Early Bird could carry one relatively low quality black and white TV signal. Although this pales into insignificance compared with the dozens of high definition colour channels handled by today's satellite transponders, it indicates the inherent 'transparency' of the satellite transponder to the information carried.

During the launch of Early Bird on 6 April 1965, Arthur C. Clarke was at Comsat HQ, observing the spectacle on a TV monitor. In his book, *The Promise of Space*, he reports in a footnote on how political enthusiasm often seems to overshadow common sense:

> The ... Thor Delta booster was still on the way up when Vice-President Humphrey started to give us one of his little speeches. The circuit to Cape Kennedy was switched off, and it occurred to me that if anything went wrong now, everyone in the United States would know it *except* the staff of Comsat [71].

Intelsat's commercial service, via Early Bird, began on 28 June 1965. Just ten days earlier it had proved the usefulness of satellite communications by taking over from a transatlantic submarine cable which had failed. It did this repeatedly throughout its lifetime and the restoration of services due to cable outages is still one of the largely unsung applications of the geostationary satellite.

Amongst Early Bird's more newsworthy feats were the medical diagnoses that became possible with patient and doctor some 6500 km miles apart and its services which allowed collectors in New York to take part in Sotheby's auctions in London. It also made possible the arrest of a Canadian gangster, thus earning itself the title 'Public Satellite No.1' [72].

Although Early Bird was primarily a technology demonstrator with a design life of only 18 months, it remained in service for four years [70]. It was switched off in January 1969, by which time it had been replaced by the Intelsat II series, but was reactivated for a couple of months later that year to provide additional capacity to meet the growing demand for satellite communications. In fact, it was still capable of TV transmissions some ten years after launch and was reactivated as recently as 1984 [73].

Later birds

As with many aspects of space exploration, the early years were dominated, as far as the general public was concerned, by American developments. This was largely because of the secrecy that shrouded most developments, technical or otherwise, within the Soviet Union. Of course, this does not mean that the USSR was not conducting satellite communications experiments with as much vigour as the Americans; after all, they had to develop the systems necessary to communicate with their scientific satellites and manned spacecraft which provided experience in all the relevant techniques.

As it turned out, the Soviet Union's first communications satellite, Molniya 1-1 (the first of the Molniya 1 generation), was launched on 23 April 1965, just two weeks after Early Bird. A further Molniya first-generation satellite was launched in October 1965, followed by two more in 1966 (Figure 5.14).

As might be expected from what has been said about the other Soviet satellites, the Molniya ('lightning') was quite different from American communications satellites. Structurally, it was a cone-topped cylinder with six solar panels deployed in a windmill pattern at the base and two 90cm-diameter parabolic communications antennas deployed from the sides. The three-axis stabilised spacecraft was 3.4 m long, 1.4 m in diameter and weighed over 800 kg, while the arrays spanned about 5 m in the open position. They produced around 700 W which was used, among other things, to power a payload of three 40 W transponders, for TV, telephony and telegraphy services, and a system of TV cameras for meteorological observations [64]. The satellites were built by the Applied Mechanics NPO in Krasnoyarsk, now NPO-PM, a major Russian spacecraft manufacturer [74].

A further key difference was the orbit used for the Molniya, which stemmed from the fact that much of Soviet territory is at high latitudes. Partly because of the penalty of launching payloads into geostationary orbit from high latitudes, but also because the high latitude coverage of the geostationary satellite is limited, the USSR exploited an elliptical, high inclination orbit for its first satellite communications network. Now known as a Molniya orbit, after the satellites, it has an inclination of about 63°, an apogee and perigee of approximately 40,000 km and 500 km, respectively, and a period of 12 hours.

The Molniya orbit's high inclination allows satellites to provide coverage of high latitudes, while its 12 hour period produces two coverage areas on opposite sides of the Earth, as a result of its two apogees in a 24 hour period. Full-time coverage of a given

Figure 5.14   Soviet Molniya satellite display at Le Bourget Air and Space Museum, Paris [Mark Williamson]

area is typically provided by three satellites synchronised in similar orbital paths so that each one contributes about eight hours of continuous service as it approaches and recedes from its apogee. A disadvantage is the orbit's low perigee, which increases atmospheric drag on the spacecraft and increases radiation exposure, owing to its passage through the Van Allen belts. Although the Soviets later placed direct broadcasting satellites in GEO, the preferred orbit for telecommunications for many years was the Molniya orbit which, according to one author, was chosen from the various possibilities as early as 1950 (seven years before Sputnik 1) [75].

Despite the impressive size and performance of the Molniya, the Soviet Union has lagged behind the United States in satellite communications throughout the application's history. Initially, this was because the USSR lacked the commercial imperative of the United States, or its culture of communications. After all, a nation which keeps its technological developments under wraps and closets its population away from outside influences is hardly likely to make satellite communications a high priority. Later, the technological lag was due more to lack of investment – commercial or otherwise – and consequent quality control and reliability issues (even in later years, Russian-built payloads and other subsystems were unlikely to match the longevity of their western counterparts).

Despite this, Molniya 1 was the first comsat to feature the now-universal parabolic communications antenna, which allows the satellite's transmitted energy to be focused into a beam and generally improves the gain, or amplification factor, of the communications system. Strangely, the first commercial US comsat to benefit from the parabolic antenna was Intelsat IV, launched in 1971, although other types of spacecraft had used them well before that.

Perhaps because American satellite communications was married to the Hughes 'spinner', which dominated the industry throughout the 1960s, it failed to develop the high gain paraboloid sooner. NASA's Applications Technology Satellites ATS-1 and ATS-3 demonstrated mechanically 'de-spun' antennas, which pointed constantly towards the Earth, in 1966 and 1967. However, they were not used operationally until TRW incorporated them into the Intelsat III satellite series, launched from 1968 [76]: the solution in this case was a simple microwave horn projecting from the top of the spinning satellite with an angled reflector to direct its beam towards the Earth.

As described in the previous chapter, America's first three-axis stabilised remote sensing satellite, Nimbus 1, was launched in August 1964, so it was not as if the technology was alien to the United States. Perhaps if General Electric, which built Nimbus, had won the contract for the early Intelsat satellites, the spinner would have died an early death and Hughes would not have become the world's leading satellite manufacturer.

Certainly, by the time Europe entered the satellite communications scene with the French/German Symphonie 1, launched in December 1974, it had already decided that three-axis stabilisation was the route to take. Thus Symphonie was the world's first three-axis stabilised, geostationary communications satellite – the progenitor of the modern-day comsat.

Not least because of the difficulties involved in mounting large antenna reflectors on a spin-stabilised platform, the three-axis stabilised satellite has become an industry standard. As antenna diameters increased – to provide smaller and smaller spotbeams – they graduated from solid deployables to much larger unfurlable reflectors (similar in appearance to a wire-mesh umbrella). NASA's ATS-6, launched in May 1974, led the development with its 9m-diameter unfurlable antenna, which allowed it to broadcast educational TV to the rural populations of India via 3 m chicken-wire antennas placed in thousands of villages. As a result of this application, which later turned into an industry for satellite TV, ATS-6 could be considered the first direct broadcasting satellite (Figure 5.15).

The military angle

Not surprisingly, the military services were also interested in satellite communications, not least because it offered the potential of reliable communications with ships

Figure 5.15   *The 9m-diameter ATS 6 unfurlable antenna under test at Lockheed Missiles and Space Co., Sunnyvale, CA. Note mould at lower left on which antenna was constructed (aluminium ribs covered in copper-plated, silicone-coated Dacron mesh) [NASA]*

at sea and with other mobile units in areas not covered by terrestrial communications. As with remote sensing applications, the military's activities in satcoms on both sides of the Iron Curtain began early in the Space Age, as shown by the US Army's development of Courier described above. Meanwhile, as already mentioned, the Soviet Union's developments were conducted clandestinely under its Cosmos programme: Cosmos 4, launched on 26 April 1962, was the first Soviet military satellite, since its task was to measure radiation before and after US nuclear tests [77], while Cosmos 41, launched on 22 August 1964, was a test vehicle for the Molniya, which of course could be used for military communications as easily as civil or commercial.

Among the earliest dedicated military satellites was the USAF's LES 1, the first of a series of Lincoln Experimental Satellites named after MIT's Lincoln Laboratory which designed and built them. LES 1 was launched to LEO on 11 February 1965, and was followed by LES 2 in May and LES 3 and 4 in December of the same year. These satellites pioneered the use of the X-band (8–12 GHz) part of the spectrum [78], which became the 'sovereign territory' of the military comsat; they were also the first 'all solid state' communications satellites, in that solid state amplifiers, rather than TWTAs, were used in their communications payloads; and LES 6, launched on 26 September 1968, was the first military comsat to reach geostationary orbit [79].

In the same way as their civilian cousins, these experimental military satellites were followed closely by more advanced, operational versions. Thus, on 16 June 1966, a Titan IIIC launch vehicle placed seven satellites of the Initial Defense Satellite Communications System (IDSCS) in orbit. The eventual system of 26 satellites was the first satellite constellation and, among other things, the programme demonstrated the technology of the electronically, as opposed to mechanically, de-spun antenna (on IDSCS 19/DATS).

Following an invitation from the United States to cooperate in the IDSCS programme, the UK installed an earth station which was in place to receive signals from the satellites as soon as they were launched. Shortly thereafter, Britain began its development of the Skynet military satellite communications programme, which continued through various spacecraft generations into the twenty-first century. The first satellite, Skynet 1, was launched on 21 November 1969: a spinner built by Ford Aerospace in the United States, it was only the second military satellite to be stationed in GEO. Its sibling, Skynet 2A, launched on 19 January 1974, was the first comsat built outside the United States and USSR: it too was a spinner based on Ford's design for Skynet 1, but manufactured by Marconi in the United Kingdom. Unfortunately, its American launch vehicle placed it in the wrong orbit and it re-entered six days later [80]; Skynet 2B was successfully launched by another Delta the following November and placed in GEO.

Another related development of satellite technology which attracted the attention of the military early on, in both the United States and the USSR, was its application to navigation, which used techniques similar to those of the communications satellite insofar as it involved the transmission and reception of satellite signals. The first navigation satellite system was developed by the US Navy, primarily to allow its Polaris missile submarines to establish their positions within about 150 m, but it soon became evident that it would be useful for all types of shipping. Following an initial launch failure, Transit 1B was placed in LEO on 13 April 1960, where it operated for three months, and by 1968 a total of 23 satellites had been launched to form a constellation providing global coverage [81]. The continued development of what was later called the Navy Navigation Satellite System (NNSS) led eventually to the USAF-managed Navstar Global Positioning System (GPS), which by the late 1990s was in use by car drivers and hill walkers around the world.

Technology development

Although communication between individuals, companies and nations was imperfect, for all manner of personal, corporate and political reasons, the technologies for the communications satellite were being developed in the same arena as those for science and remote sensing satellites. Although different technologies were adopted for different applications at different times, the technologies of spacecraft structures, materials, attitude control, propulsion, power, thermal control and TT&C were essentially the same for all types of spacecraft.

What made the communications satellite special, however, was the fact that its application – telecommunications – was the first commercial space application. This provided an important spur to the development of satellite technology, since it created an identifiable financial advantage to the pursuit of research and development by satellite manufacturers and space agencies. Once communications satellites moved from the experimental to the operational realm, and customers began to realise their potential, an inexorable drive towards improved capability and increased complexity began.

The most obvious development was to increase the number of transponders a satellite could carry, since this relates directly to the amount of communications traffic that can be carried. Increasing the size of the payload means increasing the size and mass of the satellite, supplying more power from larger solar arrays and improving the satellite's thermal subsystem to deal with the additional heat produced by a larger, more powerful payload. As a direct consequence, it also requires the development of larger, higher-capacity launch vehicles, and so a symbiosis between satellite design and launch vehicle design became a permanent feature of space technology development.

In addition, the identification of geostationary orbit as the preferred location for the communications satellite encouraged not only the development of launch vehicles capable of delivering them there, but also the development of increasingly reliable subsystems and components. Satellites in GEO are well above the Van Allen radiation belts and therefore naturally disposed to longer operational lifetimes; and they avoid the continual thermal cycling of moving in and out of the Earth's shadow every hour or so. As a result, it was only logical to attempt to ensure that once a satellite had reached GEO, it remained in operation for as long as possible. Apart from building more reliable components, this meant providing redundant subsystem and payload equipment to replace any failed units, larger propellant loads for longer-term station keeping and oversized arrays to account for the unavoidable degradation of the solar cell in the space environment. As a result, satellite lifetimes doubled between the mid-1960s and the mid-1970s, to an average of about seven years, then doubled again by the 1990s to about 15 years.

Conclusions

In the 1960s and 1970s, the by-line 'Live via Satellite' often marked a special television event. The fact that the label was seen on our screens only rarely from the 1980s onwards shows how the technology has been absorbed into our everyday lives.

Today, millions of people across the world rely on communications satellites to deliver TV entertainment, news and other services direct to their homes, while others make phone calls via satellite without even realising it. Businesses use satellites for data transfer, financial services and staff training, to name but three applications, while military organisations are increasingly dependent on space-based systems for a wide range of defence-related applications. Meanwhile, satellites provide distance education services to millions who would otherwise remain uneducated and medical services to some who might otherwise die. In financial terms, the worldwide satellite industry is thought to be worth around $100 billion per annum [82], but its impact on society is invaluable.

It is unlikely that the communications satellite will ever be truly recognised by the general public since, unlike passenger aircraft and consumer appliances, it cannot be experienced directly. Even in 1962 the lack of public appreciation, of both the technology and those who create it, was clear:

> Telstar has definitely proved the efficacy of communications satellites. The attendant pub-
> licity on both sides of the Atlantic has given the lay public an inkling of what the future holds.
> However, as is usually the case, the attention of the layman has not been directed to the
> behind the scenes, everyday [responsibility] of the engineer and scientist for the reliability of
> operation of the communications satellite. So long as the initial trial broadcast is a dramatic
> success, the press and the public is satisfied with this technological achievement [22].

The arrival of the hand-held satellite telephone may have prompted a few to ponder on the usefulness of the communications satellite, likewise the availability of GPS receivers, but the appreciation probably faded after a few weeks. History has shown that successful new technologies tend to become adopted, and then taken for granted, within a relatively short time. One apocryphal tale has a cab driver complaining 'What do I need satellite technology for? I've got GPS!'. In the final analysis, the most successful service attracts the least attention, and the unquestioning acceptance of a technology is a testament to its maturity.

References

1 Ley, Willy: 'Our Work in Space' (The Macmillan Company, New York, 1964) p. 102

2 Gatland, Kenneth W.: 'Astronautics in the Sixties' (Iliffe Books Ltd, London, 1962) p. 57

3 Weston, M. A. and Pearson, J. W.: 'Experiments in Moon Reflection Communications...', Proceedings of the International Conference on *Satellite Communications*, IEE, Savoy Place, London (22–28 November 1962) pp. 53–9

4 Gatland, Kenneth W.: 'Astronautics in the Sixties' (Iliffe Books Ltd, London, 1962) p. 58

5 Taylor, John W. R., and Allward, Maurice: 'Eagle Book of Rockets and Space Travel' (Longacre Press Ltd, London, 1961) p. 130

6 Williamson, Mark: 'The Cambridge Dictionary of Space Technology' (Cambridge University Press, Cambridge, 2001) pp. 140–2

7 Ley, Willy: 'Our Work in Space' (The Macmillan Company, New York, 1964) p. 106

8 Gatland, Kenneth W.: 'Astronautics in the Sixties' (Iliffe Books Ltd, London, 1962) p. 60,145

9 Johnson, Nicholas L. (personal communication, 25 March 2004)

10 Ley, Willy: 'Our Work in Space' (The Macmillan Company, New York, 1964) p. 107

11 Thompson, Tina D. (Ed.): 'TRW Space Log 1996' (TRW Space and Electronics Group, Redondo Beach, Volume 32, 1997) p. 69

12 Williamson, Mark: 'The Cambridge Dictionary of Space Technology' (Cambridge University Press, Cambridge, 2001) p. 399

13 Taylor, F. J. D.: 'Telstar', Proceedings of the International Conference on *Satellite Communications*, IEE, Savoy Place, London (22–28 November 1962) pp. 179–81

14 Bray, W. J.: 'Some Technical Aspects of the Design of Worldwide Satellite Communications Systems', Proceedings of the International Conference on *Satellite Communications*, IEE, Savoy Place, London (22–28 November 1962) pp. 7–17

15 Pardoe, G. K. C.: 'World Communications Satellite System', Proceedings of the International Conference on *Satellite Communications*, IEE, Savoy Place, London (22–28 November 1962) pp. 18–9

16 Ringereide, T.: 'Some System Approaches to Multiple Access Communication Systems', Proceedings of the International Conference on *Satellite Communications*, IEE, Savoy Place, London (22–28 November 1962) pp. 19–25

17 Kington, C. N., and Pearson, H. E.: '...Goonhilly', Proceedings of the International Conference on *Satellite Communications*, IEE, Savoy Place, London (22–28 November 1962) pp. 112–120

18 Welber, I.: 'Telstar', Proceedings of the International Conference on *Satellite Communications*, IEE, Savoy Place, London (22–28 November 1962) p. 168

19 Abrahamson, B. N., and Brady, M. E.: 'Relay', Proceedings of the International Conference on *Satellite Communications*, IEE, Savoy Place, London (22–28 November 1962) pp. 168–78

20 Shennum, R. H.: 'Telstar', Proceedings of the International Conference on *Satellite Communications*, IEE, Savoy Place, London (22–28 November 1962) p. 198

21 Bond, D. S.: 'Performance Characteristics of a System for Direct TV Broadcasts from Earth satellites', Proceedings of the International Conference on *Satellite Communications*, IEE, Savoy Place, London (22–28 November 1962) pp. 259–64

22 Feingold, S. K.: 'Decommutation Equipment for Communications Satellites', Proceedings of the International Conference on *Satellite Communications*, IEE, Savoy Place, London (22–28 November 1962) pp. 311–14

23 Solomon, Louis: 'Telstar: Communication Break-through by Satellite' (Constable Young Books Ltd, London, 1963) p. 33

24 Williamson, Mark: 'The Communications Satellite' (Adam Hilger, Bristol, 1990) pp. 243–66

25 Perrotta, Giorgio, and Somma, Roberto: 'Reliability of Space Systems: Need for New Tools', *Space Communication and Broadcasting*, **1** (2), July 1983, pp. 189–98

26 Solomon, Louis: 'Telstar: Communication Break-through by Satellite' (Constable Young Books Ltd, London, 1963) p. 37

27 Gatland, Kenneth W.: 'Astronautics in the Sixties' (Iliffe Books Ltd, London, 1962) pp. 78–9

28 Von Braun, Wernher, Ordway, Frederick I., and Lange, Harry H.-K.: 'Science of the Seventies: History of Rocketry and Space Travel' (Encyclopaedia Britannica Educational Corp, 1969) p. 185

29 Hansen, James W.: 'Hughes TWT/TWTA Handbook' (Hughes Aircraft Company, Electron Dynamics Division, Torrance, USA, 1993) p. 1

30 Kompfner, Rudolph: *Wireless World*, **52** (November 1946) pp. 369–72

31 Pierce, John: *Proceedings I.R.E.* **35** (February 1947) pp. 111–23

32 Gittins, J. F.: 'Power Travelling Wave Tubes' (The English Universities Press Ltd, London, 1965) p. 1

33 Hansen, James W.: 'Hughes TWT/TWTA Handbook' (Hughes Aircraft Company, Electron Dynamics Division, Torrance, USA, 1993) p. 2

34 Solomon, Louis: 'Telstar: Communication Break-through by Satellite' (Constable Young Books Ltd, London, 1963) p. 47

35 ibid., p. 43

36 ibid., p. 49

37 Lynch, Arnold C. (personal communication, 5 July 1991 at Nineteenth Week-end Meeting on the History of Electrical Engineering (IEE PG S7), Lancaster, UK: Lynch suggested that the French horn had cost five times that of Goonhilly Aerial 1)

38 Solomon, Louis: 'Telstar: Communication Break-through by Satellite' (Constable Young Books Ltd, London, 1963) pp. 15–18

39 ibid., p. 10

40 Taylor, F. J. D. (personal communication, 9 August 1991)

41 Williamson, Mark: 'The Cambridge Dictionary of Space Technology' (Cambridge University Press, Cambridge, 2001) p. 285

42 Solomon, Louis: 'Telstar: Communication Break-through by Satellite' (Constable Young Books Ltd, London, 1963) pp. 53–4

43 ibid., p. 55

44 Taylor, F. J. D. (personal communication, 27 July 1991)

45 Pritchard, W. L., and Puente, J. G.: *Journal of the British Interplanetary Society (JBIS)*, **23**, 1970, pp. 125–36

46 Thompson, Tina D. (Ed.): 'TRW Space Log 1996' (TRW Space and Electronics Group, Redondo Beach, Volume 32, 1997) p. 68

47 ibid., p. 75

48 Ley, Willy: 'Our Work in Space' (The Macmillan Company, New York, 1964) p. 99

49 McAleer, Neil: 'Odyssey: The Authorised Biography of Arthur C. Clarke' (Victor Gollancz, London, 1992) p. 61

50 Clarke, Arthur C.: 'Extra-Terrestrial Relays', *Wireless World*, October 1945, pp. 305–7

51 Williamson, Mark: 'The Communications Satellite' (Adam Hilger, Bristol, 1990) pp. 235–6

52 Clarke, Arthur C.: 'How the World Was One: Beyond the Global Village' (Victor Gollancz Ltd, London, 1992) p. 166

53 McAleer, Neil: 'Odyssey: The Authorised Biography of Arthur C. Clarke' (Victor Gollancz, London, 1992) p. 62

54 Clarke, Arthur C.: 'How the World Was One: Beyond the Global Village' (Victor Gollancz Ltd, London, 1992) pp. 163–6

55 Clarke, Arthur C.: 'The Promise of Space' (Harper and Row, New York, 1968) p. 102

56 Pierce, J. R.: 'Orbital Radio Relays', *Jet Propulsion*, **25** (4), pp. 153–7

57 Heppenheimer, T. A.: 'Countdown: A History of Spaceflight' (John Wiley and Sons, New York, 1997) p. 274

58 Solomon, Louis: 'Telstar: Communication Break-through by Satellite' (Constable Young Books Ltd, London, 1963) p. 31

59 Clarke, Arthur C.: 'Voices from the Sky' (Harper and Row, New York, 1965)

60 Martin, J.: 'Communications Satellite Systems' (Prentice-Hall, 1978) p. 18

61 Clarke, Arthur C.: 'The Promise of Space' (Harper and Row, New York, 1968) p. 105

62 'A World Without Boundaries' (Hughes Aircraft Company, Space and Communications Group, Los Angeles, US, 1989 – booklet no. 890333/30M/89) p. 3

63 Gatland, Kenneth W.: 'Astronautics in the Sixties' (Iliffe Books Ltd, London, 1962) p. 72

64 Caprara, Giovanni: 'The Complete Encyclopedia of Space Satellites' (Portland House, New York, 1986) p. 65

65 'A World Without Boundaries' (Hughes Aircraft Company, Space and Communications Group, Los Angeles, US, 1989 – booklet no.890333/30M/89) p. 5

66 Heppenheimer, T. A.: 'Countdown: A History of Spaceflight' (John Wiley and Sons, New York, 1997) p. 275

67 Von Braun, Wernher, Ordway, Frederick I., and Lange, Harry H.-K.: 'Science of the Seventies: History of Rocketry and Space Travel' (Encyclopaedia Britannica Educational Corp, 1969) p. 186

68 Clarke, Arthur C., and Editors of Time Life Books: 'Man and Space' (Time-Life International, The Netherlands, 1970) p. 100

69 Heppenheimer, T. A.: 'Countdown: A History of Spaceflight' (John Wiley and Sons, New York, 1997) p. 273

70 Sharpe, Mitchell R.: 'Satellites and Probes' (Aldus Books, London, 1970) p. 125

71 Clarke, Arthur C.: 'The Promise of Space' (Harper and Row, New York, 1968) p. 107

72 Allward, Maurice (Ed.): 'The Encyclopedia of Space' (The Hamlyn Publishing Group Ltd, Feltham, 1968–9) p. 223

73 Bennett, S. B.: 'Intelsat News', **6** (2), 1990, pp. 4–5

74 Harvey, Brian: 'The New Russian Space Programme' (John Wiley and Sons, Chichester, 1996) p. 197

75 Smolders, Peter L.: 'Soviets in Space' (Lutterworth Press, Guildford, 1973) pp. 90–1

76 Heppenheimer, T. A.: 'Countdown: A History of Spaceflight' (John Wiley and Sons, New York, 1997) p. 276

77 Turnill, Reginald (Ed.): 'Jane's Spaceflight Directory, 1987' (Jane's Publishing Co Ltd, London, 1987) p. 205

78 Thompson, Tina D. (Ed.): 'TRW Space Log 1996' (TRW Space and Electronics Group, Redondo Beach, Volume 32, 1997) p. 96

79 ibid., p. 119

80 ibid., p. 157

81 Turnill, Reginald (Ed.): 'Jane's Spaceflight Directory, 1987' (Jane's Publishing Co Ltd, London, 1987) p. 355

82 Pelton, Joseph N., Oslund, Jack, and Marshall, Peter: 'Communications Satellites: Global Change Agents' (Lawrence Erlbaum Associates, New York, 2004)

Chapter 6

Probing the Moon – the development of the lunar science spacecraft

I for one do not want to go to sleep by the light of a Communist Moon

Lyndon B. Johnson (US Vice-President)

From the late 1950s onwards, the field of space science developed quickly from simple observations of the Earth's magnetic field and micrometeoroid flux in low Earth orbit to observations of extraterrestrial radiations and investigations of the nearby planets. But space science from Earth orbit has always had its limitations and while the only way to collect data on distant stars and galaxies is by the remote sensing techniques of space astronomy, there is another alternative for the planetary bodies of our own solar system. Indeed, it was evident from the early days that by far the best way to conduct planetary astronomy was to transport the observing instruments, on board spacecraft commonly known as space probes or planetary probes, to the vicinity of the body in question.

Thus a subsidiary phase of the space race began: the race to land a spacecraft on the surface of the nearest planetary body, the Earth's own Moon. The fact that, less than a year after Sputnik 1, both nations were aiming for the Moon indicates the depth of their respective desires to be first in the various heats of the race.

It is interesting to note that spacecraft were sent to both the Moon *and* Venus even before the first orbiting observatory, OSO 1, reached orbit in March 1962. The pragmatism of remote sensing was apparently outweighed by the desire to 'get there' physically, but mainly, as with most aspects of space exploration at the time, by the political need to score points against the opposition.

The exploration of the planetary bodies, including the Earth's Moon, was typically pursued in phases – fly-by, impact, landing and orbital injection – which represented increasing accuracy in guidance and control (see Table 6.1). In the early years, however, 'accuracy' was a word not often heard in reports of the missions.

Table 6.1   Lunar spacecraft targeting accuracy – progression from fly-by and impact to orbit and landing (selected missions)

Launch date	Name	Nation	Comments
11 October 1958	Pioneer 1	USA	Distance from Earth: 113,854 km (Earth–Moon distance: 400,000 km approx.)
2 January 1959	Luna 1/Lunik 1	USSR	Passed within 6000 km; escaped Earth's gravity
3 March 1959	Pioneer 4	USA	Passed within 60,000 km
12 September 1959	Luna 2/Lunik 2	USSR	First spacecraft to impact Moon (13/9/59)
4 October 1959	Luna 3/Lunik 3	USSR	First lunar farside images
11 March 1960	Pioneer 5	USA	First 'deep space' probe: returned data from 36.2 million km from Earth
26 January 1962	Ranger 3	USA	Passed within 36,800 km; spacecraft inoperative
23 April 1962	Ranger 4	USA	Farside impact (26/4/62); spacecraft inoperative
28 July 1964	Ranger 7	USA	First Ranger success: 4,308 photos; impact (31/7/64)
31 January 1966	Luna 9	USSR	First hard landing (3/2/66); transmitted photos for 3 days
31 March 1966	Luna 10	USSR	First spacecraft to enter lunar orbit (3/4/66)
30 May 1966	Surveyor 1	USA	First soft landing 2/6/66; >11,000 photographs
10 August 1966	Lunar Orbiter 1	USA	First US orbiter/lunar mapper; crashed onto Moon 29/10/66
21 December 1966	Luna 13	USSR	Landed 24/12/66; first soil density measurements
7 November 1967	Surveyor 6	USA	Landed 10/11/67; first spacecraft to lift off from lunar surface (~4 m on 17/11/67)
21 December 1968	Apollo 8	USA	First manned spacecraft to orbit Moon
16 July 1969	Apollo 11	USA	First manned lunar landing: 20/7/69
12 September 1970	Luna 16	USSR	First automated lunar sample return mission. Orbital insertion 17/9/70; landing 20/9/70; lift-off 21/9/70; re-entry 24/9/70
10 November 1970	Luna 17	USSR	First automated rover: Lunokhod 1 (landed 17/11/70)

Phase 1: fly-by

America was first to attempt to launch a spacecraft to the Moon, in August 1958, with a spacecraft in its Pioneer series. Unfortunately, it failed when the Thor-Able launch vehicle's first stage exploded after just 77 s of flight. Close on the heels of its triumphs with the Sputniks, the USSR made its first attempt the following month, but it and two subsequent launches failed even to reach Earth orbit.

Although America's second attempt, on 11 October 1958 with Pioneer 1, failed to reach the Moon, it did set a distance record – or, in geocentric terms, a record height above the Earth – by reaching an apogee of 113,854 km [1]. However, the first spacecraft to reach the distance of the Moon (about 400,000 km) was the Soviet Luna 1 – also called Lunik 1[1] or *Mechta* (Russian for 'dream') (Figure 6.1). It was launched on 2 January 1959 and, two days later, passed within 6000 km of the lunar surface. In doing so, it was also the first man-made object to escape from the Earth's gravitational field, which could be regarded as a more fundamental achievement. In fact, since it continued into an orbit around the Sun, it was sometimes known as 'Planet 10' (there being nine natural solar planets) [2].

America's best effort, with Pioneer 4 on 3 March 1959, managed to pass within 60,000 km of the Moon some 41 h 13 min after launch [3]. An article in the UK's *Daily Telegraph and Morning Post* of 6 March 1959 reflected the contemporary fascination with large numbers: its main title was 'U.S. Satellite 305,000 miles out in space' while its subtitle read simply '4,300 m.p.h.' [4]. In fact, Pioneer 4 reached 650,000 km (407,000 miles) before its battery expired, while the signal level at the Goldstone tracking station in California suggested that, had the battery lasted, it could have been tracked to more than 1.1 million km [3].

Beyond this fascination with numbers, the newspaper story itself was about Jodrell Bank's success in tracking Pioneer 4 and its 'puzzling' failure to track Luna 1. Professor Bernard Lovell, the astronomer behind the building of the 75 m radio telescope, was quoted as saying that Pioneer 4 was 'an impressive achievement by the Americans' which had 'increased the early possibility of sending a human being into space'. Although he could not account for Jodrell Bank's inability to detect the beacon of 'the Russian Lunik', he said 'they had no reason to doubt Soviet claims' and had 'always found our Russian colleagues very genuine'.

With the proviso that this report appeared in a national daily newspaper, which always leads one to doubt its absolute accuracy, the quoted language provides an interesting snapshot of an independent nation's view of unfolding events. While US–Soviet relations were frosty to say the least, and both US and Soviet newspaper coverage was understandably biased in favour of the home team, Britain was attempting to give credit to both sides. Certainly, scientists at Lovell's level were pursuing a professional dialogue with both US and Soviet scientists to every possible extent and would not wish to alienate either side, but as the space race moved on, Britain would come to ally itself more and more with America – not least in order to get

---

[1] The first three Luna spacecraft were originally called Lunik, but later renamed Luna.

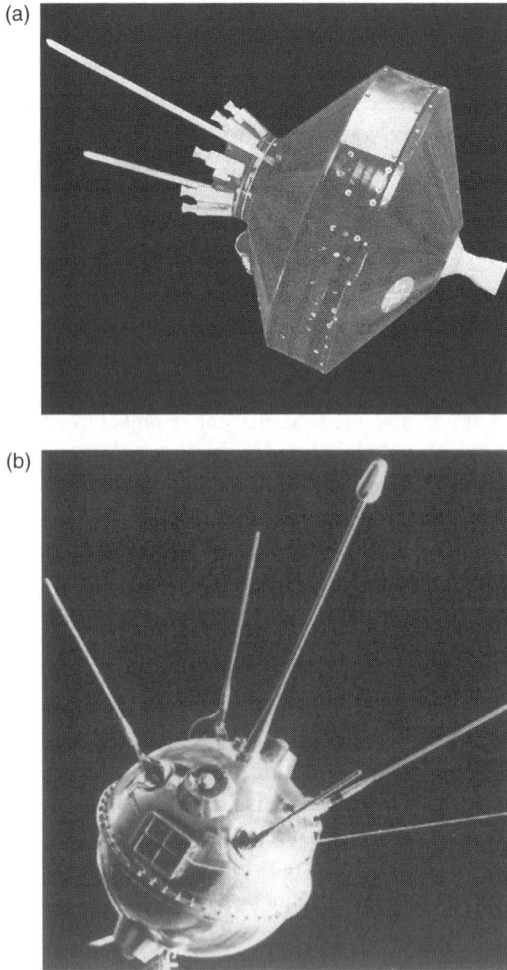

Figure 6.1   First lunar probes: (a) Pioneer 1; (b) Luna 1 [NASA]

its satellites launched. France, meanwhile, would favour the Soviets as partners in space exploration, leading to joint scientific missions and, ultimately, to the flight of Jean-Loup Chrétien as the first French cosmonaut – not astronaut – to Russia's Salyut 7 space station in June 1982.

Pioneer 4 was, however, an entirely American undertaking. The spacecraft was spin-stabilised and conical in shape, with a height of 50 cm – including a 7 cm antenna spike – and a diameter of 23 cm at the base. The thin fibreglass cone section was gold-plated to make the surface electrically conducting, allowing it to be used in conjunction with the spike as an unsymmetrical dipole antenna [3]. For thermal control, the outside of the spacecraft was painted in black and white stripes. DC power

Figure 6.2   Pioneer 3 (similar to Pioneer 4): note 'yo-yo' de-spin mechanism at left-hand end [NASA]

was supplied by a battery of 18 mercury cells mounted around the squat cylindrical base section of the spacecraft, which could be despun to some extent (for the benefit of some experiments) by a simple but effective mechanism. Sometimes referred to as a 'yo-yo', it consisted of a set of two 1.5 m weighted wires wrapped in the opposite direction to the spin. When the weights were released they would fly outwards and, like an ice skater stretching out her arms, reduce the spin speed. The yo-yo device was released at the end of its travel and succeeded in reducing the spin from 415 to 11 rpm (Figure 6.2) [3].

Its payload included two Geiger tubes, designed to provide a 'space radiation profile' between the Earth and the Moon and confirm the discovery by Pioneer 3 of a second, higher Van Allen belt. The satellite also carried a transistorised (0.5 kg) transmitter which relayed 'operational and environmental data' on three telemetry channels. The transmitter had a total effective radiated power of 180 mW, some three times that of America's first satellite Explorer 1 but still a whisper in the cosmos [3].

In common with the early, low-powered satellites used for communications and remote sensing, the onus was on the ground segment to detect, receive and amplify these cosmic whispers. In the case of spacecraft heading off towards the Moon, as opposed to circling the Earth in low-altitude orbits, the decrease in power flux density resulting from the spreading of the antenna's beam was of serious concern. Spacecraft signals were typically measured in fractions of a billionth of a watt by the time they reached the Earth. Effectively, this meant that it was a hundred times as difficult to

Figure 6.3   Pioneer 5 with its 'paddle wheel' solar arrays [NASA]

detect a signal from the Moon as it was from geostationary altitude and, incidentally, 56 million times as hard from Mars [5].

The eventual solution for NASA was the development of its Deep Space Network (DSN), an international antenna network designed to transmit commands to spacecraft, acquire telemetry from them, determine their positions and velocities and measure variations in radio waves for science experiments. However, at the time of the early Pioneer missions only the Goldstone 26 m station was operational, along with a smaller, 3 m diameter antenna in Puerto Rico [3], which is why the cooperation of Jodrell Bank was so useful. Indeed, it was even used to transmit the signal which triggered the separation of Pioneer 5 (Figure 6.3) from its launch vehicle on 11 March 1960 and, thereafter, to repeatedly switch its transmitters on and off [6].

Whereas some of the early spacecraft provided little more than a temporary entry in the space record book, Pioneer 5 proved conclusively that there was a scientific aspect to long distance travel into the solar system. In addition to becoming the first of the true deep-space probes by returning data from a distance of 36.2 million km from Earth, it confirmed the existence of the interplanetary magnetic field and provided warning of solar storms which could cause electromagnetic interference and other effects here on Earth. This early-warning function later became an important space application in its own right under the banner of 'space weather'.

Phase 2: impact

Having made a number of lunar fly-bys, the two competitors moved towards the second phase of planetary exploration – impact. On 13 September 1959, some

(a)   (b)

Figure 6.4   (a) Luna 2; (b) Luna 3 [NASA]

34 hours after launch [7], Luna 2 became the first spacecraft to impact the Moon.[2] In addition to its serious scientific intentions, the mission delivered a political message in the form of a 26 kg sphere made from tiny hammer-and-sickle medallions. According to one contemporary account, it disintegrated on impact, scattering the medallions over the surface of the Moon [2], though in fact the force of the impact is just as likely to have buried the sphere under several metres of lunar crust.

Two days later, Premier Khrushchev arrived at Andrews Air Force Base to begin a tour of the United States. One of his first official acts was to present President Eisenhower with a replica of one of the pentagonal plates of which the metal sphere was composed. Apart from the familiar Russian emblems, a ribbon wound round stalks of wheat bore the Communist slogan 'Workers of the world, unite!' [8].

Luna 3 was arguably more useful. Launched two years to the day after Sputnik 1, on 4 October 1959, it was placed in a large elliptical Earth orbit which caused it to loop around the Moon every 16.2 days, thus granting the Soviet Union a 'first' in that it was able to photograph the lunar farside for 40 minutes on its first orbit. This was a significant technical achievement because, until then, one could only guess at the appearance of the lunar farside. Since the Moon is in 'captured rotation' about the Earth, it presents only its familiar face to Earthbound observers. Luna 3's 29 historic pictures showed that there were many more craters and fewer maria than on the nearside.

---

[2] Only successful probes were given a 'public serial number', although they all had manufacturers' product designations.

(a)   (b)

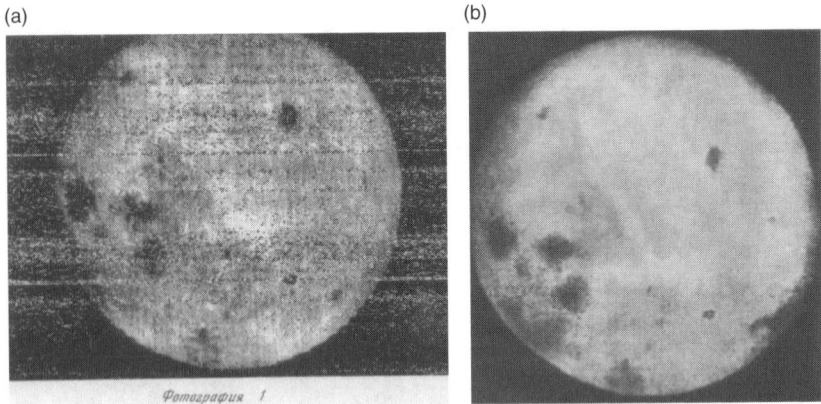

Figure 6.5   First image of lunar farside (from Luna 3 on 7 October 1959): (a) noisy and low resolution, (b) higher resolution. Dark spot at upper right is Mare Moscoviense; small dark circle at lower right with white dot in centre is the crater Tsiolkovsky and its central peak [NASA]

Techniques for imaging in the early days of the Space Age, when electronic imaging was not available, were based on photographic film and now seem archaic. Luna 3 was, in effect, a space-based photoprocessing laboratory required to operate under the weightless conditions and wide temperature extremes for which space is renowned. Following exposure, the film was automatically developed, fixed and dried, then placed in front of a TV scanning system, which converted the photographs into a stream of telemetry data [2]. Once received on Earth, this was converted back to a hard-copy photograph. Today, the intermediate (non-electronic) film stage seems cumbersome, inefficient and prone to mechanical failure, but before the vidicon tube became available for spaceflight, it was the only method available.

Nevertheless, the USSR made the most of its technology by arranging a 'repeat showing' of the data two weeks later. According to one contemporary author, the orbit of Luna 3 was the result of a 'complex mathematical analysis ... begun in 1953 at the Mathematics Institute of the Academy of Sciences' using early digital computers [9]. The orbit ensured that the spacecraft would loop around the Moon and return to the vicinity of the Earth 'after a two week flight covering almost 800,000 miles'. As it passed over Soviet territory, the spacecraft was commanded to televise its stock of 35 mm film, this time, because of the reduced range, with greater quality of reproduction (Figure 6.5) [10].

Again, with today's understanding of orbital dynamics, the immensity of this achievement is not self-evident. Only two years after the first man-made object had orbited the Earth – a relatively trivial dynamical problem – the Soviet Union had succeeded in utilising the gravity of the Moon to retarget a spacecraft back towards the Earth.

America's next lunar mission, which followed a month after Luna 3's 'TV show', was a Pioneer launch (designated Atlas-Able 4B), which failed when the payload shroud broke away from the vehicle about 45 s after lift-off [7]. NASA had to wait until March 1960 for its Pioneer 5 success, mentioned above. Unfortunately, this was itself followed by another two launch failures: Atlas-Able 5A in September, the result of a second stage malfunction, and Atlas-Able 5B in December, due to an explosion 70 s after launch [11].

Of course, the space agency had a lot of other things on its plate at the time – including communications and weather satellites – but they were no help to America's lunar programme. Whereas the Soviet Union had reached the fly-by and impact phases, and was moving inexorably towards the landing phase, America had still to hit the target.

Then, on 12 April 1961, the Soviets delivered another body blow to American pride by placing Yuri Gagarin in orbit. It was one of the events which led President John F. Kennedy to make a speech, the following month, which was to chart America's path in space for the best part of a decade. Apart from galvanising American industry into developing a system that would land men on the Moon under the Apollo programme, Kennedy's declaration also reinvigorated the unmanned lunar programme in support of Apollo. The main beneficiaries were the Ranger impact and Surveyor soft-lander projects, which are discussed below.

Kennedy's backing of the lunar exploration programme was good news for NASA, since political uncertainty had already threatened America's lunar probes: according to one author, 'by the summer of 1960, the fiscal reality of NASA's limited budget began to erode the funds for the Ranger effort and 100 project engineers had to be laid off'. Following Kennedy's speech, however, 'All stops were pulled out and the NASA budget began to grow dramatically' [12].

Ranger

The Ranger programme was an important one for NASA, since it offered the opportunity to test a variety of spacecraft subsystems, such as guidance, propulsion and communications systems, which would – at least in essence – be applicable to the nascent Apollo programme. The missions themselves were designed to prove that a spacecraft could be targeted accurately towards the Moon and to characterise the lunar surface in preparation for a manned landing. However, the Ranger programme did not go according to plan and many lessons were learned as a result; for this reason, it is worth looking at it in some detail.

Design studies for Ranger, a far more complex spacecraft than the early Pioneers, had begun at NASA's Jet Propulsion Laboratory (JPL) in 1960 [13]. The spacecraft was to weigh approximately 300 kg, a significant increase on Pioneer 4's 5.9 kg [14]. This was made possible because of the parallel development of launch vehicles, the payload capacity of which was increasing steadily with each new vehicle or vehicle generation.

According to one author, the design mass of the initial Ranger probes, known as the Block I spacecraft, was frozen at 306 kg at a time when the capability of their Atlas-Agena launch vehicle was 'still following a roller-coaster path' [15].

Obviously, spacecraft designers had to err on the side of caution; otherwise they would risk developing a spacecraft that was too heavy to launch. However, it was unfortunate that they could not take advantage of the 74 kg of additional capacity that the booster later acquired, since this would have allowed much-needed redundancy to be incorporated [15].

Ranger was a relatively large spacecraft for the time, measuring 3 m in height and 4.5 m across its deployed solar panels. The spacecraft was made in two main parts, a 1.5 m diameter hexagonal box section containing most of the support or service subsystems surmounted by a tall cylinder which housed the payload (Figures 6.6 and 6.7). While the distinction between subsystems and payload had always been clear to those involved, the earliest spacecraft were relatively simple devices and housed the majority of equipment together in the sphere or cylinder of the spacecraft body. However, when it became possible to envisage a series of similar spacecraft that would grow in capability throughout the series, it made sense to develop a separate service module or 'bus' which would be more-or-less standard on each spacecraft. This service and payload module philosophy was discussed in Chapter 3 in relation to the Orbiting Solar Observatory (OSO) and was also applicable to Ranger.

Ranger's standard 'bus' contained the spacecraft's propulsion, attitude control, power and telemetry subsystems. In common with the majority of satellites destined for orbits higher than LEO, Ranger carried an integral rocket engine to provide propulsion additional to that of the launch vehicle, which in Ranger's case was used for mid-course corrections to its flightpath, rather than placing it in orbit. It was capable of providing 23 kg of thrust for up to 98 s [16].

As far as attitude control was concerned, engineers faced a challenge with spacecraft like Ranger, since its camera payload had to point towards the Moon while its communications antenna remained 'locked onto' the Earth. For some spacecraft, it was possible to take pictures or other data, record them and transmit them to an earth station at a later time, but Ranger's mission made this impossible. It was designed to dive headlong onto the lunar surface, so there would be no 'later time' in which to download images. Attitude information was supplied by sun sensors, earth sensors and a set of gyroscopes controlled by an onboard computer. Attitude control itself was provided by nitrogen gas jets [17].

Electrical power was provided by a two-panel solar array hinged from a tubular 'truss structure' which surrounded the bus section; the array was secured against the payload cylinder for launch and deployed once in space. The 9800 cells produced about 200 W and the spacecraft carried two silver-zinc batteries for use during the launch phase and mid-course corrections, when the arrays would not be facing the Sun [17]. A steerable high gain communications antenna was also deployed from the tubular structure at the appropriate time. The spacecraft's sun sensors ensured that the array panels remained perpendicular to the solar direction to ensure maximum output from the solar cells, while the infrared earth sensors enabled the antenna to be pointed in the right direction.

Now that spacecraft can be targeted to enter a planet's atmosphere halfway across the solar system, or land on the surface of a comet, it is hard to imagine the difficulties involved in hitting a target the size of the Moon, which, after all, is an easily

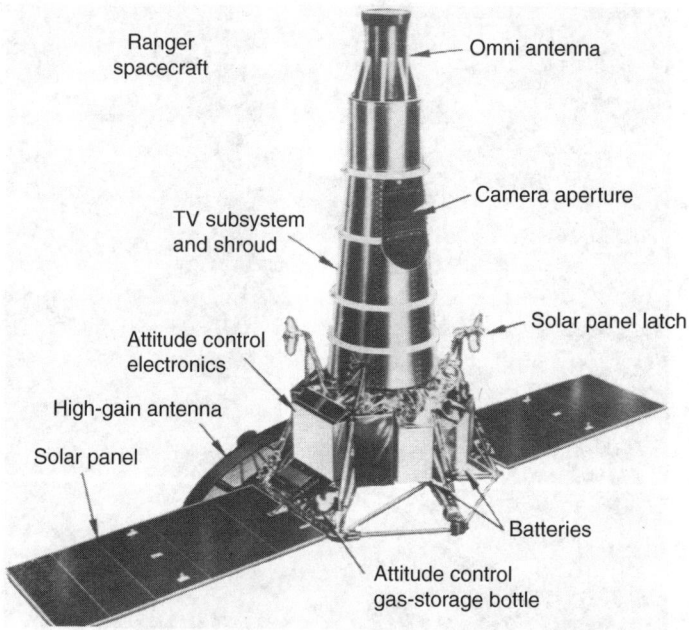

Figure 6.6   Main features of Ranger [NASA]

measurable object in the Earth's night sky. In the early 1960s, however, targeting was a significant challenge.

To impact in the specified target area, Ranger had to enter a 16km-diameter 'corridor' within about 25 km/h of the desired injection velocity (which varied from 39,360 to 39,400 km/h dependent on the launch date). This challenging specification had to be met by the Atlas-Agena launch vehicle if Ranger's mission – primarily one of lunar photography – was to be successful.

After the spacecraft separated from its launcher, its onboard computer caused it to tumble so that its sun sensors could search for and lock onto the Sun; this would allow its solar panels to produce power for its various subsystems and to recharge its batteries [18]. The spacecraft was then aligned with the Earth using an infrared sensor to maintain a communications link through its high-gain antenna (prior to this, contact was maintained via a low-gain omni-directional telemetry and command antenna). This sequence is much the same as that performed by a modern-day communications, weather or science satellite once it reaches orbit. The main difference with Ranger, and the other lunar probes, was that, having oriented themselves, they had to navigate towards the Moon.

Tracking of the spacecraft's trajectory was performed by the Goldstone station, from which commands could be transmitted, as necessary, to adjust Ranger's orientation and perform other housekeeping tasks. The propulsion, attitude control and antenna pointing systems were required to work together in a relatively complex

Figure 6.7　Ranger spacecraft [NASA]

process for mid-course manoeuvres. First the antenna had to be rotated away from the rocket plume, which would issue from the base of the service module; then the attitude control subsystem would turn the spacecraft to allow the engine to fire in the right direction; after the burn, the spacecraft would be reoriented so that its array faced the Sun; and finally the antenna would be realigned with the Earth.

The manoeuvre itself was capable of adjusting the spacecraft's trajectory at the Moon's distance by as much as 12,000 km in any direction. It goes without saying that mid-course manoeuvres were nail-biting times, since if only one of these actions failed the mission could be over.

The first two Rangers – the Block I spacecraft – were intended to validate the spacecraft and mission design while carrying their payloads beyond the Moon in highly elliptical orbits, from which they could study solar radiation and other interplanetary phenomena over extended lifetimes. The spacecraft carried a package of science experiments including a magnetometer, cosmic dust detector, ion chamber, Lyman Alpha telescope, X-ray scintillation counters and electrostatic analysers.

The spacecraft were launched in August and November 1961. Unfortunately, neither one succeeded in getting beyond low Earth orbit because their launch vehicles' Agena upper stages failed to re-ignite as designed and their orbits decayed rapidly.

The setbacks came at a time when NASA, and America itself, desperately needed a measure of success in its space programme. Even editors of British hobby magazines were well aware of the apparent Russian superiority. An editorial in the January 1962 issue of *Practical Mechanics* records that 'Russia has already announced that they plan to orbit a man around the moon and bring him safely back to earth. Subsequently a rocket will be landed on the moon's surface, equipped with all the necessary instruments to send back comprehensive details of conditions that exist there. We take it for granted that Russia will take the initiative in these projects ... the race to space between nations is now virtually non-existent [19].' Certainly, the race into space had already been won by the USSR – with its launch of Sputnik 1 in 1957 – but, as the editorial itself implies, it had been replaced by a race to land a spacecraft on the Moon.

Under the original Ranger programme, having tested the 'bus' on the first two missions, subsequent spacecraft were to take close-range photographs and deposit an instrument package – called a Lunar Rough Landing Capsule – on the lunar surface. Thus in 1960, the Aeronutronic Division of the Ford Motor Company obtained a contract from JPL for the development of a 44 kg capsule that would be mounted on the three Block II Rangers in place of some of the scientific instruments (Figure 6.8) [18,20].

According to a JPL contractor list published in 1962, Aeronutronic was one of 28 subcontractors that 'provided instruments and hardware for Ranger 3 [amounting] to $8 million'. They included a number of well-known firms such as RCA, which provided the lunar impact TV camera, Bendix, Lockheed, Motorola and Texas Instruments. In addition, there were some 1200 lesser-known industrial firms, whose supplies added more than $5 million to the bill [21]. One of the smaller subcontractors for the later Rangers was the lyrically named 'Ace of Space Inc, Pasadena' which provided an 'electronic chassis'.

The spherical instrument capsule was of unusual construction: its payload of seismometer, amplifier, battery and radio transmitter was housed in a metal sphere enclosed in a balsa wood outer casing, with a thin layer of oil between the metal and the balsa so that the sphere was 'free-floating in the cavity'. The 63 cm capsule was designed to have its centre of gravity about half an inch (1.25 cm) below its geometric centre, so that, on landing, it would right itself 'after the fashion of heavy-bottomed toys' [22]. This placed its antenna on the top, pointing towards Earth.

Despite being equipped with a retro-rocket, which would do much to reduce the effect of the mother-probe's 'kamikaze' approach, it by no means promised a soft landing. As one author put it, 'under the influence of lunar gravity ... the capsule will strike the surface at between 80 and 120 mph, bouncing and rolling until it comes to rest' [22].

The capsule design was complicated by the fact that, once stable, the oil would have to be vented off to prevent it acting as an insulator to the seismic shocks the capsule was designed to detect; this was to be accomplished by blowing out two pyrotechnic slugs. The capsule would also carry 1.7 kg of water as a thermal control system: the idea was that it would be heated by the dissipation of electrical energy in the capsule until it reached its boiling point under ambient lunar conditions. This was designed to place an upper limit on the capsule's operating temperature, while ensuring that the temperature would not fall below freezing during the 14-day lunar night [23].

*Figure 6.8   Model of Block II Ranger showing spherical instrument capsule
[NASA]*

Unfortunately, neither the capsule's unusual thermal subsystem, nor any of its other systems, had the chance to prove themselves; ironically, the culprit was the temperature regime to which the probe was subjected even before it left the ground. Following a NASA edict made in 1959, the entire spacecraft was sterilised to prevent contamination of the lunar environment. However, since alcohol wipes and ethylene oxide gas could only reach parts of Ranger's internal mechanisms, it was decided to heat the entire spacecraft to 125°C for 24 hours, despite evidence gathered in late 1961 that this would have 'detrimental effects on Ranger's electronic systems' [24].

As a result, following its launch on 26 January 1962, Ranger 3 failed to accomplish its mission. This was ostensibly because a booster and guidance system error had conspired to cause it to miss the target by some 37,000 km. However, as was shown later, its central computer and sequencer had failed from the effects of pre-flight heat sterilisation.

In contrast, Ranger 4's flightpath was so accurate that it would strike the Moon – on 26 April 1962 – without even a single mid-course correction. Sadly, heat sterilisation had claimed another victim and the master clock in the spacecraft's central computer and sequencer had stopped. Sixty-four hours after launch, Ranger 4 acquired the dubious honour of being the first spacecraft to crash on the farside of the Moon [25].

One can only imagine the frustration of the engineers and scientists engaged in the project: now that the launch vehicle and guidance system had finally been persuaded to work in unison, the spacecraft had died from 'heat stroke'. Apart from the effects of heat sterilisation, a formal investigation into the failures suggested that the Ranger missions relied too much upon the successful completion of a long chain of events and were doomed to failure by a lack of redundancy and a lack of flexibility to work around a single point failure [26].

The final Block II Ranger, Ranger 5, failed an hour after launch on 18 October 1962, this time reportedly because of a short circuit in the computer's power-switching module which directed the flow of energy from the solar panels [26]. As a result, the unusual instrument capsule, which never had a chance to prove itself, is now barely a memory in the annals of spaceflight.

Subsequent Ranger missions that were already scheduled were delayed for a year to redesign the Ranger spacecraft, solve its reliability problems and develop four 'Block III' spacecraft dedicated to high-resolution close-up lunar photography. It was not, however, equipped as a lander and would simply crash into the lunar surface in the ultimate close approach.

Ranger revamped

Following a politically unendurable litany of failed missions, the new-look programme offered Ranger a chance to succeed by supporting the much larger goals of the Apollo programme. Some saw this as an indication of how Apollo had come to dominate other NASA space programmes in the 1960s, often to their own detriment [27]; and looked at in isolation it was easy to see why. Ranger had begun as an explorer of the interplanetary medium, failed in its role as the delivery system for the first lunar seismic laboratory, and ended up as a sacrificial camera-carrying platform designed to bury itself ignominiously beneath the lunar surface as its Luna and Pioneer forbears had a few years earlier. However, the revamp could also be characterised as 'joined-up thinking': there was no hiding the fact that Apollo was the most important programme in NASA's portfolio, so it made sense to utilise existing programmes to support it wherever possible.

To increase the probability of success, the upgraded Ranger had redundant attitude control components, redundant timer, spare battery and double-sized solar panels; and its control electronics were not sterilised [28]. Its primary payload was a camera system designed by RCA, the company which provided the imagers for the Tiros weather satellites. The Ranger system comprised six vidicon-based TV cameras, two 'full-scan' devices with wide-angle lenses and four 'partial-scan' units with telephoto lenses. It was designed to transmit about 300 pictures a minute during the final 15 to 20 minutes of the spacecraft's descent towards the lunar surface.

Ranger 6 was launched on 30 January 1964, a full 15 months after the ill-fated Ranger 5. Under the redesign programme, Ranger 6 had been stripped of all scientific payloads to allow it to concentrate on the 'surface characterisation' mission in support of Apollo. The performance of its camera was therefore crucial. As it neared the

Moon – on target – its telemetry carrier was strong, but it lacked the all-important video signal. As controllers watched helplessly, the spacecraft crashed onto the Moon, having returned no useful data.

The culprit, it was discovered, was the Atlas launcher, which had vented its unburnt kerosene propellant as the booster section was jettisoned about two minutes into the launch phase. The sustainer engine had ignited the vented fuel creating 'a flash wave which momentarily engulf[ed] the entire rocket' [29]. Although the phenomenon was known about and of no consequence to the Atlas, Ranger engineers were apparently unaware of its potential to damage their spacecraft. As a result, umbilical connector pins wired from the Agena upper stage to the Ranger were shorted out by the plasma from the burning fuel vapour which damaged the control electronics for the camera system.

The Ranger programme did not celebrate its first success until July 1964, when Ranger 7 returned more than 4,300 photographs of the lunar surface before impact. Its signals were received and amplified by the Goldstone station and recorded on magnetic tape. They were also displayed on two specialised TV monitors known as kinescopes, whose screens were photographed using 35 mm cameras, the shutters of which were set to remain open long enough to record a full sweep of the electron beam across the screen, giving a 'clean' still photograph (Figure 6.9) [30]. The initial images had a similar resolution to that of Earth-based telescopes of the time (somewhat less than a kilometre), while the final images showed features as small as about 25 cm across. In most cases, the very last image was incomplete and bordered by a band of noise – a striking indication that the spacecraft had stopped transmitting as it hit the Moon.

In recognition of the achievement, the International Astronomical Union (IAU) gave the area within Mare Nubium (the Sea of Clouds), which Ranger 7 had targeted, the name Mare Cognitum (the Known Sea). However, even a well-meant and politically correct endorsement from the IAU could do little to eclipse the relatively inconsequential nature of the Ranger programme.

By this time, to put the long-awaited event in perspective, space technology had moved on significantly in other areas: communications satellites, weather satellites and environmental research satellites were in continual operation and several astronauts and cosmonauts had orbited the Earth. Moreover, the race to land a man on the Moon was in full swing, and it was deeply embarrassing to NASA that it had only just been able to target a probe to destroy itself on the lunar surface.

Since the idea of a lunar landing had been mooted, there had been various theories as to the surface composition, which predicted anything from solid basalt to 'dust bowls' that would swallow a lunar lander whole. The pressures of the schedule for Apollo meant, in the end, that redesigning the Ranger programme had been a largely pointless exercise: by early 1964, the design of the Apollo lunar module had already been frozen, so all Ranger 7 could do was confirm the design decisions that had already been made. Nevertheless, as contemporary photographs show, America over-reacted with elation at this measure of success: as one author reported, 'Some cheered and leaped for joy; others openly wept' [31].

Figure 6.9   First image of the Moon taken by a US spacecraft (taken by Ranger 7 on 31 July 1964 about 17 min before impacting the lunar surface): Mare Nubium is at centre left; large crater at centre right is the 108 km diameter Alphonsus [NASA]

Some were, however, more pragmatic. Even after the successful Ranger 8 mission in February 1965, the unconventional astrophysicist Thomas Gold of Cornell University lamented 'the Ranger pictures were like mirrors: every scientist saw his own theories reflected in them' [32]. If the disagreements over the constitution of the lunar surface were ever to be settled, it would be necessary to land there.

Phase 3: landing

For NASA, with its new focus on Apollo, it became crucial to prove that a spacecraft could be soft-landed on the Moon, and this was the aim of the Surveyor programme. It began conservatively enough, in August 1965, with the launch of a model of the spacecraft to a 'simulated Moon' somewhere in space [33], thus avoiding the possibility of a failed soft landing. Unfortunately for America, when a landing did occur, it was performed by the Soviet Union's Luna 9, launched on 31 January 1966.

Luna 9 was a fairly sophisticated spacecraft comprising two elements, a mother-craft and a landing capsule. The 1.5-tonne mother-craft was cylindrical with a large course correction and retro-rocket engine at the lower end and the 100 kg encapsulated lander at the other [34]. In addition to the main engine, there were four 'vernier' thrusters for attitude control, and a radio altimeter to measure the height above the lunar surface (Figure 6.10).

*Figure 6.10   Luna 9 mothership display at Le Bourget Air and Space Museum, Paris
[Mark Williamson]*

According to one author, 'By American standards Luna 9 did not, in fact, make a soft landing, but a semi-soft one'; the capsule (complete with shock absorbers) separated from its retro-rocket and 'fell onto the moon' [35]. Certainly, the capsule deployment process was novel: as the spacecraft approached the surface with its retro-rocket firing, a hinged arm was deployed; when it touched the surface, the capsule was ejected from the top of the vehicle and fell towards the surface, where it bounced and rolled to a stop [36].

The lander itself was a sphere with four conformal petal-shaped panels which opened after landing to right the spacecraft (Figure 6.11). The probe carried an instrument package which included a TV camera and an ingenious deployable rod with mirrored surfaces which acted as a 'rear-view mirror' for the camera. It returned panoramic views of its surroundings from the bottom of a shallow crater in the Ocean of Storms for three days [37,38].

Among its measurements on the surface, Luna 9 recorded a temperature of 300°C and a radiation level of 30 millirads per day [38]. Perhaps, more importantly, it proved

(a)

(b)

1 m

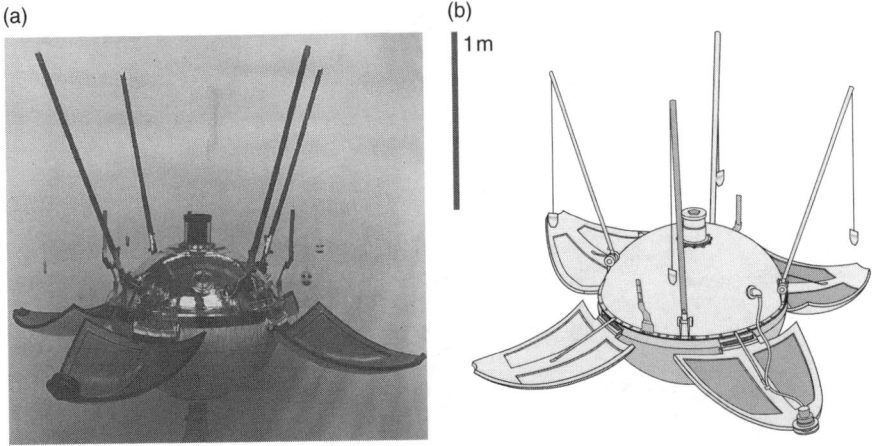

Figure 6.11   (a) Luna 9 lander [NASA] and (b) line drawing [James Garry]

once and for all that a lunar lander would not sink beneath a sea of dust, as some had hypothesised. Ironically, therefore, it was a Soviet spacecraft, and not NASA's Surveyor, that provided the first true characterisation of the lunar surface desired by the Americans for their Apollo landing mission.

Luna 13, like Luna 9, landed in the Ocean of Storms and, from December 1966, probed the surface with a radiation density meter and mechanical soil meter (Figure 6.12). The former involved bombarding the surface with gamma rays from a caesium 137 radioactive source and detecting the radiation returned; in this way it was possible to probe the lunar soil, or 'regolith', to a depth of some 15 cm. The soil meter measured the bearing strength of the surface by driving a titanium–tipped hammer into it using an explosive charge. The experiments indicated that the Moon's topsoil was physically similar to the Earth's but less dense (an average of 0.8 g/cm^3 compared with Earth's 3.3 g/cm^3) [39].

Despite the successes, by the mid-1960s, the Moon's surface was littered not only with the debris of intentionally impacted spacecraft, but also with that of a number of failed landing attempts. For example, Lunas 5, 7 and 8, all launched in 1965, crashed for a variety of reasons, including the apparent inability to fire retro-rockets at the correct time. On Luna 7, the retros fired too early and the probe fell uncontrolled to the Moon, whereas on Luna 8 they fired too late and it smashed itself into the lunar surface. Luna 4 and Luna 6 were also connected with the lunar landing mission, but both missed the Moon and ended up in solar orbit. Moreover, there were other unnumbered Luna missions which never even made it into space.

What this indicates is that both sides in the 'Moon race' had similar difficulties, first in launching spacecraft successfully, and second in targeting them towards the Moon. On balance, it is difficult to assess whether one nation's technology was better than the other's: when it worked, it was obviously better. In the early years, both sides suffered reliability problems and both were prone to human error in design,

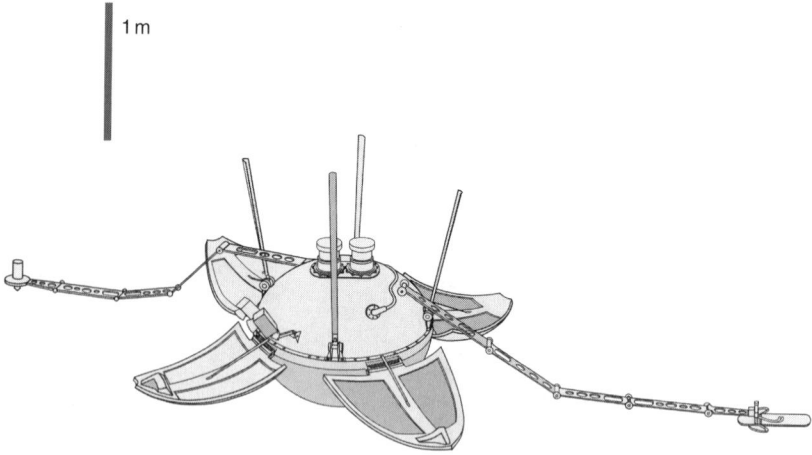

Figure 6.12   Luna 13 [James Garry]

manufacturing and testing of their respective spacecraft. The reason the Soviets scored so many 'firsts' in the early years of the Space Age was largely a result of their ability to make a large number of attempts. Disparagingly, this might be characterised as 'success by trial and error', but in engineering terms it was nothing more than a commitment to in-flight testing.

Surveyor

Though characteristically late in the game, NASA showed with its Surveyor programme that the agency and its contractors had made significant improvements in reliability compared, for example, with the early Ranger missions. Surveyor spacecraft succeeded in executing a soft landing on five out of seven attempts, providing a reasonable level of confidence that an Apollo lunar module could be guided and controlled to a safe and accurate landing.

Surveyors 1, 3, 5 and 6 confirmed the potential landing sites for the Apollo flights leaving Surveyor 7 to carry out a purely scientific mission in the lunar highlands, considered too risky for a manned landing. Together the landers produced a library of 86,589 pictures from cameras which could distinguish particles as small as 1 mm across [40]. Surveyor 1, alone, transmitted 11,237 photographs [41].

Surveyor 1, launched on 30 May 1966, was thus the first spacecraft to execute a controlled landing on another planetary body (Figure 6.13). Indeed, in the context of the history of technology, it is interesting to note that by mid-1966, when slide-rules and rotary-dial telephones were still the norm, both Russia and America had succeeded in landing spacecraft on the Moon. The Space Age was less than nine-years-old, and already terrestrial technology was operating, not only in an extraterrestrial environment, but also on the surface of another planetary body. Although most laymen at the time, as now, would not appreciate it, this was a staggering cultural achievement.

(a)

(b)

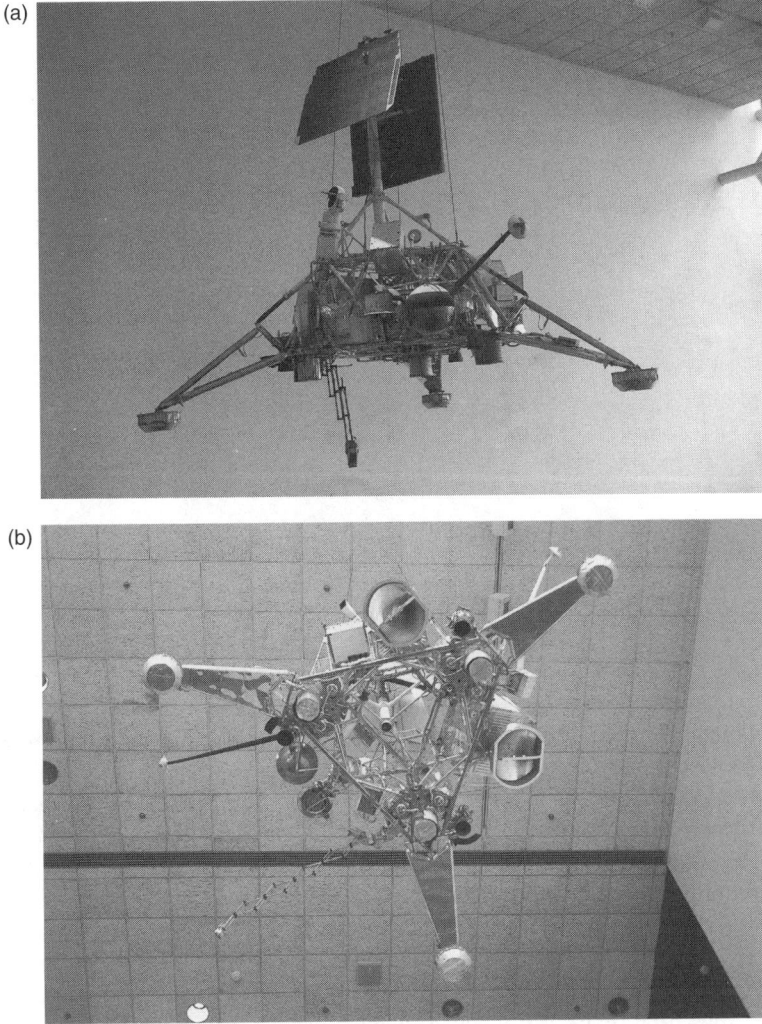

Figure 6.13   Two views of Surveyor ground test article in National Air and Space Museum, Washington DC [Mark Williamson]

Surveyor was a spindly tripod of a spacecraft, about 3 m high and 4.25 m wide across the deployed landing legs (Figure 6.13). It weighed about a tonne at launch, complete with its solid propellant retro rocket, and about 290 kg without it [41]. The entire spacecraft was based on an aluminium truss structure of mechanically strong triangular forms, at the centre of which was the retro rocket. Housed around this, within the rest of the primary structure, were propellant and pressurant tanks for the vernier thrusters, while fixed outside the structure were a pair of thermally controlled compartments housing payload and subsystem equipment. A deployable solar array

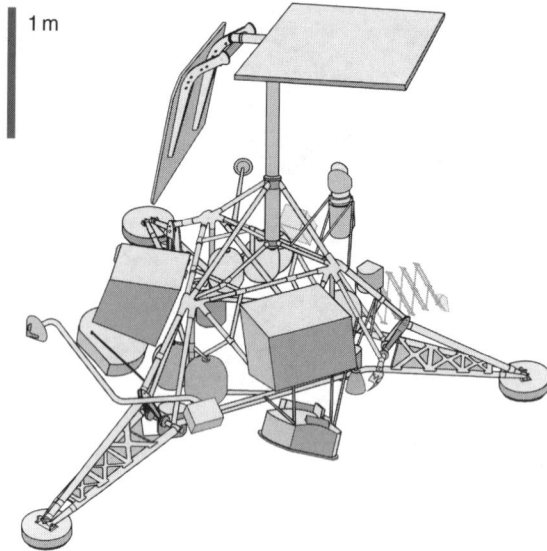

Figure 6.14　Surveyor [James Garry]

panel and a steerable flat-plate high gain antenna for communications with Earth were attached to a central mast.

Although additional equipment, such as a TV camera and the all-important soil sampler (carried on later missions), might appear to have been attached to the truss entirely at random, it was carefully placed to ensure the balance of the spacecraft about its centre of mass (Figure 6.14). This aspect of spacecraft design is common to all spacecraft, which need to maintain their stability, whether for spin-stabilisation in a transfer orbit or for mid-course corrections and landings.

The Surveyor spacecraft were designed and built for JPL by Hughes Aircraft Company, an industrial contractor better known for its Syncom and Early Bird communications satellites. Although most of the early space probes had been designed and built entirely in-house by JPL scientists and engineers, it soon became clear – especially for programmes with multiple spacecraft – that space exploration had evolved beyond a laboratory exercise to an industrial undertaking. So, while JPL took the role of programme architect and customer, the role of prime contractor was increasingly devolved to industry.

Moreover, by this point in the American space programme, spacecraft subcontractor lists contained some familiar names associated throughout the world with their own particular products: for example, Surveyor's tape recorder was supplied by Ampex, its camera lens by Bell and Howell, check-out computers by Univac and system test-stand by Scientific–Atlanta. Among the familiar space companies were Thiokol for the main retro-engine and Eagle-Picher for the batteries, while among the less obvious was the National Water Lift Co. of Kalamazoo, Michigan, which provided the landing shock absorbers.

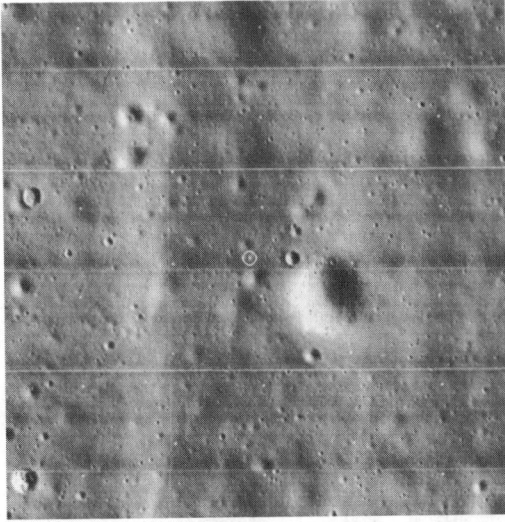

Figure 6.15   *Surveyor 1 landing site in Oceanus Procellarum imaged by Lunar Orbiter 3. The spacecraft is the bright spot in circle at dead centre (shadow of 1 m wide solar panel can also be seen on original). Framelet width (spacing between horizontal lines on image) is about 220 m [NASA]*

Surveyor's landing sequence was designed to help confirm techniques planned for Apollo. It travelled to the Moon in a 'cruise attitude' with its solar panel oriented towards the Sun, then, thirty minutes before touchdown, was rotated to align its retro-rocket with the direction of flight. The rocket was fired at an altitude of about 80 km, as measured by Surveyor's Doppler radar altimeter, when the spacecraft was travelling at around 9500 kph; by the time it shut down, at an altitude of some 11 km, the spacecraft's speed had decreased to around 650 kph. The main retro was then jettisoned to reduce the overall mass and the system of liquid-propellant verniers took over. The verniers continued to fire until the craft was about 4 m above the surface and travelling at 6 kph, resulting in a touchdown at 8–16 kph on its three spindly legs. The force of the impact was absorbed by crushable footpads, shock-absorbing pistons at the tops of the legs and crushable blocks on the lander structure.

One of the most important subsidiary goals of the Surveyor programme was to determine whether the lunar surface could support the weight of a lander. Although the design of the Apollo lunar module had been frozen and the Luna 9 landing had taken the burden of proof from Surveyor, the programme showed, at least near the landing sites chosen for Apollo, that the Moon had a solid surface covered with a shallow layer of fine dust.

The mechanical soil sampler – first carried on Surveyor 3 in April 1967 – was used to dig trenches to uncover the subsurface soil and was also pressed down and dropped onto the surface to determine its bearing strength and impact resistance.

Figure 6.16   Apollo 12 astronauts visit Surveyor 3 in November 1969 [NASA]

This confirmed the density measurements of Luna 13 and showed that the bearing strength of the regolith increased by an order of magnitude at a depth of about 5 cm, providing further confidence that the Moon would support the weight of the Apollo lunar module [42]. The sampler arm, its operations monitored using a TV camera, had a reach of about 1.5 m over a 112° arc, giving a coverage of some 2.2 square metres. The camera was also used to observe a number of small bar magnets, placed to attract any magnetic soil particles, but the iron content of the soil appeared to be low [42]. The final three Surveyors (5–7) carried an alpha particle scattering experiment, which used a curium 242 source to analyse the chemical composition of the regolith. Detectors measuring the reflected alpha particles and the energy of protons split off from the nuclei of soil atoms were used to conclude that the regolith was basaltic in nature, in other words similar to the Earth's most common igneous, or volcanic, rocks.

On 20 November 1969, in an unusual adjunct to the second manned landing mission, Surveyor 3 was visited by the Apollo 12 astronauts, Charles Conrad and Alan Bean, who had landed only 180 m away (Figure 6.16). They took photographs of the unmanned pathfinder, some showing clearly that the spacecraft had bounced on landing. Apparently its descent radar had been confused by highly reflective rocks and its verniers had been late in switching off, causing it to execute a couple of unscheduled (10 m and 3 m) bounces (Figure 6.17) [41].

The astronauts also brought some parts of the lander back to Earth for analysis. Apollo 12's accurate landing – on a planetary body 400,000 km from Earth, but within easy walking distance of another spacecraft – proved that America had finally

Figure 6.17   Surveyor 3 footpad showing how it bounced on landing. Waffle pattern of 30 cm footpad impression indicates no significant change in lunar environment since landing in April 1967 [NASA]

cracked the lunar targeting problem. It is significant that less than six years earlier, its spacecraft could not even hit a body measuring 3,500 km across!

Interestingly, the Surveyor programme was also designed to test a second, and less well-known part of President Kennedy's specification for the Apollo missions: not only would America land a man on the Moon, but also return him safely to the Earth. On 17 November 1967, having touched down seven days earlier, Surveyor 6 became the first spacecraft to lift off from the surface of the Moon. Its thrusters burned for just 2.5 s, generating some 70 kg of thrust to lift the spacecraft about 4 m above the surface [43], but it provided a useful demonstration of what was to follow in 1969 when the Apollo 11 astronauts would begin their return to Earth.

The Soviet Union made its first true soft landing in September 1970 with Luna 16 (Figure 6.18), some four years behind America's Surveyor 1. Luna 16 also conducted a lunar lift-off and succeeded in returning 100 g of lunar soil samples to the Earth. It was an extremely sophisticated unmanned mission, but failed to create much of an impression since, by then, American astronauts had landed and returned more than 22 kg of samples. In fact, Luna 16 was a second attempt: its forerunner, Luna 15, had crashed in the Mare Crisium (the aptly named Sea of Crises) on 21 July 1969, the day after Apollo 11 landed safely in the adjacent Mare Tranquillitatis (Sea of Tranquillity). It is thought that the Luna 15 mission was a failed attempt to upstage Apollo 11.

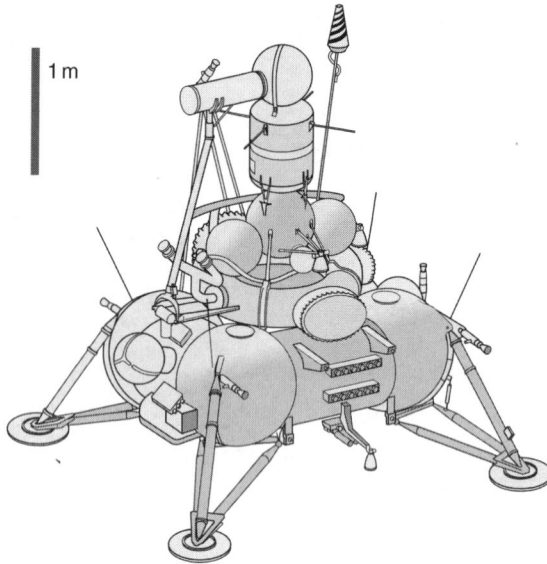

Figure 6.18   Luna 16 [James Garry]

Phase 4: orbiting

Just two months after Luna 9 became the first spacecraft to land, rather than crash, on the Moon (and two months *before* the landing of Surveyor 1), Luna 10 became the first spacecraft to enter lunar orbit, thus completing the four phases of unmanned lunar exploration (fly-by, impact, soft-landing and orbital injection).

While both superpowers were still in the midst of a political game, using space technology as a pawn, the USSR seemed particularly adept at capitalising on its abilities to beat America into the space record books. Although Luna 10 carried no camera, the orbiter was equipped with a tape recorder, which enabled it to clock up yet another first by broadcasting the Soviet national anthem, the 'Internationale', from the Moon. Apparently it was played to the twenty-third Congress of the Communist Party of the Soviet Union, which was in session at the time, reportedly bringing delegates to their feet [44].

Among other, arguably more significant achievements, Luna 10 determined that the micrometeoroid flux near the Moon is 100 times greater than in cislunar (Earth–Moon) space, an important finding for any spacecraft destined for lunar orbit. In addition, by means of a gamma ray spectrometer, it was found – well before the Surveyor landers confirmed the observation from the surface – that the lunar rocks resemble volcanic basalt [39].

Luna 10 was followed into orbit later in 1966 by Luna 11 and Luna 12, proving that the dynamical techniques of orbital insertion had been mastered – at least by the Soviet Union. This was something NASA would also have to do before a manned

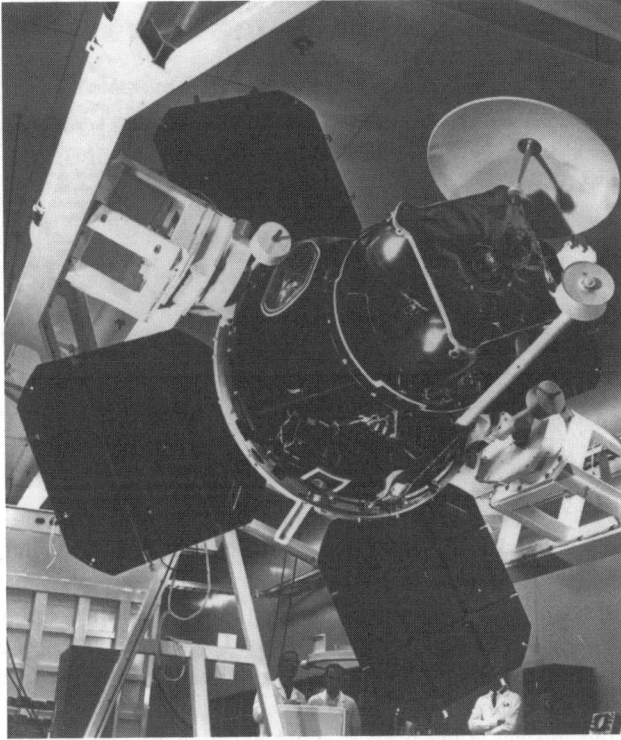

Figure 6.19   Lunar Orbiter [NASA]

landing could be attempted, since the Apollo mothership (the command and service module) was required to orbit the Moon until the lunar module returned from its landing mission. The spacecraft designed for the task were the aptly named Lunar Orbiters (Figure 6.19).

Between August 1966 and August 1967, as usual only slightly behind the Soviets, America succeeded in placing five spacecraft in lunar orbit. The Lunar Orbiters, built by Boeing, were operated in parallel with the Surveyor landing missions and returned a wealth of high-quality imagery from orbits between 1600 and 40 km above the surface. As with earlier spacecraft, the Lunar Orbiters took pictures on film, developed them automatically, then scanned them into a TV system for transmission to Earth. The imaging system, supplied by Eastman Kodak, had both wide angle and telephoto lenses which, by use of a mirror, could focus their images onto different parts of the same film loop at the same time and were capable of producing images with lunar surface resolutions of 8m and 1m respectively [45]. Although remote imaging of the Earth and Moon are not entirely comparable, because the Earth has an atmosphere, it is interesting to note that 1m resolution images of the Earth did not become available outside military circles until the late 1990s.

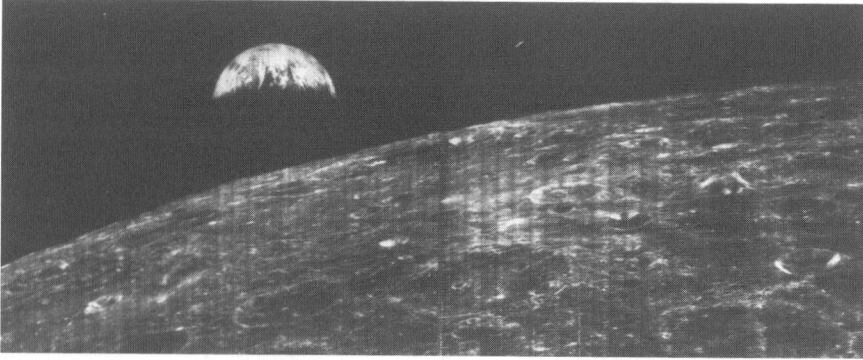

Figure 6.20   First view of Earth from Moon captured by Lunar Orbiter 1 on 23 August 1966 [NASA]

Again, the fact that the best photographs from Earth-based telescopes could only resolve features about 800 m across proved the advantages of sending spacecraft to the Moon. The catalogue produced by the Lunar Orbiters was fundamental to the planning of the Apollo missions. Although of lesser scientific interest, one of the 'firsts' was Lunar Orbiter 1's portrait of 23 August 1966 (Figure 6.20), showing Earth above the lunar horizon, a tantalising image of human perspective repeated, more famously, in 1968 by the Apollo 8 astronauts. The Lunar Orbiters also discovered mysterious large concentrations of mass (dubbed 'mascons') below the lunar maria, which proved to be extremely important for predicting the orbital paths of the later Apollo spacecraft.

NASA ended the successful series of Lunar Orbiter missions by commanding the spacecraft to crash onto the Moon or, in the case of Lunar Orbiter 4, allowing it to de-orbit [46]. This was done to 'prevent interference' with the subsequent Apollo missions. Although the policy seemed sensible at the time, considering that astronauts' lives were potentially at stake, it added to the growing catalogue of lunar surface debris. In fact, more than 80 spacecraft, propulsion units and other objects, including five Apollo Saturn V third stages, have either landed or crashed onto the Moon, creating more than 100 tonnes of debris. This is one of the lesser known and less productive outcomes of the first decade of lunar exploration.

Conclusions

The progress made in the technology of unmanned spacecraft in the 1960s was little short of remarkable, as was the fact that so many different applications were being addressed in parallel. While some teams were designing, building and launching communications satellites, remote sensing satellites and astronomical observatories into Earth orbit, others were dispatching their creations beyond the influence of the Earth's gravitational field.

This chapter has concentrated on the application of space technology to lunar exploration, but at the same time still other teams of scientists and engineers were developing spacecraft to fly to our other near neighbours in the solar system, Mars and Venus. In fact, the first attempt to send a spacecraft to Mars was made by the Soviet Union as early as 10 October 1960; unfortunately, it and a subsequent attempt four days later failed to reach space, as a result of launch vehicle failures [11]. Likewise, the Soviet Union's first attempt to launch a probe to Venus – made on 4 February 1961 with a spacecraft designated Sputnik 7 – was a failure [47]. However, the launch of a 6.5 tonne spacecraft into Earth orbit, which led to much speculation about the possibility of a manned flight, was, in itself, a demonstration of Soviet prowess in launching heavy payloads.

The first partially successful planetary mission came just eight days later, on 12 February 1961, with the launch of Sputnik 8. From an initial low Earth orbit it boosted its payload, Venera 1, on a trajectory designed to take it to Venus [48]. Sadly, radio contact was lost when the spacecraft was 7.6 million km from Earth [49], about a tenth of the way to Venus, and it raced silently past Venus into a permanent orbit round the Sun. However, following a number of successful atmospheric entry missions, on 15 December 1970 Venera 7 made the first radio transmission (lasting 23 minutes) from the surface of another planet.

In addition to further Mars missions, which eventually met with success, between 1964 and 1970 the Soviet Union made various attempts at planetary exploration under its Zond programme, which was also a cover for its plans to place a Soviet cosmonaut in orbit around the Moon.

America, of course, had its own planetary programme. Indeed, the first successful fly-by of Venus was made by NASA's Mariner 2 on 14 December 1962, when it made observations of the planet with its infrared and microwave radiometers. Interestingly, the spacecraft was based heavily on the design of the early Ranger probes, which had failed in their missions to impact the Moon. It now seems strange that the Ranger bus made a fly-by of Venus some 19 months before its first success as a lunar probe. Likewise, the first successful fly-by of Mars was made by NASA's Mariner 4 on 14 July 1965, when it transmitted the first close-up pictures of the red planet. Once again, by a strange quirk of fate, the agency was still ten months away from landing Surveyor 1 on the Moon.

Nevertheless, the techniques and technology of the era laid the foundations for later triumphs in planetary exploration and although the early fly-bys could more accurately be placed in the 'near-miss' category, targeting techniques became finely honed for later missions. Thus began a continuing effort to extend the reach of mankind from the confines of Earth orbit to the other planets of the solar system and beyond.

References

1 Thompson, Tina D. (Ed.): 'TRW Space Log 1996' (TRW Space & Electronics Group, Redondo Beach, Volume 32, 1997) p. 66

2   Taylor, John W. R. and Allward, Maurice: 'Eagle Book of Rockets and Space Travel' (Longacre Press Ltd, London, 1961) p. 143

3   'Juno' brochure (Jet Propulsion Laboratory, California Institute of Technology, undated though post-May 1961)

4   'Daily Telegraph Reporter': 'U.S. Satellite 305,000 Miles Out in Space', *Daily Telegraph and Morning Post*, Friday, March 6, 1959, p. 15

5   Deutsch, Les (JPL): quoted in Williamson, Mark: 'Extreme Contact', *IEE Communications Engineer*, **2** (5), October/November 2004, pp. 12–15

6   Gatland, Kenneth W.: 'Astronautics in the Sixties' (Iliffe Books Ltd, London, 1962) p. 93

7   Thompson, Tina D. (Ed.): 'TRW Space Log 1996' (TRW Space & Electronics Group, Redondo Beach, Volume 32, 1997) p. 67

8   Lapp, Ralph E.: 'Man and Space: The Next Decade' (Secker and Warburg, London, 1961) p. 73

9   ibid., p. 74

10   ibid., p. 75

11   Thompson, Tina D. (Ed.): 'TRW Space Log 1996' (TRW Space & Electronics Group, Redondo Beach, Volume 32, 1997) p. 69

12   Reeves, Robert: 'The Superpower Space Race' (Plenum Press, New York, 1994) p. 54

13   Gatland, Kenneth W.: 'Astronautics in the Sixties' (Iliffe Books Ltd, London, 1962) p. 173

14   Turnill, Reginald (Ed.): 'Jane's Spaceflight Directory, 1987' (Jane's Publishing Co Ltd, London, 1987) p. 91

15   Reeves, Robert: 'The Superpower Space Race' (Plenum Press, New York, 1994) p. 56

16   'Project Ranger', *NASA Facts* **2** (6), US Government Printing Office (OF-753-205), Washington DC, USA, 1964, p. 10

17   ibid., p. 9

18   Gatland, Kenneth W.: 'Astronautics in the Sixties' (Iliffe Books Ltd, London, 1962) p. 175

19   'Fair Comment' (Editorial), *Practical Mechanics* (George Newnes Ltd, London, 1962), **XXIX** (333), January 1962, p. 149

20   Reeves, Robert: 'The Superpower Space Race' (Plenum Press, New York, 1994) p. 52

21   'Subcontractors' (Rangers 3-5), 124-12/61 (Office of Public Education and Information, CalTech JPL, Pasadena, CA), dated January 18, 1962

22   Gatland, Kenneth W.: 'Astronautics in the Sixties' (Iliffe Books Ltd, London, 1962) p. 181

23   ibid., pp. 182–3

24   Reeves, Robert: 'The Superpower Space Race' (Plenum Press, New York, 1994) pp. 56–7

25   ibid., pp. 57–8

26   ibid., p. 58

27 Baxter, Stephen: 'How NASA Lost the Case for Mars in 1969', *Spaceflight*, **38** (6), June 1996, pp. 191–4

28 Reeves, Robert: 'The Superpower Space Race' (Plenum Press, New York, 1994) p. 64

29 ibid., p. 66

30 'Project Ranger', *NASA Facts* **2** (6), US Government Printing Office (OF-753-205), Washington DC, USA, 1964, p. 8

31 Reeves, Robert: 'The Superpower Space Race' (Plenum Press, New York, 1994) p. 68

32 ibid., p. 69

33 Thompson, Tina D. (Ed.): 'TRW Space Log 1996' (TRW Space & Electronics Group, Redondo Beach, Volume 32, 1997) p. 93

34 Woods, David: 'Probes to the Moon' in Gatland, Kenneth (Ed.): 'The Illustrated Encyclopedia of Space Technology' (Salamander Books, London, 1981) p. 130

35 Smolders, Peter L.: 'Soviets in Space' (Lutterworth Press, Guildford, 1973) pp. 216–7

36 Woods, David: 'Probes to the Moon' in Gatland, Kenneth (Ed.): 'The Illustrated Encyclopedia of Space Technology' (Salamander Books, London, 1981) p. 131

37 Thompson, Tina D. (Ed.): 'TRW Space Log 1996' (TRW Space & Electronics Group, Redondo Beach, Volume 32, 1997) p. 97

38 Sharpe, Mitchell R.: 'Satellites and Probes' (Aldus Books, London, 1970) p. 146

39 ibid., p. 147

40 ibid., p. 148

41 Yenne, Bill: 'The Encyclopedia of US Spacecraft' (The Hamlyn Publishing Group, Twickenham, 1985) p. 146

42 Sharpe, Mitchell R.: 'Satellites and Probes' (Aldus Books, London, 1970) p. 150

43 Yenne, Bill: 'The Encyclopedia of US Spacecraft' (The Hamlyn Publishing Group, Twickenham, 1985) pp. 147–8

44 Lewis, Richard S.: 'The Moon' in Lewis, Richard S. (Ed.): 'The Illustrated Encyclopedia of Space Exploration' (Salamander Books, London, 1983) pp. 136–7

45 Neumann-Ezell, Linda: 'Unmanned Lunar Missions – American Programmes' in Rycroft, Michael (Ed.): 'The Cambridge Encyclopedia of Space' (Cambridge University Press, Cambridge, 1990) p. 162

46 Johnson, Nicholas: 'Man Made Debris in and from Lunar Orbit' (IAA-99-IAA.7.1.03), IAA/IISL Scientific–Legal Round Table on Protection of the Space Environment, 50th IAF Congress, Amsterdam, 1999

47 Reeves, Robert: 'The Superpower Space Race' (Plenum Press, New York, 1994) p. 171

48 ibid., p. 172

49 Thompson, Tina D. (Ed.): 'TRW Space Log 1996' (TRW Space & Electronics Group, Redondo Beach, Volume 32, 1997) p. 70

Chapter 7

Man in space – the development of the manned capsule

It is of course difficult to imagine future space travel without manned spacecraft. Space cannot possibly be explored by automatic luniks and interplanetary probes only

Yuri Gagarin

The formative years of the Space Age set a hectic pace with the development of satellites for both scientific and propaganda purposes, and the parallel development of increasingly powerful rockets to place them in orbit. Although the low payload capacity and poor reliability of early launch vehicles meant that only relatively simple, unmanned spacecraft could be carried, it was only a matter of time before one nation or another attempted a manned launch.

The impetus for manned spaceflight was both cultural and political. Ever since the early poets and satirists dreamt of eagles flying to Venus and sailing vessels swept on a whirlwind to the Moon, the passengers had been human. So whether space was considered 'the next Everest' or, later, 'The Final Frontier', it was inevitable that people would want to make that ultimate journey of exploration. It was equally inevitable, in the era of Cold War rivalry, that manned space exploration should be an integral part of the 'space race' (Table 7.1).

Technology requirements

Many of the technological requirements for manned space travel are the same as for unmanned missions, so developments made to produce everything from Sputnik 1 onwards had a bearing on the techniques, and the confidence in those techniques, that would be required for the first manned mission.

Table 7.1   Early manned spaceflights and capsule development flights

Launch date	Name	Nation	Comments
15 May 1960	Sputnik 4/KS1	USSR	Korabl Sputnik 1 (KS1); Vostok prototype
29 July 1960	MA-1	USA	Abort compatibility test (failed); Mercury capsule #4
19 August 1960	Sputnik 5/KS2	USSR	KS2 carried dogs Belka and Strelka
21 November 1960	MR-1	USA	Suborbital Mercury–Redstone qualification test flight (failure); capsule #2
1 December 1960	KS3	USSR	Flight failure
19 December 1960	MR-1A	USA	Mercury–Redstone qualification test flight (success); capsule #2A
31 January 1961	MR-2	USA	Suborbital flight carried chimpanzee Ham; capsule #5
21 February 1961	MA-2	USA	Abort compatibility test (success); capsule #6
9 March 1961	KS4	USSR	Vostok test flight with dog and dummy
24 March 1961	MR-BD	USA	Redstone development flight; boilerplate capsule
25 March 1961	KS5	USSR	Vostok test flight with dog and dummy
12 April 1961	Vostok 1	USSR	First man in space: Yuri Gagarin (one orbit)
25 April 1961	MA-3	USA	Single orbit test of Mercury–Atlas and tracking network; capsule #8
5 May 1961	MR-3	USA	First American in space: Alan Shepard (suborbital flight); capsule #7
21 July 1961	MR-4	USA	Virgil Grissom (suborbital flight); capsule #11
6 August 1961	Vostok 2	USSR	Gherman Titov (17 orbits)
13 September 1961	MA-4	USA	Repeat of MA-3; capsule #8A
29 November 1961	MA-5	USA	Mercury orbital test flight (chimpanzee Enos); capsule #9 (2 orbits)
20 February 1962	MA-6	USA	First American in orbit: John Glenn; capsule #13; first manned Atlas flight (3 orbits)
24 May 1962	MA-7	USA	Orbital flight: Scott Carpenter; capsule #18 (3 orbits)
11 August 1962	Vostok 3	USSR	Orbital flight: Andrean Nikolayev (64 orbits)

Table 7.1 Continued.

Launch date	Name	Nation	Comments
12 August 1962	Vostok 4	USSR	Orbital flight: Pavel Popovich; came within 6 km of Vostok 3 (48 orbits)
3 October 1962	MA-8	USA	Orbital flight: Walter Schirra; capsule #16 (6 orbits)
15 May 1963	MA-9	USA	Final Mercury flight: Gordon Cooper; capsule #20 (22 orbits)
14 June 1963	Vostok 5	USSR	Orbital flight: Valeri Bykovsky (81 orbits)
16 June 1963	Vostok 6	USSR	First woman in space: Valentina Tereshkova (48 orbits)
8 April 1964	Gemini 1	USA	Gemini unmanned flight test: 4 orbits and recovery
12 October 1964	Voskhod 1	USSR	First multi-person crew (three cosmonauts on 24 h, 16-orbit mission)
19 January 1965	Gemini 2	USA	Gemini unmanned flight test: suborbital
18 March 1965	Voskhod 2	USSR	First spacewalk in LEO by Alexei Leonov (1 day 2 h mission with Pavel Belyayev)
23 March 1965	Gemini 3 (GT-3)	USA	Orbital flight: Virgil Grissom; John Young (5 h mission)
3 June 1965	Gemini 4 (GT-4)	USA	Orbital flight: James McDivitt; Ed White (first US spacewalk); 4 days 2 h
21 August 1965	Gemini 5 (GT-5)	USA	Orbital flight: Gordon Cooper; Charles 'Pete' Conrad; 7 days 23 h
4 December 1965	Gemini 7 (GT-7)	USA	Orbital flight: Frank Borman, Jim Lovell; 13 days, 18 h
15 December 1965	Gemini 6 (GT-6)	USA	First close orbital rendezvous (with Gemini 7); Walter Schirra, Tom Stafford; 1 day 2 h
16 March 1966	Gemini 8 (GTA-8)	USA	First successful Gemini-Agena docking; Neil Armstrong, Dave Scott; 11 h in orbit
3 June 1966	Gemini 9 (GTA-9)	USA	Tom Stafford, Gene Cernan; 3 days in orbit
18 July 1966	Gemini 10 (GTA-10)	USA	Gemini-Agena docking; John Young, Michael Collins; 2 days, 23 h in orbit

Table 7.1   Continued.

Launch date	Name	Nation	Comments
12 September 1966	Gemini 11 (GTA-11)	USA	Gemini-Agena docking; Pete Conrad, Richard Gordon; 2 days, 23 h in orbit
11 November 1966	Gemini 12 (GTA-12)	USA	Gemini-Agena docking; Jim Lovell, Edwin 'Buzz' Aldrin; 3 days, 22 h in orbit
23 April 1967	Soyuz 1	USSR	First manned Soyuz flight: 1 day, 2 h in orbit; Vladimir Komarov died on re-entry

The first requirement was a launch vehicle capable of lifting a heavy payload, the second was a spacecraft large enough to hold an astronaut and the necessary life-support systems, and the third was the ability to return the astronaut safely to the Earth. The Soviet Union, with its more powerful boosters, immediately had the upper hand in being able to launch large and heavy payloads. Beyond that, the basic structural design of the manned spacecraft was essentially an extension of those developed for the early unmanned satellites, where the protection of a delicate payload was already a priority. However, apart from a few capsules designed to return film to Earth for processing, the majority were not required to re-enter the Earth's atmosphere intact. This meant that a reliable form of heat shield – based largely on warhead re-entry work – had to be developed to protect returning astronauts or cosmonauts.

Experience was being gathered through the development of unmanned satellites with most of the other subsystems: propulsion, attitude and orbital control, power and communications. A manned spacecraft would require a propulsion system to change its orbital parameters and adjust its attitude, just like an unmanned spacecraft, and some sort of control system to oversee its operation. Especially important for a manned spacecraft would be the 'de-orbit burn', the thruster firing that would (hopefully) return its occupant to Earth.

As with other spacecraft, a manned spacecraft would also require a supply of electrical power, whether derived from batteries, solar arrays or some other means, and a communications system, not only for spacecraft telemetry and command, but also for maintaining contact with the crew. The reliability of the hardware was of paramount importance since lives were at stake. Although there was every political advantage to be gained from sending the first man into space, there would be an equal, if not greater, disadvantage in becoming the first nation to kill one of its citizens in the attempt.

Finally, in contrast to the unmanned spacecraft, there was a requirement for a life support system, which encompassed an oxygen supply, sustenance by way of food and drink (depending on the duration of the mission), and control of the cabin environment in terms of temperature, pressure and humidity. So, although a fair amount of engineering experience could be transferred directly from the unmanned spacecraft, a good deal of research and development remained.

The Soviet approach

Once the Soviet Union had placed Sputnik 1 in orbit, in October 1957, it was inevitable that a manned mission would follow as soon as the technology was available. Although Sputnik 1 itself weighed only 83.6 kg, much less than a manned spacecraft, its R-7 launch vehicle had placed it in orbit without an upper stage, the addition of which would make the rocket much more capable. When the half-tonne Sputnik 2 was launched a month later, the R-7's potential as a manned launch vehicle became obvious. The 3 m diameter of its sustainer section was more than enough to support a manned capsule, and the injection of the launch vehicle's 6-tonne core stage into orbit provided a striking demonstration of its capabilities.

As discussed in Chapter 3, the signposts for Soviet intentions to launch a satellite were conspicuous – for those who wished to read them – well before the launch of Sputnik 1. The same was true for a Soviet manned mission, with articles published for both domestic and foreign consumption. For example, a report on the ramifications of Sputnik's launch appeared in *Pravda* on 11 October 1957: written by an academician at the Byurukan Astrophysical Observatory, it ended with the sentence: 'Without question, future technical progress will enable us to launch manned vehicles into outer space'[1]. Another, unattributed *Pravda* article published ten days after the launch of Sputnik 2 stated: 'Without a doubt, these investigations ... will pave the way for evolving the means to guarantee safety to human beings in space travel ... [allowing us] to foresee a day when all circumsolar space will be accessible to direct investigation by man' [2].

By 1958, Soviet interest in manned spaceflight was readily accessible to Western readers with the publication of a book entitled 'Sputnik into Space' (translated from an Italian version which was previously published in Russian). Among other things, it stated: 'The first manned satellite will not differ in any great respect from Sputnik 1 which carried only instruments. Naturally it will be larger and hermetically sealed ... A protective shield against direct action of cosmic rays, a minimum supply of food and water – all this will be installed for the comfort of man. Sputnik II, which carried the dog Laika, conformed to these principles.'[3] Although this would have read like much of the science fiction of the time, for those with vision, the potential for Soviet manned spaceflight was clear.

Meanwhile, back in the USSR, Sergei Korolev was working on possible designs for a manned space capsule, since this was the next obvious stage following the success of the Sputniks. Although he had started his design work in 1958, it was November 1959 before the Council of Chief Designers approved the project for the first manned flight. That same month, a meeting of the Academy of Sciences discussed the options for that first flight: their initial preference was for a suborbital or ballistic trajectory

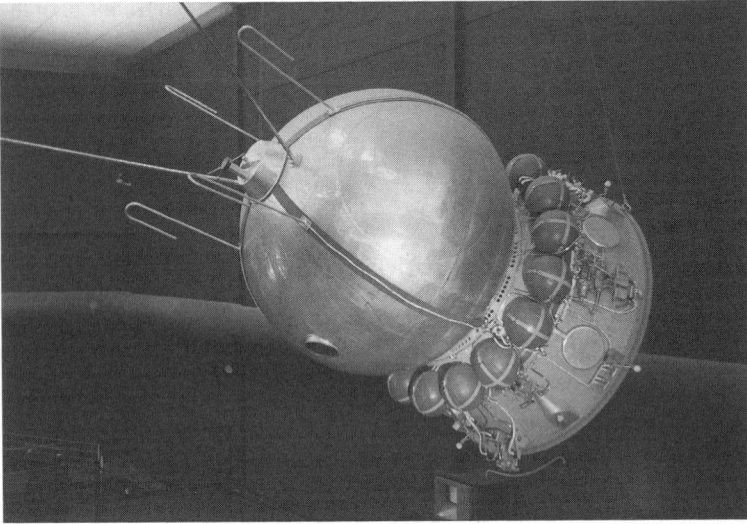

Figure 7.1   Vostok capsule display at Le Bourget Air and Space Museum, Paris [Mark Williamson]

that would allow the cosmonaut a brief sojourn in the microgravity environment of space and return him automatically to Soviet territory; it would also avoid the hardware risks associated with orbital insertion and retro-fire. However, arguments by Korolev and Valentin Glushko led to a final recommendation for *orbital* flight, to which Premier Khrushchev agreed [4]. This led to Yuri Gagarin's historic mission of 12 April 1961.

Gagarin's spacecraft, called Vostok or 'East', comprised a 2.3 m diameter sphere – the 'descent module' in which the cosmonaut travelled – mounted on an 'instrument module' housing the retro-rocket, life-support consumables and other support systems (Figure 7.1). The overall length of the spacecraft was 4.4 m and it weighed about 4.7 tonnes, including the 2.5 tonne descent module. Although it was not revealed until much later, the descent module was too heavy to be landed safely with a cosmonaut inside and included an ejection seat which was fired at an altitude of about 7 km (Figure 7.2) [5]. Since there are no convenient large ocean expanses adjacent to the Soviet Union, all Soviet manned spacecraft have been designed to return to solid ground.

As an indication of how difficult it was to gain detailed and accurate information on the Soviet space programme, the use of the ejection seat was not confirmed until the dissolution of the Soviet Union allowed Soviet authors to write authoritative papers and books on the programme. Even an author with an 'excellent knowledge of the Russian language' and access to Soviet documents could write, with impunity, in 1971: 'Although there are conflicting reports on this point, it seems probable that Gagarin landed in the capsule ... ' [6]. As it turns out, the ejection cover-up was an example of deliberate Soviet obfuscation to ensure that they entered the record

Figure 7.2   Vostok model at Le Bourget Air and Space Museum, Paris, showing ejection seat [Mark Williamson]

books: since the rules of the Fédération Aéronautique Internationale (FAI) required a pilot to be in – if not *in control of* – a craft from take-off to landing, admitting to the ejection could have nullified the record [7].

The first successful launch of a Vostok prototype occurred on 15 May 1960 [8]. It was designated Sputnik 4 or Korabl Sputnik 1 (KS1), translated as 'Spaceship-Satellite 1'. It was fortunate that there was no one aboard, since an attempted de-orbit burn placed the spacecraft in a higher orbit instead of returning it to Earth, and the descent module did not re-enter until October 1965 [9].

Following another failed launch attempt in July 1960, Sputnik 5 (or KS2) carried the dogs Belka and Strelka into orbit on 19 August, along with a menagerie of rats, mice, flies and other biological specimens. Later that day, they were returned to Earth in the first successful test of Vostok's de-orbit and re-entry technology, including the ejection seat [10]. Although Belka had the dubious honour of being the first animal to suffer from space sickness, both dogs survived their ordeal and the way was clear for the first manned launch, which the mission's medical director recommended be limited to a single orbit [11].

However, before the manned Vostok launch could take place, other test flights were required: unfortunately, the next two flights, in December 1960, were also failures – the first because of misfiring retro-rockets, the second due to low thrust on ascent, which forced an abort [12]. This led to some careful redesign and two updated versions of the Vostok were flown in March 1961, just two weeks apart, apparently in an attempt to regain ground on the schedule which had previously foreseen a manned launch in December 1960. Both flights carried a dog and a spacecraft test dummy, or instrumented manikin, unofficially named Ivan Ivanovich by journalists. Apart from accelerometers and radiation detectors, the dummy contained radio equipment to qualify the spacecraft communications system, though to avoid any confusion about Ivan being a real cosmonaut, this was done by retransmitting popular folk songs rather than speech. In addition, should local people find the capsule before the recovery team arrived, or should Ivan's (removable) head be found after an ejection test, a cloth with the word 'MAKET' ('puppet' or 'mock-up') was placed over his face [13].

Not surprisingly, the graduation from unmanned to manned spacecraft brought about an increase in complexity, while at the same time demanding improved reliability (a difficult balance to ensure). An indication of the complexity of the spacecraft is given by various published parts-counts: according to one author, 'Each Vostok contained 15,000 m of electrical wires, 240 valves, and 6000 transistors' [14]. Another reported 'Vostok had about 240 electric bulbs, 56 electric motors … and about 800 relays and switches' [15].

Amidst all this technology, Gagarin was little more than a passenger on the first manned spaceflight, and had very little to do except look out of the window and make the occasional report. Indeed, he was not even in control of the spacecraft, which was being operated remotely from the ground because officials were unsure whether a cosmonaut would be able to operate controls in this new and unknown environment. However, as a precaution, Gagarin carried a three digit code in a sealed envelope, which he could use to unlock the manual controls if the radio link with ground control was lost [16].

Gagarin made a circuit of the Earth in 108 minutes, speeding across the Pacific Ocean in just 20 minutes, and landed safely back in the Soviet Union following a successful ejection. The world's press had been informed of the flight while Gagarin was still in orbit, not only for the propaganda value, but also to avoid him being arrested as a spy should he land on foreign soil; ironically, on his return to Earth, he was greeted with pitchforks and stakes when local farmers suspected exactly that. Although they were aware of Gary Powers' U-2 crash the previous year, they – like the rest of the world – had not been informed of Gagarin's flight until the radio announcement [17]. It may have helped that 'CCCP' (Cyrillic for USSR) had been painted onto his helmet just before the flight – a last-minute decision made at the launch site after Gagarin had already donned his spacesuit (Figure 7.3) [18].

The history of spacesuit design is almost as long as the Space Age itself, with suit development for the Vostok programme beginning at the Zvezda Research Development and Production Enterprise ('Plant No. 918') in 1959. Zvezda already had experience in developing full pressure suits for aircraft pilots and designing the pressurised cabin for Laika. Interestingly, despite this early experience, the first publication on the design and use of Soviet spacesuits did not appear in the Soviet Union itself until 1970 [19].

Figure 7.3   Yuri Gagarin in the bus to the launch pad on 12 April 1961. Seated behind him is his back-up, Gherman Titov, while cosmonauts Grigoriy Nelyubov and Andrian Nikolayev stand [NASA/Asif Siddiqi]

The specifications for the first cosmonaut suit, the SK-1 full pressure suit, were approved in September 1960. Rather than integrate the suit's systems with the Vostok cabin's life support system (LSS), it was decided to design a self-contained system, which was seen as more reliable. According to one source, it also made it possible to distribute responsibility for the respective systems between two contractors, Zvezda for the suit and Nauka for the LSS, so that 'no single entity could be found responsible for any possible failure' [20]; however, others might translate the assignment of two separate contractors as a mechanism for *assigning* responsibility.

As a result of Zvezda's previous experience, the SK-1 was designed, manufactured and tested in only six months. Among other things, it was intended to protect a cosmonaut for five hours in a depressurised cabin; on ejection at altitudes up to 8 km; and for 12 hours in cold water, should this be necessary. It was a 'soft suit' made in two layers with separate, pressure-tight closures: an inner layer of 0.6 mm thick natural rubber and an outer one composed of a strong polyester fabric, 'Lavsan'. Under the pressure suit, the occupant wore underwear and a thermal protection overall, and over it an orange coverall[1] which incorporated a flotation collar inflated from a bottle of carbon dioxide gas. The suit also featured boots adapted for a parachute landing.

The helmet had a double-glazed visor, which was useful not only in terms of redundancy but also to eliminate misting. It also incorporated an automatic closure

---

[1] Later suits, both Soviet and American, were coloured white to reflect the Sun's heat when their occupants were outside the spacecraft.

device, triggered electrically by a drop in cabin pressure or imminent ejection. The visor could be opened and the gloves removed in orbit, when the cosmonaut would rely on the cabin system [21].

Although the SK-1 suit (and the SK-2 variant for female cosmonauts) was considered a self-contained system, since it could operate autonomously if necessary, it was connected to the spacecraft in several ways. The main components of the life support system were built into the ejection seat, in which the cosmonaut would sit throughout the mission, while tanks containing compressed air (to maintain suit pressure) and oxygen were installed on the instrument module. Nominal flight conditions simply required the ventilation of the suit 'to remove heat, sweat, CO_2 and other gas discharges of the human body', as one source puts it [22]. A fan supplied the cosmonaut with air via separate feeds to the helmet and the rest of the suit. A system of pipes and valves ran from the helmet, via a reservoir, to the cabin, which meant that under normal conditions cabin air was provided by the suit system. In an emergency, such as cabin depressurisation, the helmet visor would be closed and the suit isolated from the cabin. For fire-safety reasons, the oxygen content of the air was limited to 40–45 per cent in all operating modes [23]. All Soviet/Russian spacecraft since have used a similar oxygen/nitrogen atmosphere at sea-level pressure.

Obviously, it was also necessary to control the temperature and humidity of the air in the cabin, and this was done automatically. Cabin air was circulated through a heat exchanger which transferred any excess heat to a liquid circuit connected to a radiator in the instrument module. The cosmonaut was able to set the initial temperature, which was then maintained by a thermostat, even during the period of re-entry heating [24].

Temperature, pressure and other data could be monitored by the cosmonaut on an instrument panel. For example, there were gauges showing the oxygen and CO_2 content of the air and radiation levels in the cabin, and a globe which indicated the spacecraft's position in orbit. One interesting addition was a counter showing the number of orbits completed, a celestial version of a car's odometer [6].

More than 40 years later, it is difficult to fully appreciate the enormity of the achievement of Gagarin's flight. The catalogue of different technologies that had to work perfectly was immense: from the turbopumps and stage separation devices of the launch vehicle to the guidance system and heat shield of the descent module, there was precious little margin for error. Only three-and-a-half years had passed since the first man-made device had been placed in orbit and there had been many launch and in-orbit failures to undermine the confidence of the designers; in fact, 56 out of 106 space missions conducted before Vostok 1 failed in some manner.

It is interesting to note that several decades later – by which time space technology had reached a certain level of maturity – an average commercial communications satellite would take about the same time (three to three-and-a-half years) to graduate from drawing board to orbit. There can be little better illustration of how times have changed: in the early, highly competitive years of the Space Age, there was little time for the cautious, conservative approach of modern space technology development.

The Soviet success with Gagarin's flight had a predictable effect on America, but whereas President Eisenhower had attempted to belittle Sputnik, President Kennedy's reaction to Vostok was quite different: he was irritable, and impatient for an equitable response. At a meeting with advisors and senior NASA officials two days after Gagarin's mission, he implored 'Just tell me how to catch up. Let's find somebody. Anybody. I don't care if the janitor over there has the answer, if he knows how' [25]. The solution, of course, was to make sure that the first man on the Moon was an American, but first America had to put a man in space.

The American approach

In common with the Soviet Union, America pursued many aspects and applications of space technology in parallel, one of which was manned spaceflight. However, as with America's attempts to launch its first satellite (discussed in Chapter 3), its efforts were divided – in this case between a space capsule and a space plane.

Following its success in breaking the so-called 'sound barrier' with the X-1 rocket plane, America had embarked on a decade of high-speed flight research with the X-15 rocket-propelled aircraft, three of which conducted 199 missions between 1959 and 1968 (Figure 7.4). Released at altitude from beneath the wing of a converted B-52 bomber, they flew at speeds of up to Mach 6.7 and set an altitude record of close to 108 km, which by modern definition is 8 km above the boundary to space.

By the end of the 1950s, the X-15 was regarded by many as the natural route to orbital flight and the US Air Force was planning its X-20 'DynaSoar' as a space plane launched by a conventional rocket booster. Had it been developed, the DynaSoar

Figure 7.4   X-15 test pilot Neil Armstrong with X-15 number 1 following a flight. X-15-1 is now in the National Air and Space Museum, Washington DC [NASA]

Figure 7.5   Maxime A. Faget [NASA]

(an acronym for Dynamic Soaring) would have been the natural precursor to the Space Shuttle. However, the shockwaves from Sputnik had set in motion a less logical race for space which left insufficient time to upgrade the X-15 or develop the X-20, which was still on the drawing board [26]. The result was the much simpler ballistic capsule design of Project Mercury, which became an official NASA programme on 26 November 1958 [27].[2]

Although the history of the Mercury programme has been covered in great detail elsewhere [29,30], it is worth summarising a few aspects here as examples of the efforts that were being made to boost America into the realm of manned space flight.

The designer of the one-man Mercury capsule was Maxime Faget (Figure 7.5), who in 1958 became one of 35 engineers to form the nucleus of a Space Task Group for the Mercury project based at Langley Field, Virginia (a group that would later evolve into NASA's Johnson Space Center in Houston) [27]. Faget developed and patented the 'Aerial Capsule Emergency Separation Device' (more commonly known as an escape tower), the 'Survival Couch' and the 'Mercury Capsule' itself [31]. The industrial prime contractor for the capsule was McDonnell Douglas Aircraft Corporation, as with other spacecraft contractors of the time better known for its aircraft.

The Mercury capsule would be launched by a conventional rocket and, in contrast to the space planes or 'lifting bodies', would return to Earth as a 'blunt-bodied

---

[2] The antipathy between the X-15 pilots and the Mercury astronauts is recounted in an entertaining manner in the book 'The Right Stuff' by Tom Wolfe [28], which later became a film.

projectile'. The blunt-body concept had been devised as early as 1952, by H. Julian Allen at the National Advisory Committee for Aeronautics (NACA). He realised that a blunt shape would absorb only a small fraction of the heat generated during re-entry into Earth's atmosphere, a principle which was used for missile nose cones as well as the early manned space capsules (Mercury, Gemini and Apollo) [32]. As the returning spacecraft encountered the denser layers of the atmosphere, it pushed the air molecules out of the way, forming a bow wave ahead of the capsule and carrying most of the heat away. The rest was absorbed by an ablative material applied to the heat shield, which charred and broke away, taking additional heat with it (Figure 7.6).

The capsule itself was a truncated cone measuring about 1.9 m across the base and 2.7 m tall, with a maximum in-orbit mass of about 1450 kg (Figure 7.7). It was so small inside that the astronaut was installed in a seated, as opposed to reclined, position; it was said that one didn't so much ride in the capsule as wear it! Certainly, it was too small to incorporate an ejection seat, which is why a launch escape tower – housing a solid propellant rocket powerful enough to pull the capsule clear of the rocket in an emergency – was developed.

Solid propellant motors also formed the basis of the so-called 'retro-pack', which comprised three such motors secured to the heat shield by metal straps (Figure 7.8); the pack was jettisoned following the retro-burn which initiated re-entry. In contrast to Soviet manned spacecraft, all American capsules were designed for a sea landing, or 'splashdown', and this became an iconic feature of mission success for the widely televised US manned space programme.

Prior to splashdown, the heat shield was detached from the rear pressure bulkhead to expose a rubberised canvas pneumatic cushion, known as a 'landing bag' or 'landing skirt', which was designed to absorb the impact; it was estimated that this reduced the landing loads from 45 to 15 g. Once the Mercury capsule had splashed down, the skirt filled with water to form a 'sea anchor' to maintain the capsule in an upright position [33]. On later flights, Navy divers fixed a flotation collar around the spacecraft to stop it sinking.

Arguably the most important subsystem, since this was a manned spacecraft, was the Environmental Control System (ECS), developed by Garrett Corporation under subcontract to McDonnell Douglas. Like the Vostok system, it controlled the temperature and pressure of the capsule and the spacesuit automatically, and had a manual override option. However, since Mercury had no ejection seat, the design of the system was quite different and there was no integration of components with the astronaut's couch.

The Mercury ECS comprised two largely independent systems for capsule and suit, although they used the same coolant water and electrical supply. Oxygen entered the spacesuit torso and exited from the helmet, carrying away excess heat and carbon dioxide. This was passed through a debris filter to remove any particles; then a canister of activated charcoal and lithium hydroxide to remove odours and CO_2; and then a heat exchanger to cool the atmosphere before recycling it. The heat exchanger used the vacuum outside the capsule to boil the coolant water at a low temperature and the resultant steam was exhausted into space. Interestingly, considering the high-tech nature of spaceflight, condensed water vapour from the pressure suit was

(a)

(b)

RESEARCH CONTRIBUTING TO PROJECT MERCURY

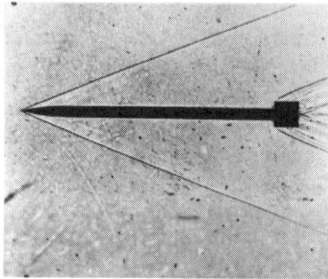

INITIAL CONCEPT

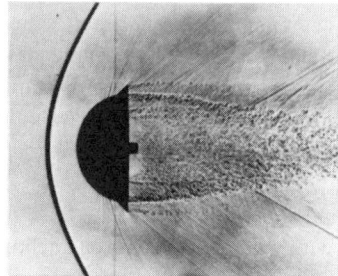

BLUNT BODY CONCEPT 1953

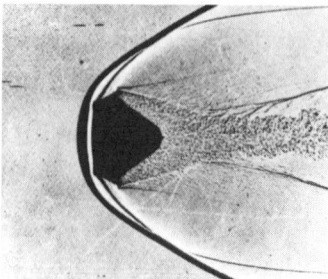

MISSILE NOSE CONES 1953-1957

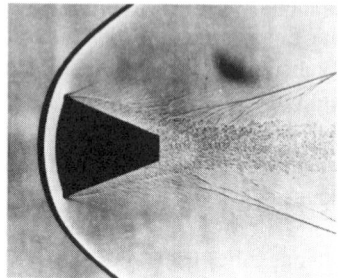

MANNED CAPSULE CONCEPT 1957

Figure 7.6   *(a) H. Julian Allen, pioneer of the blunt body concept, (b) shadowgraph images of re-entry vehicles illustrating the concept (light is refracted through fluid flow of differing densities) [NASA]*

Figure 7.7   Mercury capsule no. 2 [NASA]

collected by a 'mechanical water separator', effectively a sponge which was squeezed periodically to remove the water to a collection tank [34].

A notable difference between the Soviet and American designs was the latter's provision of a 100 per cent oxygen atmosphere, which was also used for Gemini and Apollo. However, it was supplied at a reduced pressure of about 38 kPa (5.5 psi), compared with 'sea-level pressure' (which is 101.4 kPa or 14.7 psi). In recognition of the potential fire hazard, the Mercury capsule featured a pressure relief valve which allowed the cabin atmosphere to be dumped overboard, thereby starving the flames of oxygen; and with the dump valve closed the cabin repressurised automatically [34].

The contract for the Mercury pressure suit went to B. F. Goodrich in July 1959; it was based on those worn by US Navy aviators at the time, although with several modifications and improvements. The design process was an exercise in precision tailoring which produced fitted suits for each of the seven Mercury astronauts. It involved dressing the astronauts in the long underwear they would wear beneath the suits and wrapping them in strips of wet brown paper tape. When the tape dried and hardened, it was cut off and used as a mould for the individual's pressure suit [36].

Although more tight-fitting than the cosmonaut suits, the Mercury suits had a number of layers in common: a one-piece undergarment, an inner gas-retaining layer (in this case made from a ply of neoprene and neoprene-coated nylon fabric) and an outer layer made from heat-reflecting aluminised nylon. The latter, with matching

Figure 7.8   Mercury capsule retro-pack [NASA]

white or silver boots, gave the Mercury astronauts an iconic 'spaceman' look which reflected the sleek aluminium finish of their rockets (especially the Atlas), and provided the NASA public relations machine with an enduring image of the 'Mercury Seven' kitted out for spaceflight (Figure 7.9).

Man-rating the rockets

Whereas Soviet designers had the option to choose whether to launch their first cosmonaut on a ballistic trajectory or into orbit, American designers had no choice but to plan for a suborbital flight. In 1958, the United States had no launch vehicle capable of lifting a payload as heavy as Mercury into orbit, and the only sufficiently-developed rocket with the power to launch the spacecraft onto a suborbital trajectory was the Redstone ballistic missile, developed by Wernher von Braun's Army rocket team. Thus the Redstone, which formed the first stage of America's first satellite launcher, Juno I, acquired a new role as the launch vehicle for Project Mercury.

In fact, the concept of using the Redstone to launch men into near-Earth space arose as early as 1956, when it was proposed by the US Army Ballistic Missile Agency (ABMA) as a troop carrier. It was suggested that a Redstone booster with a modified

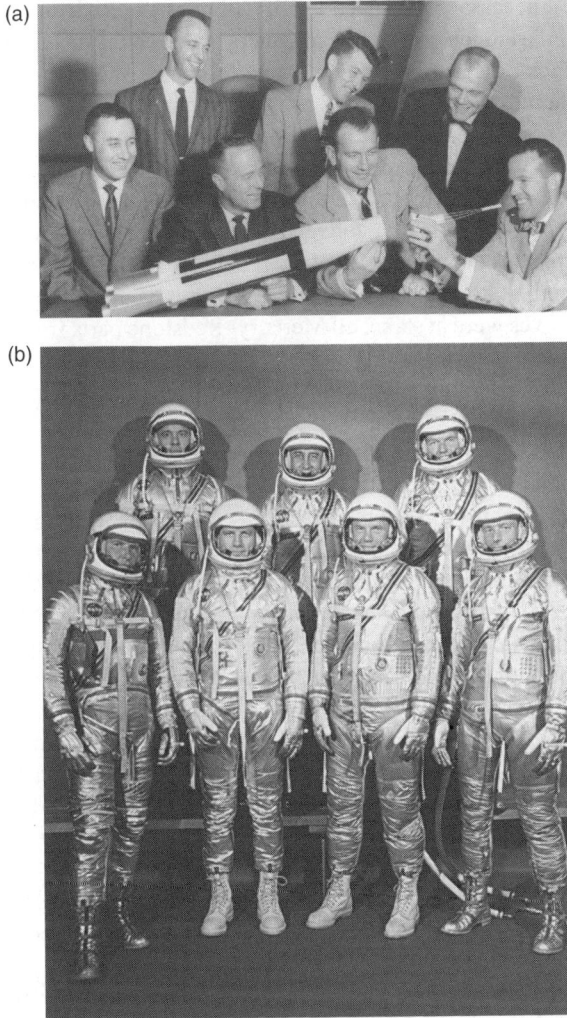

Figure 7.9   The 'Mercury Seven' in suits ... and spacesuits (a) with Mercury-Atlas model, (b) front row left to right: Walter H. Schirra, Jr. Donald K. Slayton, John H. Glenn, Jr., and Scott Carpenter; back row: Alan B. Shepard, Jr., Virgil I. Gus Grissom, and L. Gordon Cooper) [NASA]

second stage could launch 'a canister of men' to an altitude of 230 km and a range of 885 km, from where it would be lowered to Earth by a parachute and retro rockets. However, this and subsequent similar proposals came to nothing [37].

Once the decision to use the Redstone for Project Mercury had been made, von Braun's team set about improving the reliability of the Redstone and qualifying it for manned spaceflight, a procedure known as man-rating. One of the 800 engineering changes deemed necessary for space qualification of the Redstone booster was the development of the launch escape tower. Others involved the development of sensors to monitor various aspects of the vehicle's performance and trigger the firing of the escape tower should a life-threatening problem occur. This was even more difficult than at first appears, since too many sensors could give conflicting readings and, since the sensors themselves were prone to failure, could provide false information. The solution was to reduce the sensors to a few basic types to monitor important quantities such as attitude, angular velocity, voltage and chamber pressure [38].

The man-rating procedure also involved reliability studies, component testing and workforce training, a key factor in the success of any venture. To highlight the fact that astronauts' lives were at stake, all Mercury–Redstone parts from the hundreds of subcontractors were stamped with a specially designed insignia to set them apart from the ordinary Redstone components (it featured the Roman god Mercury striding over the Earth with a rocket under his arm) [39]. In addition, to personalise the endeavour and (understandably) to protect their own interests, the Mercury astronauts themselves spent time touring the various plants, meeting the workers and generally developing a good working relationship with the spacecraft contractors (Figure 7.10). A cartoon issued by Raytheon Corporation demonstrated its understanding of the concerns: it featured an astronaut climbing a rickety ladder (held on makeshift scaffolding by a nonchalant worker) to a capsule on a rocket which was lashed together with tape and string. The astronaut's speech bubble read: 'Just think, Wally, everything that makes this thing go was supplied by the lowest bidder' [40].

Famously, the Mercury astronauts also exerted their authority in an argument with NASA management over the control of their spacecraft. The control of a spacecraft's attitude in orbit had been well tested in a variety of satellites using either automated onboard systems or commands from ground controllers. While NASA managers saw no reason to change this for Mercury – no doubt feeling that the occasional firing of a few hydrogen peroxide thrusters was best handled by well-tested, redundant hardware – the astronauts were pilots and wanted to fly their own spacecraft. For the first time in the development of space systems, the human ego became part of the engineering specification.

Another bone of contention was the provision of a window, which was frowned on by the engineering fraternity for reasons of structural integrity. The astronauts rejected the suggestion of two small portholes on either side of the astronaut's head and demanded a window directly in front of the face [41]. Indeed, they also rebelled against the word 'capsule', because it implied that they were passive cargo, insisting on the use of the word 'spacecraft'; but most people – including the astronauts themselves on occasion – used the words interchangeably.

The first flight of the Mercury-Redstone, MR-1, designed to qualify the new configuration of Redstone booster and Mercury capsule (or 'spacecraft') occurred on 21 November 1960. It was a failure which has ever since provided a touch of (possibly perverse) humour in documentary films about early American rocketry. Although ignition of the Redstone engine took place correctly, the first motion of the vehicle triggered an engine shut-off signal, which in turn automatically caused the spacecraft launch escape tower to fire and the recovery parachutes to deploy.

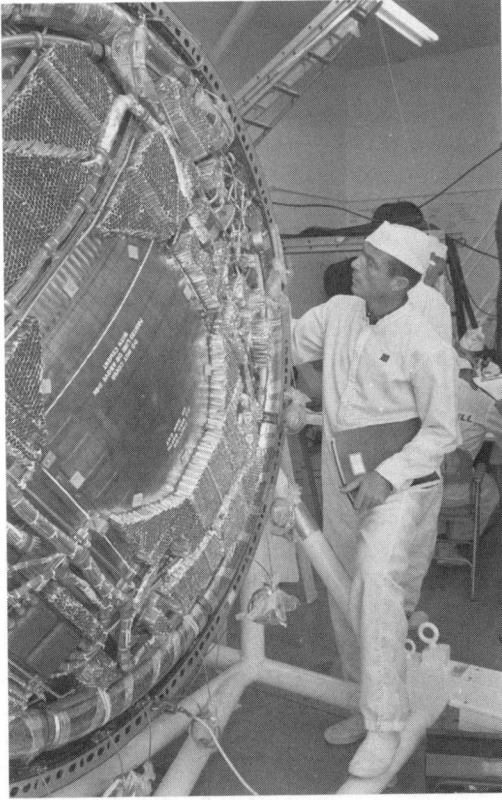

Figure 7.10   Mercury astronaut Scott Carpenter examines the honeycomb heat shield of his 'Aurora 7' capsule at Cape Canaveral, Florida [NASA]

The failure might have appeared less farcical if the capsule had actually separated from the Redstone *with* the escape tower, but it remained attached because the g-load sensing requirements for separation were not met. The parachutes were released because barometric sensors indicated an altitude of less than 10,000 ft (3048 m) [42]. The actual height achieved before the engine shut down was 'about an inch' [43] or 'a four-or-five-inch liftoff' [44], depending which official NASA source one reads, but it was almost certainly an altitude record NASA would prefer to forget. The incident provided a stark reminder that 'systems engineering' was just as important as detailed, component engineering; when a large number of components and sensors are required to work together, it is necessary to understand their interrelationships.

The cause of the embarrassing failure was traced to an anomalous impulse created when two electrical connectors disconnected from the vehicle in reverse order: the 29 ms interval caused the Redstone's thrust to be terminated and the tower to be jettisoned. It was later found that a control cable from the Redstone tactical missile had been used instead of the Mercury–Redstone version; whether or not it included a Mercury stamp is not recorded [42]. The hardware was modified by adding a longer cable to ensure that the rocket remained electrically grounded until all umbilical plugs were pulled, and the circuitry was modified to inhibit the release of the capsule until close to the end of the nominal thrust period (before that time an abort signal could only be sent manually from the ground) [45].

The second attempt on 19 December 1960 (designated MR-1A) was a success and the way was clear for the next phase of Mercury–Redstone testing: the flight of a chimpanzee named Ham. Whereas the Russians had chosen dogs as test subjects, the Americans preferred primates: for example, on 28 May 1959, Able (a rhesus monkey) and Baker (a squirrel monkey) had been launched on a suborbital trajectory by a Jupiter rocket, to test their reactions to acceleration and weightlessness (now termed microgravity). Although the Russians accepted that monkeys were physiologically closer to man, they took the view that dogs were closer in emotional terms and would 'undergo tests that nervier chimpanzees will reject' [46].

Ham made his suborbital flight on 31 January 1961 as part of the MR-2 mission. Unfortunately a valve in the Redstone's propulsion system, which controlled the mixture ratio of the propellants, failed in the open position leading to a higher thrust than planned. Thus instead of a peak altitude of 183 km and a range of 468 km, the Redstone boosted Ham to 251 km and 668 km, respectively [42], conditions which imposed a re-entry deceleration of 14.7 g, some 3 g higher than expected [47]. Despite being somewhat fatigued and slightly dehydrated, he survived the flight and lived to the ripe old age of 26 [48]. Nevertheless, Ham's unfortunate experience lent credibility to the disparaging description of the early astronauts as 'Spam-in-a-can'!

The original Mercury–Redstone launch schedule had called for the fourth launch to be manned, but the first three launches had uncovered a number of design problems, some of which remained to be flight-tested. After Ham's unintentional extension of the 'flight envelope', von Braun wanted to test the Redstone further and, following much discussion, a fourth test flight was approved.

Dr. Joachim Kuettner, the head of ABMA, was one of those against the decision. He believed that the 'small (non-critical) electronic error' discovered after the third flight could be easily corrected, adding: 'I argued that nothing statistically worthwhile would come from another unmanned flight [and showed the laboratory directors] the anticipated USSR launch schedule, [but] von Braun decided against me and another unmanned flight had to be scheduled. In this way Gagarin flew just ahead of our first manned flight' [42]. Kuettner's comments reveal not only the pressure that the race to launch the first man into space placed on those concerned, but also the internal disputes that must have raged: it takes little 'reading between the lines' to conclude that he blamed von Braun for allowing the Soviet Union to take the lead.

As a result of the decision, the unmanned MR-BD (booster development) flight was made on 24 March 1961, the same month that NASA chose Alan Shepard as, ostensibly, the first man in space. The following month, however, Yuri Gagarin became the first man to orbit the Earth ... and another race had been lost. Alan Shepard flew MR-3 ('Freedom 7') to become the first American in space on 5 May 1961, but the flight was suborbital, reaching an altitude of 187 km and a range of 486 km. It lasted only 15 minutes and 22 seconds, including just under five minutes of microgravity.

(a)　　　　　　　　　　　　　　(b)

Figure 7.11　(a) Mercury–Redstone launch of 'Freedom 7', (b) Mercury–Atlas launch of 'Friendship 7' [NASA]

In contrast, Gagarin had completed a full orbit of the Earth in a flight lasting an hour and 48 minutes.

The month after MR-4 ('Liberty Bell 7') carried Virgil Grissom on his short suborbital flight of 21 July 1961, the remainder of the planned Mercury–Redstone flights were cancelled because the programme was considered to have met its goals. The remaining Mercury flights were boosted into orbit by the Atlas launch vehicle, which like the Redstone had been adapted from a ballistic missile. The first American in orbit, John Glenn, was launched on 20 February 1962 on the third manned Mercury flight, which lasted 4 h and 55 min [49]. The final Mercury flight, launched on 15 May 1963, carried Gordon Cooper on a 34 h 20 min mission, circling the Earth 22 times (Figure 7.11) [50].

Although Project Mercury had only two official goals – to investigate an astronaut's ability to survive and perform in the space environment, and to develop the basic technology and hardware for future missions – it achieved much more from NASA's perspective. It helped develop a management system capable of organising a large number of contractors and subcontractors, while also encouraging those contractors to develop new capabilities to meet the high-tech, high-reliability requirements of spaceflight. In addition, it helped push the development of more capable and more reliable launch vehicles which could be used beyond the manned spaceflight arena, and developed a worldwide system of spacecraft tracking stations that would be used for later missions [51].

Meanwhile, the Soviet Union was continuing with its Vostok flights. After the sixth and final flight on 16 June 1963, which made Valentina Tereshkova the first woman in space, the Soviet Union was faced with a hiatus in manned spaceflight, since its first Soyuz mission was not due until 1966. As a result, Korolev devised a modified version of the Vostok, called Voskhod ('sunrise'), which would enable the demonstration of a number of manned spaceflight techniques, including the orbital rendezvous and docking of two spacecraft and the 'spacewalk' [52]. It was also decided to forego the development of a two-man vehicle, which was being pursued by the Americans, and move directly to a three-man vehicle. To do this, however, it would be necessary to remove the ejection seat and oblige the crew to wear overalls rather than spacesuits.

Voskhod 1, launched on 12 October 1964, carried a three-man crew on a politically successful 24-hour mission, predictably shocking the world once again with its audacity and apparent technical superiority. Voskhod 2, launched on 18 March 1965, pre-empted the widely-publicised first launch of America's two-man Gemini by just five days. This time, the spacecraft carried a crew of two, leaving room for a flexible telescopic airlock which enabled Alexei Leonov to make the first spacewalk, or extra-vehicular activity (EVA) as NASA would call it (Figure 7.12). Since the cosmonaut was required to leave the cabin, a new type of spacesuit (known as the Berkut suit) was designed, complete with backpack life support system. Although Leonov's ground-breaking spacewalk was successful, he reported difficulty returning to the spacecraft through the temporary airlock because his suit had expanded [53]. The final two Voskhod missions were cancelled to allow designers to concentrate on the forthcoming Soyuz spacecraft.

Bridge to Apollo

Once America had succeeded in launching a number of individual Mercury astronauts into space, it was time to take the next step towards the three-man Apollo capsule, which was already in the design phase. This was the function of the two-man spacecraft named after the constellation of Gemini ('the twins').

The Gemini missions had several aims in the overall manned spaceflight plan: to increase the number of crewmen a spacecraft could support; to increase the mission duration by demonstrating the necessary power and life support provisions; to demonstrate EVA techniques and hardware; and to prove the rendezvous and docking techniques that would be required for the Apollo missions.

The Gemini spacecraft was based heavily on the Mercury design to save both time and engineering effort (Figure 7.13). Indeed, since much of the preliminary design work had already been completed by McDonnell Douglas, NASA decided not to seek competitive bids and appointed the Mercury prime contractor for the Gemini project [54].

The Gemini was wider and taller than the Mercury, but remained a compact spacecraft: although its 2.3 m width and 3.3 m height gave it about 50 per cent more cabin space, this had to be shared by two astronauts for up to two weeks in orbit [51]. Side by side, the Gemini appeared to be a much larger vehicle, but this was because the design incorporated a tapered Adapter Module which linked the capsule

Figure 7.12 Alexei Leonov makes the first spacewalk (three frames from the Voskhod 2 movie camera) [NASA/Asif Siddiqi]

(a)

Mercury instrument panel

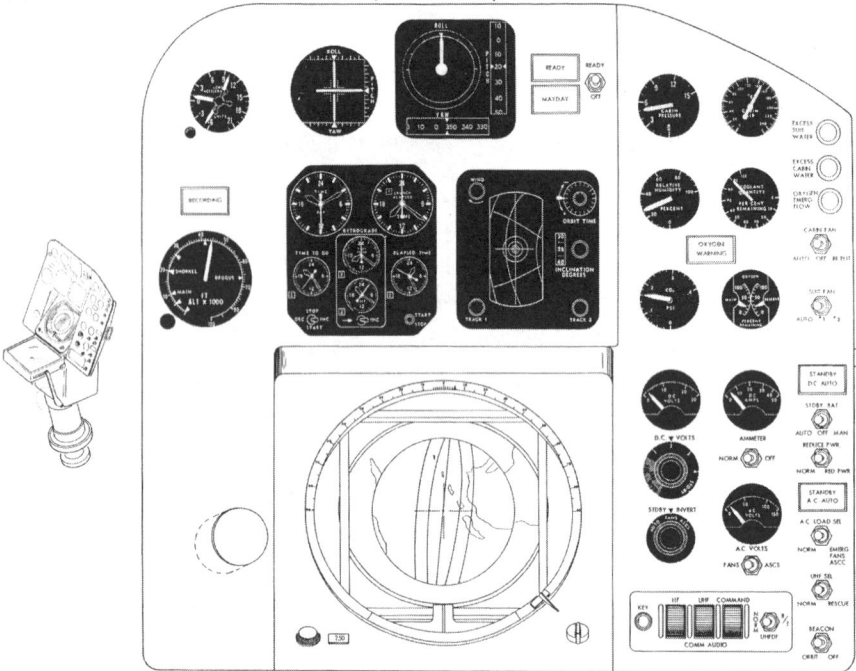

(b)

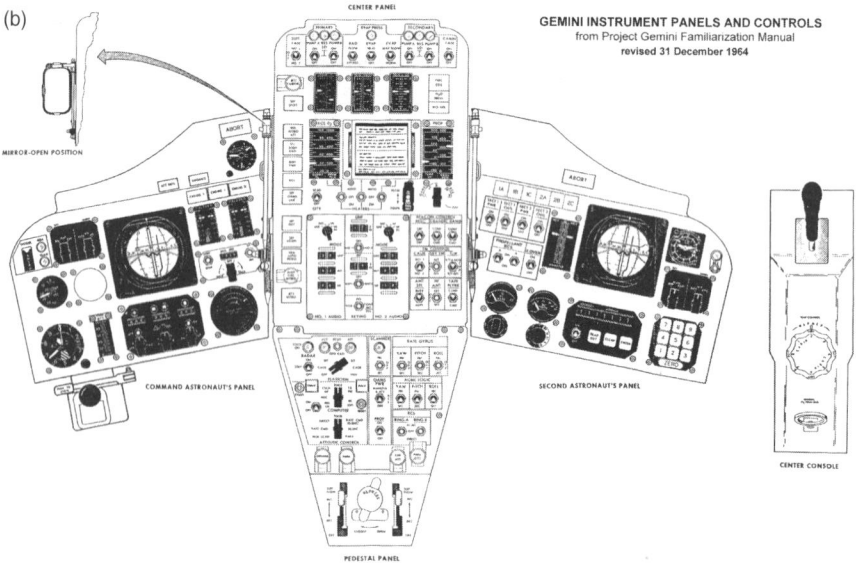

Figure 7.13   An evolution in complexity of manned spacecraft: control panels for Mercury (a) and Gemini (b) – not to same scale; compare also Figure 8.10 for Apollo command module [NASA]

GEMINI SPACECRAFT – TYPICAL INTERIOR ARRANGEMENT
from Project Gemini Familiarization Manual

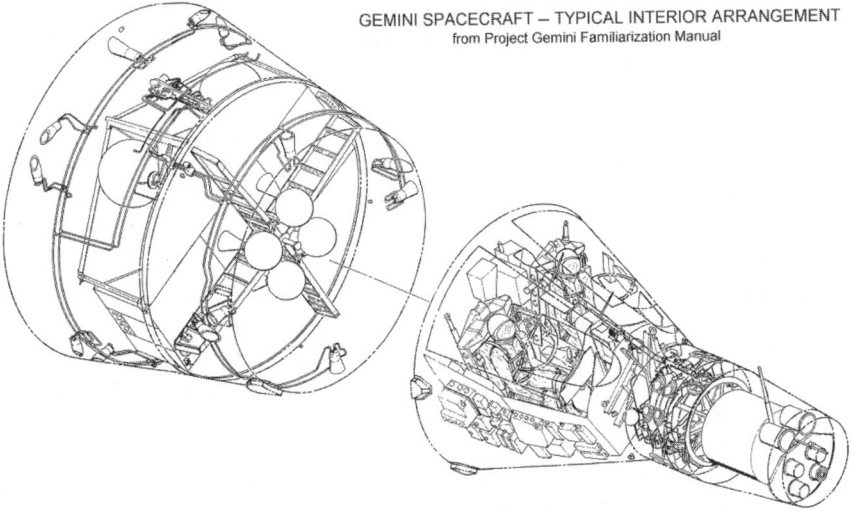

Figure 7.14   Gemini capsule (right) and adapter module (left) [NASA]

or 'Re-entry Module' to the launch vehicle (Figure 7.14). Since the combination was, at 3500 kg, more than twice as heavy as the Mercury, it could not be launched by the Redstone or the Atlas, so another converted ballistic missile, the Titan 2, became the Gemini launch vehicle. For Gemini, the concept of the launch escape tower was discarded in favour of ejection seats, which could be used during launch incidents or the latter stages of re-entry [55].

An important aspect of the Gemini missions was demonstrating that flights of the longest duration foreseen for Apollo were feasible. Apart from carrying sufficient food and oxygen for the crew, a key part of the solution was providing electrical power for up to two weeks. The duration of the Mercury missions had been limited by the size and capacity of the batteries that could be carried, but since they were only intended to be short excursions into space this was not too much of a problem. Gemini, however, was different.

The solution chosen was the fuel cell, a device used to generate power by the electrochemical combination of hydrogen and oxygen (the inverse of the electrolysis of water, which *uses* power to separate its chemical constituents). Although the invention of the fuel cell itself – by British physicist Sir William Robert Grove – dates back to 1839, it has proved extremely useful in the era of manned spaceflight, having featured in all American manned spacecraft since Gemini. A particularly useful by-product of the fuel cell reaction is water, which was used in Gemini for drinking and spacecraft cooling [56].

Gemini's fuel cells were developed by General Electric Co. and installed in the spacecraft in two groups, each of which produced about 1000 W [57]. Whereas fuel cells provided a good solution for manned spacecraft, the requirement for consumables made the technology unsuitable for unmanned spacecraft; they had much longer design lifetimes and used arrays of photovoltaic cells instead. Interestingly, Soviet designers also showed a preference for the solar array when it came to their later manned spacecraft, the Soyuz.

Though the Gemini's Environmental Control System (ECS) was based on that of Mercury, it had to be modified substantially to provide two parallel suit circuits for oxygen, as well as a cabin circuit. In addition, Mercury's bottled gas supply was replaced by a liquid oxygen system, which reduced the volume required for storage [58]. The Gemini pressure suit differed from the Mercury suit in two main ways: it had to be worn for up to 14 days in the capsule and, in a version with added protective layers, used for EVAs. The suits were pressurised by the ECS and connected to the spacecraft with an umbilical tether during spacewalks.

To allow EVAs, as well as facilitating ejections from the capsule, the Gemini hatches were also significantly different from the Mercury hatch. The latter was attached to the capsule with 70 torque bolts and incorporated two shaped explosive charges which could be triggered either by the astronaut or using an exterior lanyard [41]. Obviously, this was not the solution for a hatch that had to be used for an EVA, and replaced afterwards! The resulting Gemini hatches were comparative engineering masterpieces which could be opened, locked open, and closed by a single astronaut (because access was difficult in the confined cabin) as well as jettisoned pyrotechnically during launch or re-entry. As a result, they had to work under both vacuum and maximum dynamic pressure conditions, and under a range of temperature extremes.

The other significant aspect of the Gemini missions was the demonstration of in-orbit rendezvous and docking techniques – a fundamental prerequisite for the Apollo missions, which involved one spacecraft descending to the lunar surface while the other remained in orbit.

Following an unmanned flight test in April 1964, the first manned Gemini mission was launched on 23 March 1965: the conservative test of the new spacecraft lasted just four hours and 53 minutes (three orbits) [59]. Alexei Leonov's famous spacewalk five days earlier had prompted the Americans to accelerate their Gemini schedule and, rather than stage a 'stand-up EVA' with the astronaut not actually leaving the cabin, the next Gemini mission included a full EVA; it was made by Ed White on the first day of the four-day Gemini 4 mission in June 1965 (Figure 7.15). Before the spacewalk, pilot James McDivitt had attempted to rendezvous and fly in formation with the second stage of the Titan 2, but this proved more difficult than it had in simulations. Having used up 42 per cent of the craft's manoeuvring propellant in the first few minutes, the attempt was abandoned [60].

A proper docking attempt was planned for Gemini 6 in October 1965, using the specially designed Agena-D target vehicle, an adaptation of the Agena-B Thor and Atlas second stage which had launched many satellites and planetary probes. In addition to its standard inertial guidance system – incorporating gyroscopes, computer and associated attitude control thrusters – the Gemini carried a rendezvous radar, which was capable of detecting the Agena when they were as much as 400 km apart. It provided the range, bearing and closing speeds of the spacecraft, which were supplemented from about 80 km by visual observations of a high-intensity beacon mounted on the Agena [57]. Unfortunately, the Agena launch failed and Gemini 6,

Figure 7.15   America's first EVA conducted by Edward H. White in June 1965. In his right hand White carries a Hand-Held Self-Manoeuvring Unit (HHSMU) [NASA]

(a)

(b)

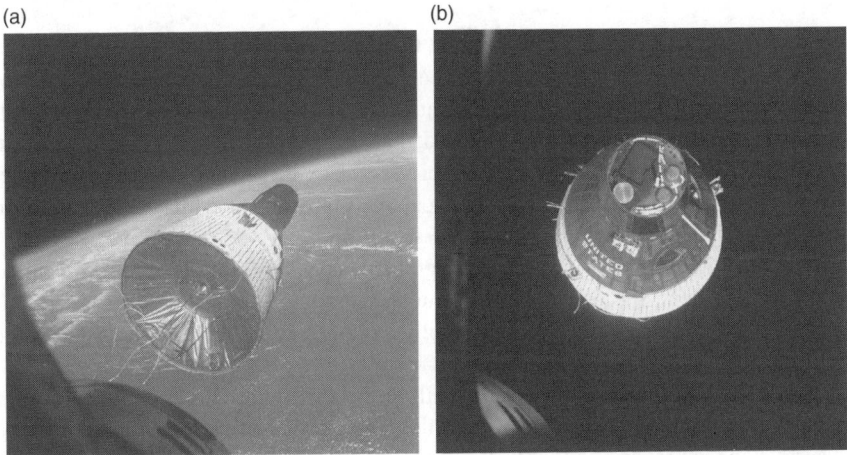

Figure 7.16   Gemini 7 in orbit on 15 December 1965 (taken from Gemini 6 during rendezvous manoeuvres which brought them within 15 cm of each other) [NASA]

having nowhere to go, was grounded. However, in December 1965, Gemini 6 and Gemini 7 performed the first close rendezvous between two spacecraft in orbit, with a closest approach of about 15 cm (Figure 7.16) [61].

The first successful docking occurred on the Gemini 8 mission of 16 March 1966, but almost led to the first in-orbit disaster. Twenty-seven minutes after astronauts

Figure 7.17   The 'Angry Alligator': the Augmented Target Docking Adapter (ATDA) as seen from Gemini 9 from a distance of 20 m. Failure of the fairings to separate prevented the docking of the two spacecraft [NASA]

Neil Armstrong and David Scott, both on their first space mission, docked with the Agena, the combination went into a spin because a short circuit in the Gemini control system had caused a thruster to fire continuously. Believing the Agena to be at fault, they separated the two craft only to find that the Gemini began to revolve at an incredible rate of up to one revolution per second. A threatened collision was only averted by using 75 per cent of the manoeuvring propellant which, under mission rules, demanded an emergency return to Earth [62].

In June 1966, the Gemini 9 crew were greeted in orbit by what Tom Stafford called 'an angry alligator', the Agena target vehicle with a half-opened payload fairing. It prevented the docking, but provided another photographic icon of the Space Age (Figure 7.17).[3]

While the Gemini missions were stealing the limelight, Soviet engineers were busy developing the Soyuz spacecraft, which in a number of variants would become Russia's standard manned spacecraft, up to and beyond the end of the century. The Soyuz comprised three modules: a spherical orbital compartment at the forward end, containing life support equipment and used for scientific experiments; a re-entry module, containing the cosmonaut's couches and flight control equipment; and an equipment module with two deployable solar array panels.

The first manned Soyuz mission was launched on 23 April 1967, carrying cosmonaut Vladimir Komarov. Tragically, he failed to return alive: as the re-entry module

---

[3] It also provided a story line for a 1967 James Bond film, *You Only Live Twice*, in which Spectre's spacecraft swallows Russian and American capsules.

descended its parachute lines became entangled and the module hit the ground at great speed, exploding on impact [63]. The incident halted the Soyuz programme, just as the Apollo 1 fire had delayed the Apollo programme three months earlier (see Chapter 8). Another Soyuz mission did not occur until October 1968, when Soyuz 2 and 3 were launched on 25 and 26 October, respectively, the former to act as an unmanned target vehicle for the latter; it was effectively a repeat of the Gemini 6/7 mission.

Conclusions

In 1965, the Gemini 7 spacecraft had spent more than 13 days and 18 h in space, proving that, although cramped and uncomfortable, an astronaut crew could survive for the time required to conduct a lunar mission. In July, September and November of the following year, Gemini 10, 11 and 12 conducted successful docking missions and completed the aims of the Gemini programme. In just five-and-a-half years of manned spaceflight experience, gained between May 1961 and November 1966, NASA had demonstrated the majority of the technology and techniques that would form the basis for a manned lunar landing mission. In addition, an interesting 'first' on Gemini 11 was the generation of artificial gravity on board a spacecraft, engineered by fixing a 30 m tether between the capsule and the Agena and rotating the two spacecraft about their centre of mass. It provided the proof of a concept mooted by Hermann Ganswindt in 1881.

Given the Soviet propensity for space records, the Western world had assumed from the early days of the Space Age that a manned lunar programme was also part of the Soviet plan, but their programme was not formally established until 3 August 1964 [64]. Having 'provoked the Americans' to compete in a Moon race, as one author puts it, they 'realized too late that there was a real race underway' [65]. In September 1968, three months before Apollo 8 circumnavigated the Moon, the USSR despatched the unmanned Zond 5 to loop around the Moon and land in the Indian Ocean six days later. This was part of its then unconfirmed attempt to land a man on the Moon before America.

As it turned out, the Soviet L-3 Moon programme was not as well organised as America's and ultimately failed to deliver in terms of technology or reliability. For example, all four launch attempts of the N-1 launch vehicle (the Soviet answer to the Saturn V) were failures and the programme was cancelled. The fact that these launches were made between February 1969 and November 1972, by which time the Apollo programme was all but over, indicates that they were nowhere near beating the Americans.

If anything, this makes the Apollo programme even more of a triumph than it undoubtedly was, since America had shown that it was the only nation capable of performing such a mission in such a short timescale. This, and the fact that Apollo was an important historical and cultural achievement in its own right, is why the next three chapters are dedicated to an analysis of the technology that made it possible.

References

1   Ambartsumyan, V.: 'Space Travel Initiated' in 'Soviet Writings on Earth Satellites and Space Travel' (MacGibbon and Kee, London, 1959) p. 158

2   'Sputnik II' (Article in Pravda, November 13, 1957) in 'Soviet Writings on Earth Satellites and Space Travel' (MacGibbon and Kee, London, 1959) p. 179

3   Vassiliev, M. (under the supervision of Dobronravov, V. V.): 'Sputnik Into Space' (Souvenir Press, London, 1958) p. 90 (published 'simultaneously in Canada' by The Ryerson Press, Toronto)

4   Harvey, Brian: 'The New Russian Space Programme' (John Wiley and Sons, Chichester, 1996) p. 34

5   Clark, Philip: 'The Soviet Manned Space Programme' (Salamander Books Ltd, London, 1988) p. 15

6   Smolders, Peter L.: 'Soviets in Space' (Lutterworth Press, Guildford, 1973) p. 114

7   Newkirk, Dennis: 'Almanac of Soviet Manned Space Flight' (Gulf Publishing Co, Houston, 1990) p. 26

8   Thompson, Tina D. (Ed.): 'TRW Space Log 1996' (TRW Space and Electronics Group, Redondo Beach, Volume 32, 1997) pp. 66–8

9   ibid., p. 68

10 Clark, Philip: 'The Soviet Manned Space Programme' (Salamander Books Ltd, London, 1988) pp. 16–17

11 Harvey, Brian: 'The New Russian Space Programme' (John Wiley and Sons, Chichester, 1996) p. 45

12 ibid., pp. 48–9

13 Abramov, Isaak P., and Skoog, A. Ingemar: 'Russian Spacesuits' (Springer-Praxis, Chichester, 2003) p. 45

14 Harvey, Brian: 'The New Russian Space Programme' (John Wiley and Sons, Chichester, 1996) p. 37

15 Stoiko, Michael: 'Soviet Rocketry: The First Decade of Achievement' (David and Charles, Newton Abbot, 1970) p. 175

16 Clark, Philip: 'The Soviet Manned Space Programme' (Salamander Books Ltd, London, 1988) pp. 19–20

17 Harvey, Brian: 'The New Russian Space Programme' (John Wiley and Sons, Chichester, 1996) pp. 57–9

18 Abramov, Isaak P., and Skoog, A. Ingemar: 'Russian Spacesuits' (Springer-Praxis, Chichester, 2003) p. 48

19 ibid., pp. xi, xvii

20 ibid., p. 37

21 ibid., pp. 41–2

22 ibid., p. 53

23 ibid., pp. 53–4

24 Stoiko, Michael: 'Soviet Rocketry: The First Decade of Achievement' (David and Charles, Newton Abbot, 1970) p. 177

25 Harvey, Brian: 'The New Russian Space Programme' (John Wiley and Sons, Chichester, 1996) p. 59

26 Harland, David M.: 'The Space Shuttle: Roles, Missions and Accomplishments' (Wiley-Praxis, Chichester, 1998) pp. 3–4

27 'Manned Space Flight: Projects Mercury and Gemini', *NASA Facts* **2** (8), US Government Printing Office (OF-758–556), Washington DC, USA, 1965, p. 2

28 Wolfe, Tom: 'The Right Stuff' (Farrar, Straus and Giroux Inc, New York, 1979)

29 Grimwood, James M.: 'Project Mercury: A Chronology' (NASA SP-4001, Washington DC, USA, 1963)

30 Swenson, Loyd S., Grimwood, James M., and Alexander, Charles C.: 'This New Ocean: A History of Project Mercury' (NASA SP-4201, Washington DC, USA, 1966)

31 Mahone, Glenn and Jacobs, and Bob: 'Legendary Spacecraft Designer Dr. Maxime A. Faget Dies At 83', NASA press release 04-350 (October 10, 2004), Washington DC, USA

32 Thompson, Elvia: 'NASA Celebrates 90 Years of Aeronautics Excellence' NASA press release 05-067 (March 3, 2005), Washington DC, USA

33 Catchpole, John: 'Project Mercury: NASA's First Manned Space Programme' (Springer-Praxis, Chichester, 2001) p. 135

34 ibid., pp. 138–9

35 ibid., p. 139

36 ibid., pp. 189–90

37 Sharpe, Mitchell R., and Burkhalter, Bettye B.: 'Mercury-Redstone: The First American Man-Rated Space Launch Vehicle' (IAA-89-740), 40th IAF Congress, Malaga, Spain, 1989, p. 1

38 ibid., p. 15

39 ibid., p. 16

40 Swenson, Loyd S., Grimwood, James M., and Alexander, Charles C.: 'This New Ocean: A History of Project Mercury' (NASA SP-4201, Washington DC, USA, 1966) p. 468

41 Catchpole, John: 'Project Mercury: NASA's First Manned Space Programme' (Springer-Praxis, Chichester, 2001) p. 161

42 Sharpe, Mitchell R., and Burkhalter, Bettye B.: 'Mercury–Redstone: The First American Man-Rated Space Launch Vehicle' (IAA-89-740), 40th IAF Congress, Malaga, Spain, 1989, p. 18

43 Grimwood, James M.: 'Project Mercury: A Chronology' (NASA SP-4001, Washington DC, USA, 1963) p. 118

44 Swenson, Loyd S., Grimwood, James M., and Alexander, Charles C.: 'This New Ocean: A History of Project Mercury' (NASA SP-4201, Washington DC, USA, 1966) p. 294

45 ibid., pp. 296-7

46 Harvey, Brian: 'The New Russian Space Programme' (John Wiley and Sons, Chichester, 1996) p. 43

47 Kennedy, Gregory P.: 'Mercury Primates' (IAA-89-741) , 40th IAF Congress, Malaga, Spain, 1989, p. 6

48 Turnill, Reginald (Ed.): 'Jane's Spaceflight Directory, 1987' (Jane's Publishing Co Ltd, London, 1987) p. 81

49 Thompson, Tina D. (Ed.): 'TRW Space Log 1996' (TRW Space and Electronics Group, Redondo Beach, Volume 32, 1997) p. 73

50 ibid., p. 78

51 'Manned Space Flight: Projects Mercury and Gemini', *NASA Facts* **2** (8), US Government Printing Office (OF-758-556), Washington DC, USA, 1965, p. 3

52 Harvey, Brian: 'The New Russian Space Programme' (John Wiley and Sons, Chichester, 1996) p. 79

53 Furniss, Tim: 'Manned Spaceflight Log' (Jane's Publishing Company, London, 1983) p. 31

54 Shayler, David J.: 'Gemini: Steps To The Moon' (Springer-Praxis, Chichester, 2001) p. 30

55 ibid., p. 34

56 ibid., p. 40

57 'Manned Space Flight: Projects Mercury and Gemini', *NASA Facts* **2** (8), US Government Printing Office (OF-758-556), Washington DC, USA, 1965, p. 5

58 ibid., p. 7

59 Thompson, Tina D. (Ed.): 'TRW Space Log 1996' (TRW Space and Electronics Group, Redondo Beach, Volume 32, 1997) p. 90

60 Furniss, Tim: 'Manned Spaceflight Log' (Jane's Publishing Company, London, 1983) p. 34

61 ibid., pp. 39–40

62 ibid., p. 41

63 Harvey, Brian: 'The New Russian Space Programme' (John Wiley and Sons, Chichester, 1996) p. 109

64 ibid., p. 91

65 ibid., p. 137

Chapter 8

First and only moonship – the development of the Apollo Command and Service Module

I believe that this nation should commit itself to achieving the goal, before this
decade is out, of landing a man on the Moon and returning him safely to Earth

US President John F. Kennedy

In April 1961, when the Soviet Union chalked up another space record by launching the first man into space, America was already deep in the political doldrums
of the 'Cuban missile crisis'. Following its ignominious defeat in the 'Bay of Pigs'
confrontation, the nation needed a boost to its collective morale. Thus it was a combination of events that led President John F. Kennedy to make one of the most famous
political directives of all time (Figure 8.1). On 25 May 1961, he made the following
declaration to the US Congress:

> I believe that this nation should commit itself to achieving the goal, before this decade is out,
> of landing a man on the Moon and returning him safely to Earth. No single space project in
> this period will be more exciting, or more impressive to mankind, or more important for the
> long-range exploration of space; and none will be so difficult or expensive to accomplish. [1]

From that point on, NASA had a goal, the political backing and the money to
turn the tables on the Soviet Union and take the lead in the race to the Moon. To say
that Kennedy's speech galvanised NASA and the nascent space industry would be
an understatement; the job of defining the mission and specifying the spacecraft to
realise that mission began in earnest.

In retrospect, the Apollo lunar programme is widely recognised as a political
imperative, designed to prove American superiority in the face of Soviet competition.
Indeed, Kennedy's speech had included the words:

> ... if we are to win the battle that is now going on around the world between freedom and
> tyranny, the dramatic achievements in space which occurred in recent weeks [Gagarin's
> flight] should have made clear to us all, as did the Sputnik in 1957, the impact of this

Figure 8.1   *President John F. Kennedy makes his famous speech to Congress on 25 May 1961. In the background are Vice President Lyndon Johnson (left) and Speaker of the House Sam T. Rayburn (right) [NASA]*

adventure on the minds of men everywhere ... Now it is the time to take longer strides – time for a great new American enterprise – time for this nation to take a clearly leading role in space achievement which, in many ways, may hold the key to our future on Earth [1].

Despite the political overtones, the Apollo programme was also a triumph of technology over seemingly insurmountable challenges. At the time of Kennedy's address, America had just 15 minutes and 22 seconds of experience in manned space-flight, accrued on just one suborbital hop which reached an altitude of 187 km; the Moon, by contrast, was about 375,000 km away and a successful lunar mission would last at least 8 days. NASA itself had been in existence for less than three years, and the technology of the day included the room-sized computer and the rotary-dial telephone. By any measure, the enormity of the challenge was unprecedented.

Kennedy himself realised this, as shown by another much quoted statement made at Houston's Rice University on 12 September 1962: 'We choose to go to the moon in this decade and do the other things, not because they are easy, but because they are hard, because that goal will serve to organise and measure the best of our energies and skills ...'. And he was right. The political decision, and the funding allocations which supported it, provided a spur to the development of space technology that has never been seen since.

It is significant not only in the scope of the history of technology, but of history itself, that only eight years after Yuri Gagarin became the first human being to orbit the Earth and less than 12 years after Sputnik 1, two men were standing on the surface of the Moon.

A key part of the system which facilitated that historic event was the spacecraft that ferried the astronauts and their lunar landing vehicle to the Moon: the Apollo command and service module (CSM). This chapter describes some of the engineering challenges of the Apollo CSM and a selection of design errors which led to well-publicised accidents. In addition to illustrating the challenge and complexity of spaceflight, they serve to illustrate an important truism: that much can be learnt from history.

Table 8.1   Apollo hardware development flights (selected) and manned missions

Launch date	Name	CM/LM	Comments
7 November 1963	PA-1	BP-6	First CM boilerplate flown from White Sands Missile Range in launch pad abort test
13 May 1964	A-001	BP-12	First high-altitude abort demonstration of Launch Escape System (LES); Little Joe 2 rocket from White Sands
28 May 1964	SA-6/A-101	BP-13	First boilerplate CM launched to temporary orbit by Saturn I
18 September 1964	SA-7/A-102	BP-15	Boilerplate CM launched to temporary orbit by Saturn I
26 February 1966	AS-201	CSM-009	First flight test of Apollo CSM; first flight of Saturn IB (Apollo-Saturn 201)
5 July 1966	AS-203	(no s/c)	Saturn IB test; first orbital flight of SIVB upper stage
25 August 1966	AS-202	CSM-011	Demonstrated LES/separation of CM from SM and operation of onboard systems, first Pacific recovery
27 January 1967 (ground test)	AS-204/Apollo 1	CM-012	Fire in CM cabin killed three astronauts: Virgil Grissom, Edward White and Roger Chaffee
9 November 1967	AS501/Apollo 4	CSM-017	First flight test of Saturn V/Block I CSM (first CSM in space; unmanned)
22 January 1968	AS-204/Apollo 5	LM-1	First flight of fully operational LM; Saturn IB (no CSM)
4 April 1968	AS502/Apollo 6	CSM-020	Unmanned flight test of Saturn V/final Block I CSM

(Continued)

Table 8.1   Continued

Launch date	Name	CM/LM	Comments
11 October 1968	Apollo 7 (AS-205)	CSM-101	First manned Apollo flight; Saturn IB; tested Block II CSM primary prop system. Crew: Walter Schirra, Don Eisele, Walter Cunningham
21 December 1968	Apollo 8 (AS-503)	CSM-103	First manned lunar orbit mission (10 revolutions); Saturn V; CSM + boilerplate LM. Crew: Frank Borman, Jim Lovell, William Anders
3 March 1969	Apollo 9 (AS-504)	CSM-104 LM-3	First manned Saturn V with full Apollo CSM and LM in LEO. Crew: James McDivitt, Dave Scott, Russell Schweickart
18 May 1969	Apollo 10 (AS-505)	CSM-106 LM-4	Dress rehearsal for lunar landing. Crew: Tom Stafford, John Young, Gene Cernan
16 July 1969	Apollo 11 (AS-506)	CSM-107 LM-5	First manned lunar landing. Crew: Neil Armstrong, Michel Collins, Buzz Aldrin
14 November 1969	Apollo 12 (AS-507)	CSM-108 LM-6	Crew: Pete Conrad, Richard Gordon, Alan Bean
11 April 1970	Apollo 13 (AS-508)	CSM-109 LM-7	Moon landing aborted 13/4/70 when SM LOX tank exploded. Crew: Jim Lovell, Jack Swigert, Fred Haise
31 January 1971	Apollo 14 (AS-509)	CSM-110 LM-8	Crew: Alan Shepard, Stuart Roosa, Edgar Mitchell
26 July 1971	Apollo 15 (AS-510)	CSM-112 LM-10	Crew: Dave Scott, Alfred Worden, James Irwin
16 April 1972	Apollo 16 (AS-511)	CSM-113 LM-11	Crew: John Young, Tom Mattingly, Charles Duke
7 December 1972	Apollo 17 (AS-512)	CSM-114 LM-12	Crew: Gene Cernan, Ronald Evans, Harrison 'Jack' Schmitt

Notes: LM-2 was donated to the Smithsonian Air and Space Museum in Washington.
LM-9 is on display at Kennedy Space Center, Florida.

Design overview

Apollo, as a programme name, dates back to July 1960, when NASA publicised its proposal for a manned space programme that would build on the experience of the one-man Mercury missions. The goal envisaged at the time for the Apollo programme – with America still 10 months away from its first manned Mercury launch – was to build a three-man spacecraft for Earth orbital and circumlunar flights [2]; there was no mention of a lunar landing. It was not until Kennedy made his speech to Congress that the press and public had any serious indication that Apollo might be developed as a lunar landing mission.

Following a number of study contracts, let in an attempt to define the mission, NASA distributed the CSM request for proposals (RFP) in September 1961 [3] and received responses from five aerospace companies: General Dynamics, General Electric, Martin Marietta, McDonnell Douglas and North American Aviation (later North American Rockwell) [4]. On 15 December, North American was selected as the CSM's prime contractor [5], setting a chain of subcontracting events in motion that would produce the first – and so far only – spacecraft capable of delivering men to our nearest astronomical neighbour.

From the very early days, several different lunar missions were considered, and the delay in choosing between them caused problems in finalising the design of the CSM. The simplest concept involved developing a large launch vehicle capable of delivering a spacecraft directly to the Moon, landing the whole spacecraft and returning it all to Earth. Although this avoided docking separate elements in space, the landing gear would need to support the weight and landing forces of the entire spacecraft and a great deal of empty tankage would have to be returned from the Moon and decelerated for re-entry. Moreover, it would require the development of a truly massive launch vehicle.

An alternative concept involved launching the spacecraft in parts, on several smaller launchers, rendezvousing the elements and constructing the spacecraft in Earth orbit; the remainder of the mission would be the same as the direct option. A third alternative was to launch several unmanned craft, including a crew return vehicle, to the Moon, where they would await a crew. As with most engineering challenges, there was more than one possible solution and the challenge for NASA's management was to choose the best one.

As it turned out, all of these options had significant disadvantages and the mission design finally chosen for Apollo featured the command and service module and a separate lunar module launched together by the newly developed Saturn V rocket (Figure 8.2). In this case, only the lunar module would land on the Moon and only the command module would return to Earth, separating from the service module about 15 minutes before re-entry. The concept was known as lunar orbit rendezvous (LOR) and the decision to base the Apollo mission on this concept was made in July 1962 [6].

The initial uncertainty in the mission design, coupled with the need to begin designing, building and testing the hardware as soon as possible, led to two parallel CSM development paths: a 26,531 kg Block I CSM without a docking system and a 30,385 kg Block II *with* a docking system. When the Block II vehicle became

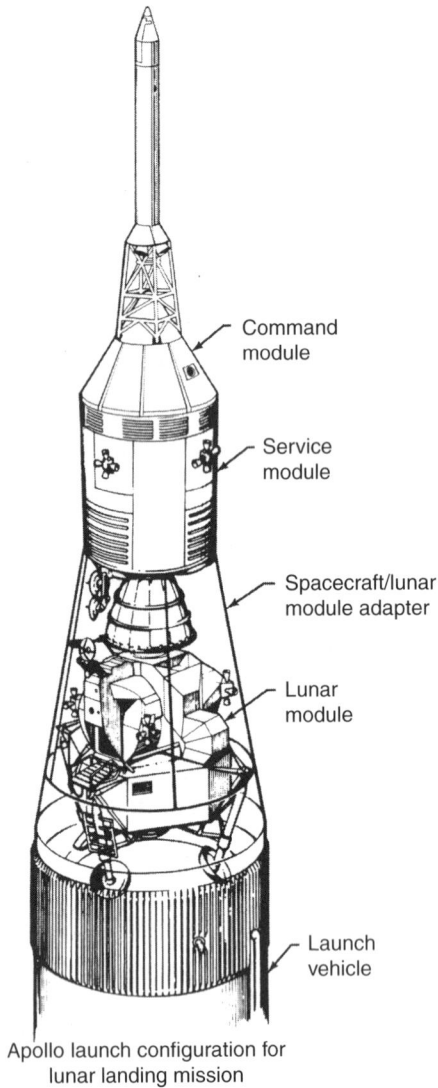

Apollo launch configuration for
lunar landing mission

Figure 8.2   Apollo design concept: CSM and LM on the Saturn V [NASA]

the favoured option, the Block I CSMs which had already been built were relegated
to unmanned flight tests of the Saturn V (Apollo 4 on 9 November 1967 and Apollo 6
on 4 April 1968). According to a NASA document produced in 1972, the CSMs for
Apollo 7 to 14 cost $55 million apiece, whereas those for the final three missions cost
$65 million each [7].

In addition to these flight models, a number of so-called boilerplate CMs were
built for use in tests of the launch vehicle stages and the launch escape system,

Figure 8.3   Launch of a CSM boilerplate on a Little Joe rocket [NASA]

a solid rocket motor designed to lift the CM away from the launch vehicle in case of catastrophic failure. A number of boilerplates were flown from the White Sands Missile Range, New Mexico throughout the mid-1960s and two were even launched to temporary orbits by Saturn IB rockets in 1964 (Figure 8.3).

The amount of spacecraft and launch vehicle hardware produced for the Apollo programme was unprecedented, as was the degree of reliability required. Although government laboratories and rocket test ranges operated by the military could handle the requirements of the early Space Age, the level of organisation required to send manned spacecraft to the Moon was far greater . . . and another challenge for NASA. Indeed, apart from any technological spin-offs, the NASA-led Apollo programme is credited with much of the development of modern industrial practice. Never before had it been necessary to coordinate so many different industrial contractors, in so many states, to produce so many leading-edge technologies on such a short timescale. The following sections describe some of the challenges encountered in the development of the CSM's subsystems.

Structure and thermal subsystems

In common with all spacecraft – whether manned or unmanned – the main limiting factor of the Apollo spacecraft was its mass, simply because of the limitations of

Figure 8.4   Apollo command and service module undergoes final integration [NASA]

the launch vehicle. This meant that the CSM, and by definition all of its subsystems and components, had to be as light as possible. On the other hand, everything had to be designed, built and tested to ensure it would survive the acceleration forces and vibrations of the launch, including stage separation shocks, and later spacecraft engine firings. Maintaining the balance between strength and weight was a challenge that design engineers would face throughout the programme.

As its name suggests, the command and service module comprised two main parts: a command module (CM) and a service module (SM) (Figure 8.4). The command module was a habitable capsule designed to support the requirements of three astronauts for almost two weeks in space, while the service module contained all the 'services' required by the command module and its crew (Figure 8.5).

The CM was conical, 3.5 m high and 3.9 m in diameter at the base; it weighed about 5830 kg and provided a habitable volume of 5.95 m^3 [8] (less than 2 m^3 for each man; about the volume of a public telephone box). It contained three couches, an instrument panel from which the operation of the craft could be monitored and controlled, and most of the electronics concerned with the control of the spacecraft and its internal environment. Ground access was provided by a main hatch on the sloping surface of the cone and access to the lunar module, when in space, was made via a docking tunnel at the apex. A total of five windows were provided for navigation observations, photography and, importantly, peace of mind. Since the CM

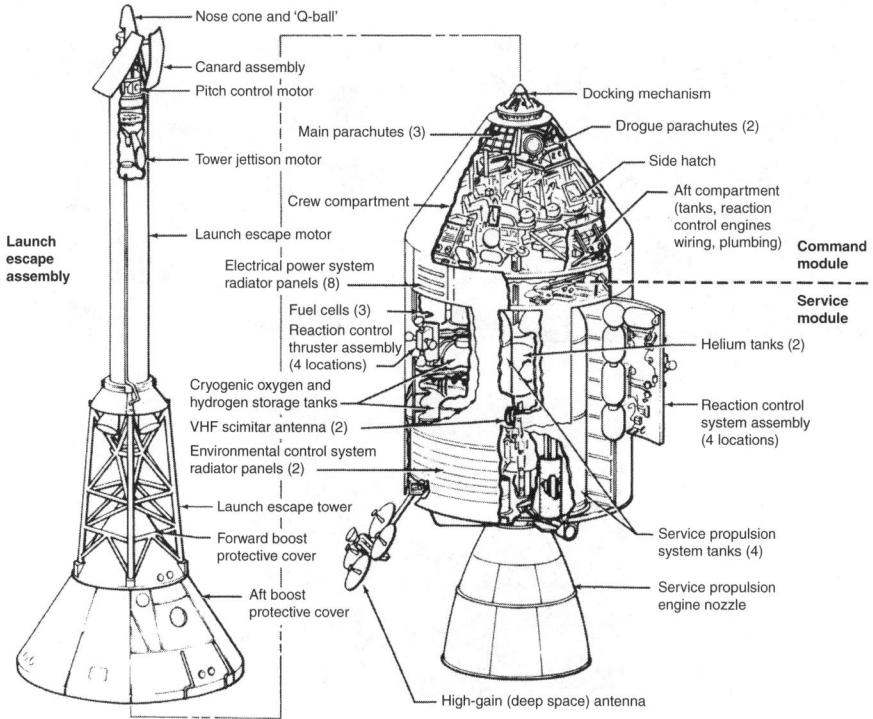

Figure 8.5   Apollo CSM and Launch Escape System (LES) [NASA]

was the only part of the Apollo spacecraft to return to Earth, it also incorporated an attitude control subsystem (pitch, roll and yaw thrusters), a heat shield and a parachute system.

The CM's shape, designed to be as simple as possible, was based loosely on that of the Mercury and Gemini capsules with their blunt heat shields and tapering cabin walls. The heat shields of all three spacecraft were designed to build up a bow shock-wave on re-entering the atmosphere which would deflect heat away from the spacecraft. The design was thus inherently a weight-saving one.

The form of the cone itself was very important and had to be determined by wind tunnel tests at NASA's Ames laboratory. For example, if the cone's taper was too shallow, the bow wave would curve in and reattach to the upper part of the capsule, causing it to heat up; of course, if it had been any steeper than the chosen design, the capsule would have been too squat to house the astronauts.

Although, in essence, the command module was a simple conical structure, bounded by various types and amounts of thermal shielding which produced its external appearance, this masked a more complex internal structure. Its central primary structure was based on a roughly conical crew compartment surmounted by a cylindrical docking tunnel. This pressurised inner shell was fabricated from aluminium

(a)   (b)

(c)

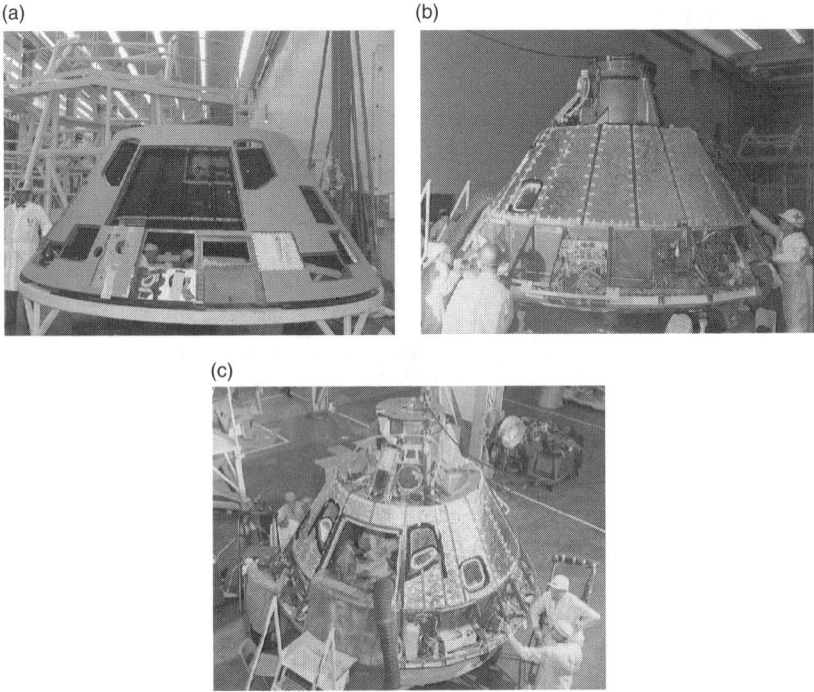

Figure 8.6   Command module CM12 structure and thermal subsystem, showing (a) basic structural shell, (b) primary structural components, (c) thermal insulation [NASA]

honeycomb panels and separated from the heat-resistant outer shell by a micro-quartz fibre insulator to help maintain internal temperatures. At the base of the cone, a circular heat shield completed the structure (Figure 8.6).

Since friction with the Earth's atmosphere would produce temperatures of around 2700°C (5000°F), the thermal design of the CM was crucial. At the time, there were no materials that would resist such high temperatures and remain intact, so an ablative material that burnt away gradually, effectively taking the heat energy with it, had to be used. The CM's outer shell was made from a 1.8 cm thick brazed stainless-steel honeycomb, comprising some 40,000 individual cells, to which an ablative layer of glass fibre honeycomb was applied in varying thickness dependent upon the expected heat load. The cells of the latter honeycomb were filled, by hand, with phenolic epoxy resin which served as the ablative material.

Protection from atmospheric heating experienced during launch, although much less extreme, was provided by the boost cover, a conical fairing attached to the launch escape system and entirely covering the CM. The boost cover was jettisoned with the escape tower once the Saturn V had left the atmosphere (Figure 8.7).

The CM itself was a marvel of packaging design, and no volume, however small, was wasted. For example, the space between the docking tunnel and the apex of

Figure 8.7   Test of Launch Escape System (LES) and boost cover in wind tunnel [NASA]

the cone housed parachutes and other equipment for the landing phase, while a small aft compartment, which ran around the base of the crew compartment, contained the manoeuvring system and electrical and fluid connections to the service module.

The command module was mounted on top of the service module, a cylinder 7.5 m high, 3.9 m in diameter and weighing about 24,550 kg [8]. The SM carried an environmental control system which provided the CM with a pure oxygen atmosphere at 34.5 kPa (5 psi); a power subsystem based on hydrogen/oxygen fuel cells; a thermal subsystem including a number of radiator panels; a propulsion subsystem for orbital injections, course changes and attitude control; and a communications system with a deployable, high gain S-band antenna.

The communications system devised for Apollo was more complex than any previous spacecraft system simply because of logistics. The CM had not only to communicate with the Earth, but also with the LM in lunar orbit and on the surface. The system supplied by Collins Radio Co (which was merged into Rockwell in 1973) was required to provide voice, television and data links, most importantly for telemetry, tracking and command data. Although an omnidirectional antenna allowed a link with Earth to be maintained irrespective of the spacecraft's orientation, the high gain S-band (2–4 GHz) antenna was crucial for the higher rate data links. Considering the state of satellite communications technology in the mid-1960s, the antenna system was advanced. It comprised four 80cm-diameter parabolic dishes on a deployable boom, which could be steered manually or automatically to track the LM or maintain the all-important link with Earth.

The cylindrical service module provided the structural backbone of the CSM combination. Its primary structure was constructed from milled aluminium radial beams forming six wedge-shaped sectors around a central cylinder. The 1.12m-diameter cylinder contained the main engine – the Service Propulsion System – and two nitrogen tanks. The main components housed within each of the six sectors are given in the Box.

Service module components by sector [9]

Sector 1: camera and science instruments (on later missions).

Sector 2: reaction control system (RCS) thruster quad and propellant tanks for all thrusters; service propulsion system (SPS) oxidiser sump tank; environmental control system (ECS) radiator.

Sector 3: RCS thruster quad and SPS oxidiser tank; ECS radiator (continued).

Sector 4: three fuel cells; two liquid oxygen and two liquid hydrogen tanks for fuel cells (LOX tanks also supplied cabin atmosphere).

Sector 5: RCS thruster quad; SPS fuel sump tank; ECS radiator.

Sector 6: RCS thruster quad; SPS fuel tank; ECS radiator (continued).

The SM had an outer skin of 2.5cm-thick aluminium honeycomb, much of which formed the access panels for the various sector compartments. The remaining area was occupied by the two large environmental control system (ECS) radiators which ran around the base of the module. Since the CM had a convex heat shield, a shallow dish structure on top of the SM was provided to support it, surrounded by a ring fairing which held eight much smaller radiators used to vent heat from the fuel cells.

When in space, the combined command and service modules were protected first by a Mylar foil coating applied to most surfaces, and second by the initiation of what came to be know as a barbecue roll (rotating the spacecraft along its roll axis to disperse the effect of solar heating). As with most other spacecraft, individual subsystems and major components were provided with their own electrical heaters to maintain their temperature within operating limits during cold periods.

Propulsion subsystem

In addition to its role as CSM prime contractor, North American also supplied the reaction control system (RCS) for the CM and was the main supplier of propulsion systems for the Saturn V launch vehicle. The Service Propulsion System (SPS) – the CSM's main engine – was, however, subcontracted to Aerojet General. It was a crucial engine, since it provided the all-important burns to inject the spacecraft into lunar orbit and bring it home again; it was also the only engine on the spacecraft capable of doing this, which in engineering terms made it a 'single point failure'.

One of the ways to improve its reliability was to use the hypergolic propellant combination of unsymmetrical dimethylhydrazine (UDMH) and nitrogen tetroxide (N_2O_4), as fuel and oxidiser respectively, since they ignite on contact. In fact,

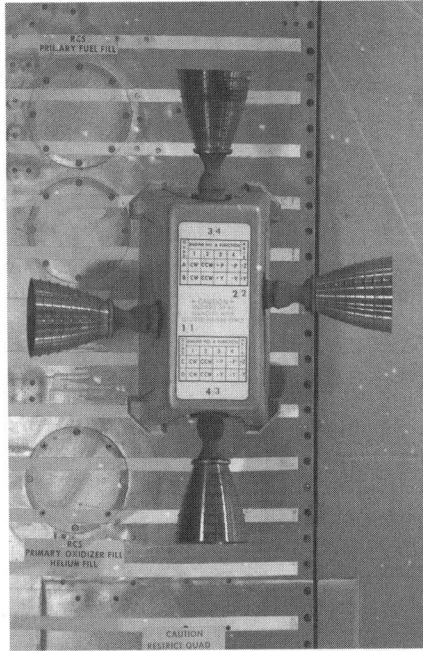

Figure 8.8   RCS thruster quad on Apollo CSM at Kennedy Space Center [Mark Williamson]

all propellant combinations used on Apollo were hypergolic to simplify the propulsion system by removing the need for an ignition system. CSM attitude and orbital control was engineered by four groups of four RCS thrusters, known as quads, positioned symmetrically around the midriff of the module. In common with the CM's thrusters, they used monomethyl hydrazine (MMH) and N_2O_4 (Figure 8.8).

Another classic method of ensuring the eventual reliability of a system was, and still is, to conduct a detailed test programme. After several years of design, development and ground-based testing, actual flight testing of the Apollo CSM began on 26 February 1966 (the same year the five final Gemini missions were conducted). On a mission designated AS-201, for Apollo-Saturn 201, a CSM was launched on the inaugural flight of the Saturn IB. That it was felt necessary to test both the CSM and the new Saturn on the same flight indicates the pressure that the schedule was placing on the respective design teams.

A key part of the flight involved a test of the SPS, which was supposed to fire for three minutes, cease for 10 s, and then resume firing to prove that it could restart. This restart capability is something a layman may take for granted in an engine, but most rocket engines, even today, are designed to fire only once before either being discarded or returned to Earth for refurbishment. If the SPS failed to restart following the lunar module's rendezvous with the CSM after the landing, the crew would have to spend the rest of their days in lunar orbit. The SPS performed each step on schedule, but halfway through the first burn the thrust level dropped about 30 per cent; telemetry showed that it was due to a leak in the oxidiser line [10].

Although another test of the Saturn IB was performed in July that year, proving the SIVB upper stage in orbit, the next CSM flight did not occur until 25 August 1966. This time the SPS fired four times, precisely on schedule, and the command module separated from the service module and returned for the first Pacific Ocean recovery of an Apollo capsule. The mission, which also checked out other onboard systems, was a success. As one author put it, 'The flight of spacecraft 11 brought everybody face to face with the moment of truth. The time had come to put people inside' [11].

It is difficult to summarise the development of such an important subsystem in a few paragraphs – indeed many books have been dedicated to various aspects of the Apollo programme – but a problem that occurred during the development of the CSM's titanium propellant tanks serves as an example. If anything, it happened *not* as a result of any incompetence, but *because* of the high standards in manufacturing, cleanliness and attention to detail that had been instilled in the Apollo workforce.

During pressure testing of one of the SM's oxidiser tanks, the tank exploded on the test stand. Despite detailed investigation by North American's engineers, they were mystified as to the cause. Then another tank undergoing a similar test exploded. Surprisingly, when pressurised with water rather than nitrogen tetroxide, the tanks retained their integrity; and to complicate matters, similar tanks being tested by the manufacturer, GM's Allison Division in Indianapolis, were not experiencing failures, even when filled with N_2O_4. North American engineers began to think the problem had something to do with the Los Angeles smog!

The cause of the problem was eventually traced to a refinery operated by the US Air Force, which supplied the propellant. An investigation of delivery records showed that the N_2O_4 supplied to Allison had a tiny amount of water in it, while the later batch sent to North American was almost perfectly pure. It turned out that the refinery operators, recognising the importance of the Apollo programme, had made a change to the process and triple-distilled the oxidiser to make it as pure as possible. The problem was that when N_2O_4 is better than 99 per cent pure it attacks titanium. The oxidiser had eaten into the SM tanks and weakened them, causing them to rupture during the pressure tests, while the relatively impure Allison oxidiser did not attack its tanks in this way [12].

This and other similar events led to a philosophy during Apollo whereby unexplained events were forbidden: every anomaly, however small, had to be investigated and explained before a unit could be passed for flight. Despite this attention to detail, the system was imperfect, as detailed later.

Power subsystem

Since the Mercury missions had been of limited duration, it had been feasible to draw power from non-rechargeable batteries. However, an Apollo lunar mission was expected to last the best part of two weeks and thus had much greater power

*Figure 8.9   Apollo fuel cell in National Air and Space Museum, Washington DC
[Mark Williamson]*

requirements. The solution was to use the fuel cell, a technology which had been demonstrated and proven on the Gemini spacecraft.

The Apollo command and service module's power source was a system of three hydrogen–oxygen fuel cells manufactured by United Aircraft Corporation's Pratt & Whitney Aircraft Division (Figure 8.9). Each fuel cell provided 1.5 kW at 28 V [13], a standard voltage for both manned and unmanned spacecraft in the 1960s. Thus, with three operating in parallel, the total power available to the CSM was theoretically around 4.5 kW. However, only two of the three were required for the lunar mission and the CM itself had a power budget of only 2 kW [13] (the fact that this is equivalent to the power rating of an average electric kettle here on Earth speaks volumes on the efficiency of spacecraft equipment).

For pre-launch check-out, DC power was derived from ground support equipment, while AC power was supplied for some functions by the spacecraft's inverters. Since the service module would be jettisoned prior to re-entry and landing, the command module also carried three zinc–silver oxide batteries, which were maintained fully charged during the mission via a battery charge regulator connected to the fuel cells [14].

Each fuel cell contained 31 separate sub-cells connected in series. Each sub-cell had a hydrogen compartment containing a nickel electrode and an oxygen compartment with a nickel/nickel oxide electrode. An electrolyte solution of potassium hydroxide and water provided electrical conduction between the two compartments and supported the chemical reaction which produced electric current. As with the Gemini cells, in addition to providing electrical energy, the chemical reactions at the electrodes produced heat and water as useful by-products. The warm water was piped to the CM where it could be used for washing or for the reconstitution of dehydrated food. The heat was used to keep the electrolyte at the correct temperature and the excess was radiated into space.

To enable sufficient amounts of oxygen and hydrogen to be carried in such a small space, they had to be held in their liquid state at cryogenic temperatures: liquid oxygen (LOX) boils at about $-183°C$ and liquid hydrogen (LH_2) at about $-253°C$. Although this, in itself, was not a problem for an industry that used LOX and LH_2 as rocket propellants – in the second and third stages of the Saturn V for example – the design of the SM's cryogenic propellant tanks presented a challenge.

In order to maintain their contents in a liquid state for up to two weeks under varying external temperature conditions, the tanks had to be very well insulated. In fact, they were so well insulated that, in a room warmed to $20°C$, they could have held water ice in its frozen state for more than eight years. They were also extremely well sealed to ensure a low leakage rate: it has been calculated that the equivalent leakage rate in a typical car tyre would flatten it in about 32 million years [15]. Despite their impressive specifications, it was one of these tanks which was the root cause of the Apollo 13 accident (see later).

Guidance, navigation and control

The early manned spacecraft were, in general, more complex than unmanned spacecraft because of the requirement for additional subsystems concerned with environmental control and life support. Coupled with enhanced redundancy requirements for safety critical systems, such as guidance, navigation and communications, this increased the component count to the point where further miniaturisation was essential.

One only has to compare the control and display panels of the first American manned spacecraft, Mercury, and the Apollo lunar spacecraft, to see how spacecraft developed in the five years that separated the final Mercury mission and the first manned Apollo mission. Whereas Mercury was a tiny one-man capsule with a few dozen dials and gauges, the Apollo command module had some 800 controls and displays (Figure 8.10). And whereas the instruments of the Mercury capsule were connected using an incredible 11 km of wire, the Apollo contained almost 24 km of wire [14].

In the interests of safety, manned spacecraft also carried a caution and warning system, giving two levels of notice to the crew. If a malfunction or 'out-of-tolerance condition' occurred in one of the subsystems, its status light would be illuminated and a master alarm circuit tripped; a corresponding warning light would be illuminated only if immediate corrective action was required [14].

The caution and warning system had a particularly severe test during the launch of Apollo 12, on 14 November 1969, when much of the electrical system – including

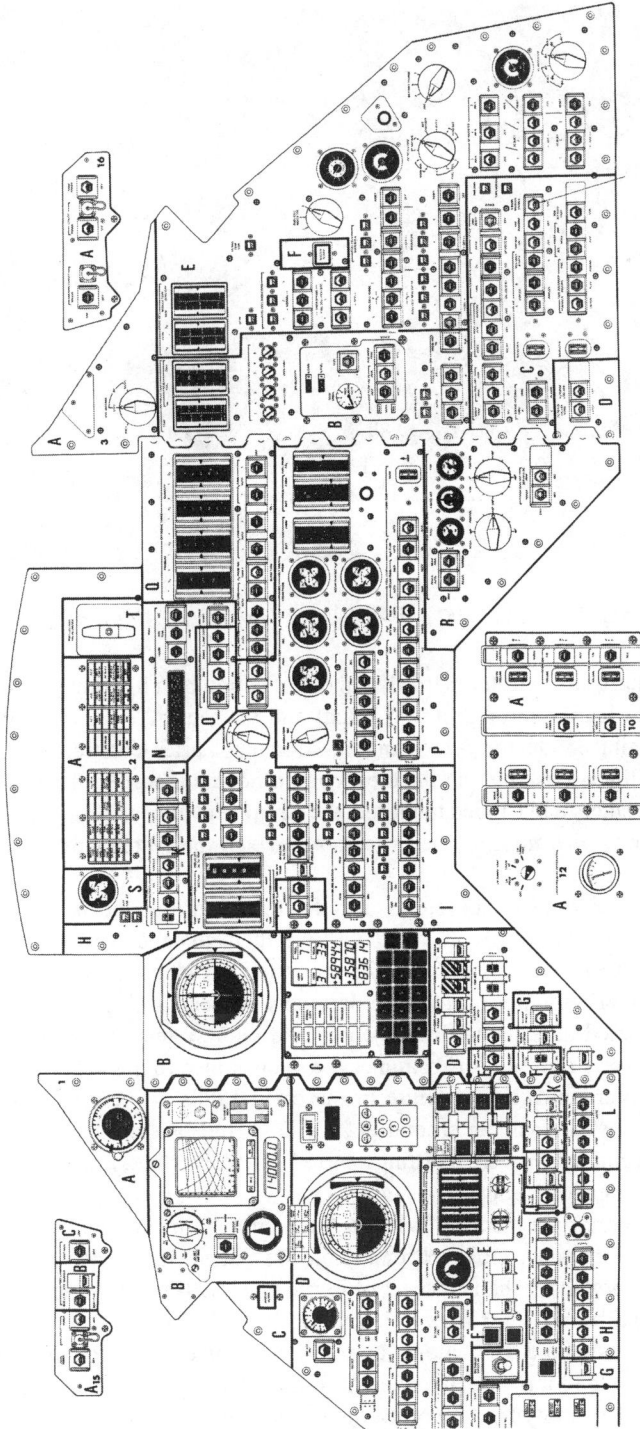

Figure 8.10   CM control panel [NASA]

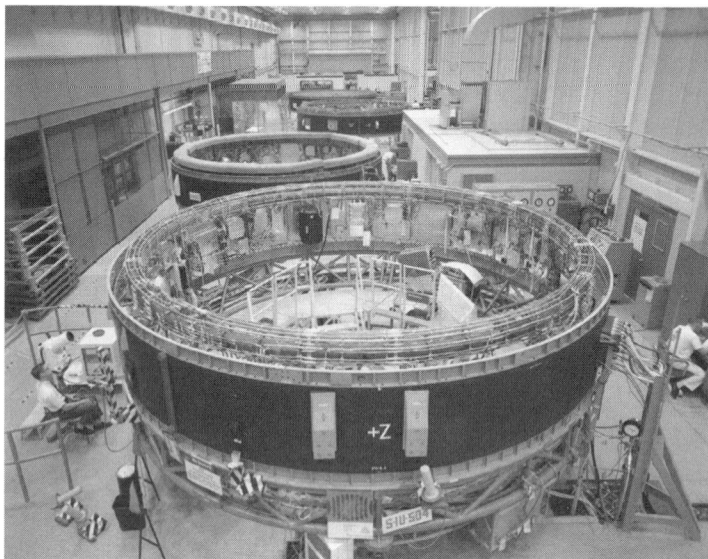

Figure 8.11   Saturn V instrument unit at IBM in Huntsville, Alabama [NASA]

the guidance platform – was temporarily put out of action by a lightning strike. It had been decided to launch through a rain squall and the Saturn V and its plume had acted as a lightning conductor. According to astronaut Charles Conrad, there were so many warning lights flashing that the crew couldn't count them. By design, the system's circuit breakers had tripped to protect the equipment and, although the incident could have aborted the mission, the crew soon revived the systems by resetting the breakers.

Of course, electronics subsystems also featured large in the launch vehicles designed to get men to the Moon, particularly in their guidance and telemetry systems. The Apollo Saturn V, for example, incorporated a guidance system with 80,000 components which could perform 9,600 operations per second (Figure 8.11). Unsurprisingly, this paled into insignificance compared with systems built just 10 years after the Saturn was retired. For example, the control electronics unit designed for the Olympus communications satellite in the early 1980s was capable of a quarter of a million instructions per second (0.25 MIPS), and by the turn of the century space-qualified computers were boasting up to 35 MIPS. But this simply reflected the general development of computer systems.

Today, the 'core rope memory' used in the Saturn V, in common with much of the technology of the day, looks extremely old-fashioned. It was constructed from hundreds of tiny doughnut shaped pieces of ferrite material (the magnetic cores) and a mass of fine signal wiring (the rope), which strung the 'microscopic washers' together. It operated by passing a current along a wire to magnetise a core ring, either in the north–south direction for a 'one' or south–north for a 'zero'. It had a storage capacity of just 460 kbit [16], a tiny fraction of the megabyte memory of a turn-of-the-century floppy disc. Despite modern-day comparisons, the Saturn V was 'state-of-the art' in the 1960s. Its guidance computer and associated data adapter weighed 125 kg, occupied a volume of 0.15 m^3 and used 438 W of power, all outstanding specifications in their day.

As discussed in Chapter 6 for unmanned lunar spacecraft, guidance and navigation was one of the key challenges for any lunar exploration programme. Although techniques are now sufficiently refined to aim spacecraft at distant planets and expect them to arrive on time and on target, to within fractions of seconds and fractions of kilometres respectively, this was not always the case. A full three years after Kennedy's rousing speech, America had still not managed to hit the visible face of the Moon with one of its unmanned spacecraft (Ranger 7 finally did so in July 1964).

In August 1961, more than three months before North American became CSM prime contractor, NASA had selected the Massachusetts Institute of Technology's Instrumentation Lab to design and develop the Apollo guidance, navigation and control (GNC) system, the key component of which was the Apollo Guidance Computer (AGC). At the time, America was still five months from placing its first astronaut in orbit which, contrary to simplistic presentations of American manned spaceflight as a serial endeavour, gives an indication of how the Mercury, Gemini and Apollo programmes were run in parallel.

Computers were in their infancy at the time of Apollo and the majority of processing for the mission was performed on the ground, by some of the most powerful computers in the world at NASA's Manned Space Flight Center in Houston. Nevertheless, all the Apollo spacecraft carried onboard computers. Both command modules and lunar modules had a primary guidance, navigation and control system computer (abbreviated PGNCS and pronounced 'pings'), which was used to calculate position and velocity (together known as the state vector), and to control spacecraft attitude, the operation of the propulsion systems and the path followed by the CM during re-entry.

The main reference for the guidance system was the inertial measurement unit, a gimbal-mounted, gyro-stabilised platform against which acceleration and rotation rates could be measured in all three axes. Although theoretically the inertial platform would remain stationary relative to the stars, it was a mechanical device and therefore prone to inaccuracy. The extent of the platform's drift was checked periodically by instructing the computer to point the CSM (and its fixed sextant) towards a given star or lunar feature, so that any errors could be read off and used to update the platform.

Using the platform as a reference, the computer was responsible, among other things, for controlling the precise duration and direction of main engine firings, particularly to inject the Apollo spacecraft into lunar orbit and onto a return path to the Earth. Some burns had to be accurately timed to avoid either crashing onto the Moon or heading off into space, as many of the early unmanned spacecraft had. Such was the degree of control necessary that the SPS engine was designed to be gimballed to allow the direction of thrust to be different from the pointing direction of the spacecraft. This was required because the CSM's centre of mass changed as the propellant in the service module was consumed. The SM's reaction control thrusters could also be controlled by the computer, for example to 'trim' an SPS burn or, during docking, to maintain a required attitude.

The flight computers themselves were housed in boxes measuring 61 cm × 32 cm × 15 cm and had a mass of about 32 kg [17]. Early designs used transistor-based circuitry, but certain disadvantages led the MIT team to consider integrated circuits (ICs), which had been developed in 1958 [18]. Since ICs did not become commercially available until 1961, they could offer little in the way of a reliability record or 'heritage' for Apollo, so one of MIT's first challenges was to convince NASA that they were worth the risk.

The ICs themselves were single logic gates, available in three classical variants: AND gates which turn 'on' if all inputs are 'on'; OR gates which turn 'on' if any input is 'on'; and NOR gates which turn 'on' if all inputs are 'off'. Although it would have been simpler to use a selection of different gates, MIT chose to make the circuits from three input NOR gates because the single-part solution improved reliability. Some 5000 gates were used in each Apollo computer and something like 60 per cent of the total US production of ICs was being used on Apollo prototypes by the summer of 1963 [17]. This is one of the reasons that the Apollo programme is credited with encouraging the development of the computer industry.

Data were entered using a relatively simple display and keyboard interface known as a DSKY ('disky'), whose keys had to be pressed some 10,000 times per mission. There were two DSKYs in the CM and another in the LM. Raytheon, the contractor which assembled the computers, increased its production staff from 800 to 2000 to build the 57 computers and 102 DSKYs for the Apollo programme (Figure 8.12) [19].

The final configuration of the machine – the Block II Apollo Guidance Computer – used a CPU with direct-coupled transistor logic (DCTL) with 36,864 words of read-only memory (ROM) and 2048 words of random-access memory (RAM); it had a 16-bit word-length. Like the Saturn V's, the ROM was a core rope memory; the RAM was a more conventional coincident-current core memory with an erasable ferrite core. The computers consumed just 55 W at 28 V and their clocks ran at 2.048 MHz [20]. This is, of course, extremely slow compared to today's computers, but compared quite well with the 4.77 MHz of the first IBM PC-XT [21].

The software – for a program called Luminary – was written in MAC (MIT Algebraic Compiler) language by some 300 people at MIT under principal designer Alan Klumpp. Luminary was stored on the read-only core rope system which, in a nod to much earlier technology, was assembled on a machine based on the Jacquard loom [21].

The computers that guided the Apollo spacecraft to the Moon are often used in a disparaging comparison with the pocket calculator to show how basic computers were in the 1960s. This is unfair to the Apollo computer's designers, however. As one of them, Eldon Hall, has written, the electronic components and digital technology of the early 1960s were 'archaic'. There were no single-chip microprocessors or integrated circuit memories, and single-chip ICs contained a single logic gate, not the many thousands we are used to today. The designers had to start from scratch, without the components required and often without the knowledge. If they had known what *was* required, writes Hall, they might have given up and declared the guidance

Figure 8.12   Apollo display and keyboard unit (DSKY) [NASA]

computer unfeasible [22]. This is an argument that must have been repeated many times, with respect to many subsystems, during the development of the Apollo spacecraft.

Everyone involved in the Apollo programme would have recognised the incredible challenge presented by President Kennedy, though few could have realised at the outset how difficult it would be to meet the deadlines, the performance specifications and the safety criteria. In fact, mistakes were made and lives were lost in the realisation of Kennedy's goal. Two examples, the Apollo 1 fire and the Apollo 13 explosion, show how critical even the smallest components and most trivial procedures were to the success of Apollo.

The Apollo 1 fire

In January 1967, the Apollo spacecraft had been in development for five years and NASA had just three years to meet Kennedy's target. Two launches of unmanned Apollos had been conducted successfully in 1966, but the first manned launch had been progressively delayed by technical problems with the Block I CSM that would be used for that mission [23].

Whereas the Mercury astronauts had developed a good working relationship with the spacecraft contractors, the Apollo crews had been frustrated in their dealings with industry. For example, the design of the spacecraft was being updated so often and so rapidly that the configuration of the simulator, that was supposed to replicate the CM exactly for training purposes, could not keep pace. And since the CM control panel contained some 24 instruments, 40 event indicators, 71 lights and 566 switches, it was important the astronauts knew where they were – especially in an emergency.

The Apollo 1 crew – Virgil ('Gus') Grissom, Edward White and Roger Chaffee – complained about the simulator at Kennedy Space Center on several occasions; indeed Grissom was apparently so disgusted with it that one morning he famously hung a lemon on it [24].

Nevertheless, the launch of Apollo 1 – also designated AS-204 because it was to use the fourth Saturn IB rocket – was set for 21 February 1967 with the astronauts on board. Before that, however, there would have to be a countdown dress rehearsal, a so-called 'plugs-out test', to make sure that the launch would go as planned. It was scheduled for 27 January 1967.

Tragically, the test itself was a disaster: a fire broke out in the cabin and the three astronauts died.

Following 10 weeks of investigation, involving some 1500 people, the Board of Inquiry determined that the most likely cause of the fire was electrical arcing from a bruised or broken wire which ran under a small trapdoor into the environmental control system beneath Grissom's couch. In fact, the spacecraft was so badly burnt that it was impossible to be categorical about it and nine other possible causes were documented [25]. All, however, involved electrical failures in the same area of the cabin (Figure 8.13).

In itself, a spark might not have proved fatal to the astronauts, but there were several aspects of the CM design that contributed to the incident:

1. The cabin atmosphere was 100 per cent oxygen at a pressure of 16.7 psi (2 psi higher than atmospheric pressure to purge contaminants from the cabin). This, and the fact that the cabin had been soaked in pure oxygen for 5 hours, allowed the fire to spread rapidly and with great intensity. Solder joints and fluid pipes melted and streams of oxygen and cooling fluid from the life support system fed the fire [26]. Ironically, in an attempt to make up for the delays, NASA had eliminated an unmanned launch-pad test in which the spacecraft was to have been pressurised with 100 per cent oxygen for the first time, instead combining this and other tests with the Apollo 1 countdown rehearsal.
2. The CM contained over 30 kg of flammable fabric including nylon couch covers, plastic foam insulation and over 5000 square inches (about $3\,\text{m}^2$) of Velcro. All were classified as hazardous in NASA reports and the astronauts themselves had expressed concern regarding the Velcro; the warnings were, however, ignored by the prime contractor. It is thought that the arcing from the wire shot sparks into nylon netting, which was used among other things to make the sleeping hammocks strung under the astronauts' couches.
3. The CM's hatch was in three parts: an inner hatch which opened inwards; an outer hatch that opened outwards; and an additional hatch in the boost cover. Under normal circumstances, it took 90 seconds to open the hatches, but the astronauts were dead within seconds. The official cause of death was 'asphyxia due to

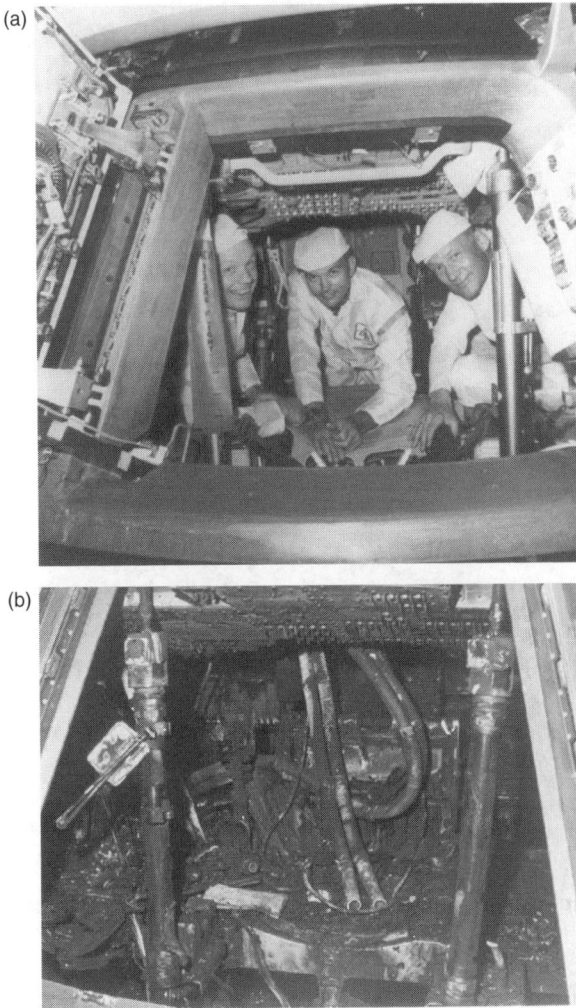

Figure 8.13   (a) Apollo 11 astronauts peer out of a command module hatch (left to right: Neil Armstrong, Michael Collins and Buzz Aldrin); (b) Apollo 1 after the fire [NASA]

inhalation of toxic gases due to fire', while 'burns' were cited as a contributory cause [26].

4. NASA's safety regime was inadequate in that 'plugs-out' tests were not considered 'hazardous' (NASA safety rules stated there had to be fuel in the rocket for it to be classed as such). Thus there were no safety personnel on the launch pad and no equipment to fight the type of fire which erupted (there were gas masks for fuel spillage, but no smoke masks). The ramifications of this were shown by the fact that heat and smoke prevented rescuers from opening the hatch for

Figure 8.14   Apollo 1 CM the day after the fire, once the boost cover had been partially removed [NASA]

five minutes following the outbreak of fire and all 27 men working on the gantry at the time had to be treated for smoke inhalation (Figure 8.14) [26].

The Accident Review Board produced a 3000-page report which, among other things, blamed NASA and North American for poor management, negligence, and poor design, installation and workmanship. As a result, top management at North American's space division was almost completely replaced, and NASA launched a $75 million programme to improve the Apollo spacecraft, detailing corrective actions, schedule changes and cost modifications [25,26].

The improvements made for Apollo 7, which became the first manned Apollo mission in October 1968, included the use of an oxygen–nitrogen atmosphere at normal atmospheric pressure on the ground (switching to pure oxygen at 5 psi in orbit); a redesign of the capsule's wiring; improved protection for fluid lines; extensive fireproofing of the spacecraft and spacesuits; and the design of a quick-release hatch which allowed escape in just three seconds (Figure 8.15) .

Some have argued that the pre-Apollo 1 command module was not safe to fly and that a disaster could easily have occurred when the astronauts were in space or on the Moon. Undoubtedly, a death in space would have had a far more devastating impact on the Apollo programme and may have led to its cancellation before the manned lunar landing.

Figure 8.15   Redesigned CM hatch in National Air and Space Museum, Washington DC [Mark Williamson]

Interestingly, despite his obvious concerns, 'Gus' Grissom is on record as saying that, if the crew died, he wanted people to accept it as worth the risk for the conquest of space. However, two years after the fire his widow filed a lawsuit against North American Rockwell and NASA. The company settled out of court in 1971 with a payment of $350,000 and, on its own initiative, made similar settlements with the wives of White and Chaffee [27].

The Apollo 13 explosion

Following the astounding successes of Apollo missions 7–11, each of which eclipsed the previous missions, the press and general public became bored with Apollo and by 11 April 1970, with Apollo 13 on the pad awaiting launch, audiences had dwindled. Although few would have admitted it, it seems likely that even the professionals had come to take success for granted.

Any complacency soon evaporated when, on 13 April, the number-2 LOX tank in Apollo 13's service module exploded. The events that followed – including the previously unplanned use of the lunar module as a 'lifeboat' – have been

(a)   (b)

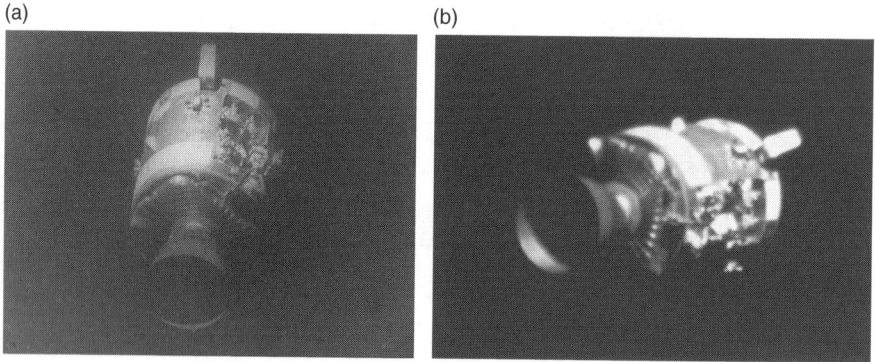

*Figure 8.16   Two views of the Apollo 13 service module after it had been separated
from the command module just prior to re-entry. The images are indis-
tinct, but the SPS engine is clearly visible, as is damage to SM sector 4
(the 'flap' is cable trunking which joined the CM and SM) [NASA]*

well-documented in print [28,29] and, in the mid-1990s, by a major cinema film
[30].[1] Among other things, the Apollo 13 accident made it abundantly clear that
transporting people to the Moon, and in the words of JFK 'returning them safely to
the Earth', was not as easy as the programme had made it appear.

Tracing the root cause of the explosion was a tortuous and complex process,
but the results show how relatively minor events or omissions can have disastrous
consequences.

The explosion occurred in sector 4 of the service module, which contained three
fuel cells in the forward compartment, two oxygen tanks in the middle, and two
hydrogen tanks in the rear (Figure 8.16). The LOX tanks, subcontracted by North
American to Beech Aircraft, were made of Inconel, a nickel–steel alloy, and were
strong enough to contain a pressure of 900 psi ($62 \, kg/cm^2$) [32]. The space between
the outer and inner tank shells was filled with insulation, some of it flammable.

The contents of the LOX and LH_2 tanks were compact and slushy at cryogenic
temperatures, but just 'warm' enough for sufficient oxygen and hydrogen to evaporate to feed the fuel cells and provide oxygen to the cabin. However, since the
pressure could potentially drop below what was required for this evaporation, the
tanks incorporated heaters and stirring systems.

The operating voltage of the CSM was fixed at 28 V early in the design process,
since this was what the fuel cells supplied, and Beech designed the heaters and
thermostats to work at this voltage. But it was later realised that while sitting on

---

[1] As is often the case with historical quotations, Commander Jim Lovell's report immediately after
the explosion is almost always misquoted as 'Houston, we have a problem'. In fact, transcripts from tape
recordings show that CM pilot Jack Swigert said 'I believe we've had a problem here'; and after the capsule
communicator replied 'This is Houston. Say again, please', Lovell said 'Houston, we've had a problem.
We've had a main B bus undervolt' [31].

the pad the SM would be running on external power at 65 V, so North American instructed Beech to uprate the components. For some reason, the thermostat switches were not changed and, despite reviews by NASA, North American and Beech, the error remained undetected. The result was that all Apollo spacecraft up to Apollo 13 flew with the wrong switches [32].

Unfortunately, Apollo 13's LOX tank 2 had a little too much 'heritage', having originally been installed in the Apollo 10 service module. During tests of that spacecraft, a problem had been detected and it was decided to remove the tank for investigation. However, a bolt securing the tank shelf to the structure was not removed and when a crane attempted to lift the unit out of the SM, the tank's drain line was damaged [32].

A year or so later, two weeks before the flight of Apollo 13, the LOX tanks in its service module – including the one removed from Apollo 10 – were filled with liquid oxygen as part of a countdown test. When, having completed the test, the ground crew found that they were unable to drain the oxygen from the tank, they activated the tank's heaters and fans in an attempt to force it out. When external power at 65 V was applied to the thermostatic switch, rated at 28 V, it overheated and fused shut [32]. The thermostat should have kept the temperature below 27°C, but the heaters remained on for 8 hours and the temperature rose, melted Teflon insulation on the fan motor cables and exposed the wires. Sadly, the associated thermometer was only calibrated to 29°C so no one realised what was happening [33].

Some 55 hours 53 minutes into the Apollo 13 mission, with the spacecraft 320,000 km from Earth, the oxygen in tank 2 had dropped to the point where the wires had become exposed; until that point, the LOX itself had prevented arcing. Following a suggestion from Mission Control that it was time to stir the LOX tanks, one of the crew did so . . . and the explosion occurred. It is thought that the oxygen heated by the arc from the exposed wires increased the pressure in the tank until the dome section on the top of the tank blew off. The Mylar insulation lining the SM's sector 4 then ignited and the gases produced blew the cover panel off into space [33].

It has been suggested that it was fortuitous that the panel blew off, since it acted as a safety valve, and that if the explosive force had been directed towards the forward bulkhead the command module could have been jettisoned into space [34]. There would then have been no hope for the astronauts. In fact, confirmation that the heat shield had not been damaged in the explosion was not received until the CM, having survived re-entry, was spotted hanging from its parachutes a few moments before splashdown.

It is also lucky[2] that the explosion occurred when Apollo 13 was on its way to the Moon rather than on the way back, since following the lunar landing the lunar module ascent stage would no longer have been available as a 'lifeboat', having already been crashed onto the Moon. Within minutes of the explosion the command module was

---

[2] Ever since the Apollo 13 incident, the superstitious have had a field day with the numbers associated with the mission, but even those who would never admit to superstitious tendencies have dined out on the fact that when Apollo 13 left the launch pad the clocks in Houston read 13:13 h. Interesting though this is, they rarely point out that, at the launch site itself, the clocks read 14:13, which is far less interesting.

without power (apart from the batteries needed for re-entry and splashdown), so the crew would have died without the LM.

On a more light-hearted note, this unplanned use of the LM led its prime contractor, Grumman Aerospace, to issue a spoof invoice to North American following the safe return of the astronauts. It was dated 17 April 1970 and requested a payment of $400,540.05. Items included a 400,000 mile 'tow' and 'fast battery charge' for $400,004 (based on four dollars for the first mile and a dollar per mile thereafter); 50 lbs of oxygen at $10/lb; 'prepaid' accommodation for two ('no TV, AC, with radio, modified American plan'); and 'extra guest $8.00/night' charged at $32.00 [35]. It is not recorded whether North American paid the invoice.

Once again, an accident review board criticised NASA and North American Rockwell (and this time also Beech Aircraft Corp) for a 'serious oversight' in the design and testing of the LOX tank. As a result, later spacecraft carried an additional oxygen tank for cabin air and a back-up battery in the SM for power. All fans and thermostatic switches were removed from the LOX tanks. Once again, lessons were learned and the remaining Apollo missions proceeded without incident.

Conclusions

The engineering development of the Apollo spacecraft – and the era of lunar exploration it facilitated – was one of the most notable achievements of the twentieth century. Given the state of technology in 1961, and America's minimal space experience, it still seems incredible today that the goal of placing a man on the Moon could have been realised in little over eight years and it has often been suggested since the end of the Apollo programme that a similar venture, if attempted today, could not achieve the same result in the same timescale. This is probably true because no one can envisage reproducing the level of political and financial support for a manned lunar mission that existed in the 1960s. Thus Apollo itself can be celebrated for its uniqueness.

By any account, the cost of the Apollo programme was high: according to a NASA fact sheet on manned spaceflight dated May 1972, the costs up to the time of the Apollo 11 landing were US$21 billion. The same document estimated the total programme cost – to completion of the Apollo 17 mission – at US$25 billion, the figure that is most often quoted today.

Whether Apollo was worth the costs – which included the lives of the Apollo 1 astronauts – is still open to debate and continues to have repercussions for the future development of manned spaceflight. Attempts to balance the cost against the weight of Moon rock returned completely miss the point, however, as do attempts to demean Apollo as little more than a jobs programme.

From a social perspective, the jobs it created were important. At its peak in 1966, more than 20,000 companies and 350,000 people throughout the United States (including nearly 50,000 at prime contractor North American Rockwell) participated directly in the Apollo programme [36]. The experience they gained helped fuel the US space programmes of the 1970s and 1980s.

In addition to refining our scientific understanding of the Moon, the Apollo programme made significant contributions to the development of electronics, computer systems, materials and many other engineering disciplines. It also inspired a generation of children to study science, engineering and mathematics, and influenced our culture by enhancing awareness of global issues. Indeed, the image of the beautiful, fragile and insignificant Earth, rising above the surface of a barren Moon, is credited with focusing the vision of the environmental movement.

From an historical perspective, there can be no doubt that the programme was also significant in terms of exploration: it created the means to place men on the surface of another planetary body, thereby transforming mankind from a single-planet species to an explorer of alien worlds.

References

1 Kennedy, John F.: 'Urgent National Needs', US President's speech to Joint Session of Congress, 25 May 1961

2 NASA Fact Sheet, Marshall Space Flight Center: Form 2914-24 (Rev. March 1980) /6F1080, p. 1

3 Dethloff, Henry C.: 'Suddenly Tomorrow Came ... A History of the Johnson Space Center' (NASA SP-4307, Lyndon B Johnson Space Center, Houston, TX, 1993) p. 64

4 Burrows, William E.: 'This New Ocean: The Story of the First Space Age' (Random House, New York, 1998) p. 334

5 Dethloff, Henry C.: 'Suddenly Tomorrow Came ... A History of the Johnson Space Center' (NASA SP-4307, Lyndon B Johnson Space Center, Houston, TX, 1993) p. 65

6 NASA Information Summaries: 'The Early Years: Mercury to Apollo-Soyuz', PMS 001-A (KSC), May 1987, p. 4

7 'Manned Spaceflight', NASA press document, May 1972, pp. 6–7

8 NASA Information Summaries: 'The Early Years: Mercury to Apollo-Soyuz', PMS 001-A (KSC), May 1987, p. 9

9 Allday, Jonathan: 'Apollo in Perspective' (IOP Publishing Ltd, Bristol, 2000) pp. 159–60

10 Gray, Mike: 'Angle of Attack: Harrison Storms and the Race to the Moon', W W Norton & Co, New York, 1992) pp. 212–3

11 ibid., p. 213

12 ibid., pp. 171–2

13 Allday, Jonathan: 'Apollo in Perspective' (IOP Publishing Ltd, Bristol, 2000) p. 161

14 Simpson, Len L.: 'Electrical Systems' in Blashfield, Jean, F. (Ed.): 'Above & Beyond: The Encyclopedia of Aviation and Space Sciences' (New Horizons Publishers Inc, Chicago, IL, 1968) pp. 681–3

15 Allday, Jonathan: 'Apollo in Perspective' (IOP Publishing Ltd, Bristol, 2000) p. 162

16 Bednarski, Gene: 'Computers' in Blashfield, Jean, F. (Ed.): 'Above & Beyond: The Encyclopedia of Aviation and Space Sciences' (New Horizons Publishers Inc, Chicago, IL, 1968), p. 558

17 Allday, Jonathan: 'Apollo in Perspective' (IOP Publishing Ltd, Bristol, 2000) p. 177

18 Hall, Eldon C.: 'Journey to the Moon: The History of the Apollo Guidance Computer' (AIAA, Reston, 1996) p. 80

19 Allday, Jonathan: 'Apollo in Perspective' (IOP Publishing Ltd, Bristol, 2000) p. 181

20 Hall, Eldon C.: 'Journey to the Moon: The History of the Apollo Guidance Computer' (AIAA, Reston, 1996) pp. 120–2

21 Fawkes, Steven: 'Apollo Lunar Module (LM)', *Spaceflight* **41** (3), March 1999, p. 128

22 Hall, Eldon C.: 'Journey to the Moon: The History of the Apollo Guidance Computer' (AIAA, Reston, 1996) p. 2

23 Gainor, Chris: 'Arrows to the Moon: Avro's Engineers and the Space Race' (Apogee Books, Burlington, 2001) p. 148

24 Shepard, Alan, and Slayton, Deke: 'Moon Shot: The Inside Story of America's Race to the Moon' (Virgin Publishing Ltd, London, 1995) p. 221

25 Benedict, Howard: 'NASA: The Journey Continues' (Pioneer Publications Inc, Houston, TX, 1989) p. 29

26 ibid., p. 28

27 Turnill, Reginald (Ed.): 'Jane's Spaceflight Directory, 1987' (Jane's Publishing Co Ltd, London, 1987) p. 40

28 Cooper, Henry S. F.: 'Moonwreck' (Granada Publishing Ltd/Panther Books, St Albans, 1975), and previously as '13: The Flight That Failed' (Angus and Robertson, 1973)

29 Lovell, Jim, and Kluger, Jeffrey: 'Lost Moon: The perilous Voyage of Apollo 13' (Houghton Mifflin Co, Boston, MA, 1994)

30 Kluger, Jeffrey: 'The Apollo Adventure: The Making of the Apollo Space Program and the Movie *Apollo 13*' (Boxtree Ltd, London, 1995)

31 Swanson, Glen E.: ' "Houston, We Have A Problem" – A History of Air-to-Ground Voice Transmissions From The US Manned Space Program' (IAC-02-IAA.2.4.08), 53rd International Astronautical Congress, Houston, TX, 10–19 October 2002

32 Allday, Jonathan: 'Apollo in Perspective' (IOP Publishing Ltd, Bristol, 2000) pp. 166–9

33 Dewaard, E. John, and Dewaard, Nancy: 'History of NASA: America's Voyage to the Stars' (Bison Books, Greenwich, 1984) pp. 87–90

34 Godwin, Robert (Ed.): 'Apollo 13: The NASA Mission Reports' (Apogee Books, Burlington, 2000) pp. 3–4

35 Allday, Jonathan: 'Apollo in Perspective' (IOP Publishing Ltd, Bristol, 2000) p. 169

36 Benedict, Howard: 'NASA: The Journey Continues' (Pioneer Publications Inc, Houston, TX, 1989) p. 153

Chapter 9

Lunar lander – the development of the Apollo Lunar Module

That's one small step for man, one giant leap for mankind

Neil Armstrong[1]

If the Apollo programme had been designed simply to transport three astronauts to lunar orbit, allow them to conduct remote observations of its surface and return them to Earth, it would have been an historic feat of technological development in the 1960s. But President Kennedy had set an even more impressive goal, which involved a landing, and this was the function of the lunar module.

The Apollo 11 lunar module, the first manned spacecraft to land on the Moon, made its touchdown in the Sea of Tranquillity on 20 July 1969. Although several decades have passed since that historic event, and many subsequent developments in space technology have taken place, the effort required to place men on the surface of the Moon and return them safely to the Earth set a precedent in engineering excellence that is hard to match even today.

Design overview

As with all other Apollo hardware, NASA put the lunar module out to contract across the United States. Nine firms submitted bids for the contract, which was worth at least $350 million and, on 7 November 1962, Grumman Aircraft Engineering Corp of Bethpage, NY, was selected to design and build 18 engineering models and flight models of the spacecraft.

The lunar module (LM) was initially known, in full, as the lunar excursion module (LEM), but the word 'excursion' was dropped after NASA's associate administrator

---

[1] Although Armstrong has insisted he said 'one small step for a man', most people who listen to the recording cannot discern the 'a'. Unfortunately, the sentence lacks the intended meaning without it.

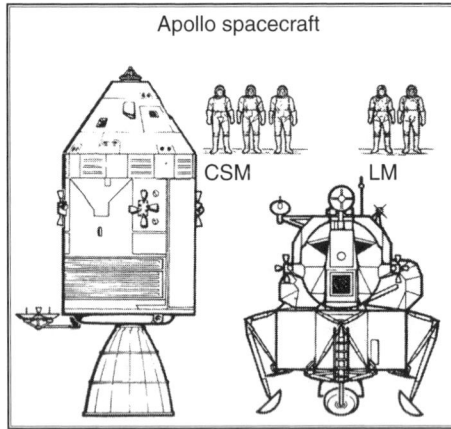

Figure 9.1   LM and CSM compared [NASA]

for manned spaceflight, George Mueller, protested that 'excursion' sounded a bit frivolous [1]. The pronunciation 'lem' was, however, retained in recognition of the spacecraft's previous acronym.

From the very beginning, the LM's design team, under Grumman's director of space projects Joseph Gavin, had some catching up to do. Whereas other parts of the Apollo programme had begun shortly after Kennedy made his famous proclamation in May 1961, contracts for the LM could not be let because NASA had not decided whether the lunar spacecraft should take a direct trajectory to the lunar surface or make the journey in several reversible stages.

Only in July 1962, when NASA announced its decision to use the orbital rendezvous technique, could LM development proceed: only then was it clear that the LM would be transported to the Moon by a separate spacecraft – the command and service module (CSM) – and would not be required to return the astronauts to Earth (Figure 9.1). Unfortunately, at the time, no one knew categorically whether the lunar surface could support the mass of a spacecraft.

It is difficult to appreciate the level of ignorance of the lunar environment at the time the lunar landing was being contemplated. Prominent astrophysicists of the mid-1960s were sincere in their views that the lunar surface was covered in a thick dust layer [2]: for example, in 1965, Tommy Gold warned Gavin to expect ten metres of dust [3]. So, at best, the dust would be blown into billowing clouds by the blast of the LM's descent engine, while at worst it would swallow the entire lunar spacecraft and its crew, never to be seen again. Luckily the detractors were misinformed and, as experience showed, only one of the six landings produced enough dust to obscure the surface in the locality of the landing area.

In fact, the first US spacecraft to soft land on the Moon, Surveyor 1, did so in June 1966, just three years before Apollo 11, while its successor, Surveyor 6, performed the first lunar lift-off in November 1967 (see Chapter 6). Although these and other missions supported the presumption that the surface was firm enough for a manned landing, their confirmation came too late to influence the design process which was, by necessity, well advanced.

Despite the unknowns, several aspects of the challenge were immediately clear to Grumman's engineers and managers: the LM design would have to adhere to the severe mass constraints imposed on it by the mission and its Saturn V launch vehicle, and meet the requirements of a unique operational environment. Indeed, the LM would have to operate in lunar orbit, where it would be 'weightless', in the one-sixth gravity environment of the lunar surface, and under dynamically varying thrust profiles between the two. These operational considerations required designers to adopt a new mindset, since all previous manned spacecraft had been engineered to survive a launch from Earth followed by microgravity operations and an atmospheric re-entry. The LM, by contrast, would be the first manned spacecraft to operate entirely 'in outer space'.

Although historians often like to bestow the honour of an idea or design on individuals, complex hardware developments rarely allow it. Unlike the design of the blunt-bodied Mercury capsule, which led to those of Gemini and Apollo and is generally attributed to Maxime Faget, the design of the lunar module cannot be traced to a single individual. Faget himself has been quoted as saying that it simply 'evolved'. Nevertheless, according to Apollo 11's command module pilot, Michael Collins, 'Caldwell Johnson and Owen Maynard of NASA and Tom Kelly of Grumman were probably the closest to being [the LM's] parents' [4]. Whatever their individual contributions, the design, development, manufacturing and test regime coordinated by Grumman was the key to producing the world's first manned lunar landing vehicle, and it is this we shall consider here.

Structure and mass constraints

As an entirely new vehicle with an unquestionably unique mission, the nub of the lunar module development programme was the design of its structure. What no one could know at the outset was the degree to which this structure would evolve throughout the programme.

The LM was built in two main sections, a larger 'descent stage' carrying a smaller 'ascent stage' (Figures 9.2 and 9.3). The descent stage, effectively a rocket engine on legs, was designed to execute a landing on the lunar surface and act as a launch platform for the ascent stage, which itself was designed to house the astronauts and return them to the command and service module in lunar orbit at the conclusion of their mission.

In its landing configuration, with its legs unfolded, the lunar module was 7 m high and measured 9.5 m across the landing gear. It was made predominantly of aluminium alloy. The descent stage was a cruciform box structure, housing the descent engine in the centre surrounded by four propellant tanks. The outer panels transformed the structure into an octagon, which provided space for pressurant tanks and other systems. The landing gear, attached to the ends of the cruciform shape, were folded to fit the spacecraft into the launch vehicle shroud and later released by pyrotechnic bolt-cutters and spring mechanisms (Figures 9.4 and 9.5).

Figure 9.2   LM-2 in National Air and Space Museum, Washington DC [Mark Williamson]

The ascent stage consisted of a stiffened cylinder, 2.35 m in diameter and 1.1 m deep, with two triangular windows and an egress hatch in the forward bulkhead (Figure 9.6). This formed the 'flight deck' of the lunar module. Directly behind this cylinder was a smaller compartment containing the environmental control system, storage space and the docking hatch. The ascent module's cramped interior had a habitable volume of only 4.5 cubic metres (less than three times the load space of the average family hatchback car). The outer skin of the module, which was about 7 cm away from the inner pressure shell, was covered almost entirely with multi-layer thermal insulation and micrometeoroid shields. In fact it was a similar covering of thermal blanket on the decent stage that gave the LM one of its many nicknames – the flying chocolate box.

The most often quoted constraint on the design of a spacecraft is its mass. But never was the mass of a manned spacecraft pared to such an extent as it was for the lunar module. Most spacecraft suffer a gradual increase in mass throughout their development which typically results in a number of swings over the amount allowed in the mass budget and subsequent efforts to bring the budget back in line. The graph (Figure 9.7) shows the variation in the overall mass of the lunar module throughout its seven-year development history. The initial proposal for the LM, which specified a mass of 22,000 lb (about 10 tonnes), was based on the lifting capability of the Saturn V launch vehicle, as predicted in the early 1960s. True to form, the LM's mass increased to nearly 13.5 tonnes within six months of the start of the design programme [5].

This preoccupation with the LM mass budget was understandable considering the effect it had on the overall Apollo–Saturn system. Each kilogram of inert mass

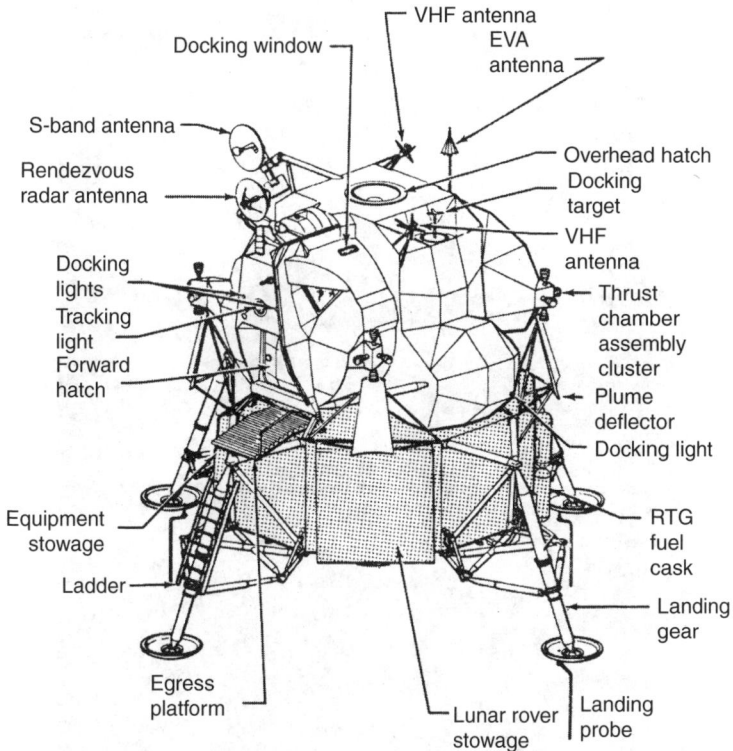

Figure 9.3   LM configuration [NASA]

lowered to the lunar surface and returned to lunar orbit required an additional 3.25 kg of propellant in the LM's tanks, so each extra kilogram of LM structure added 4.25 kg to the payload of the Saturn V. The resulting knock-on effect on the Saturn's propellant budget added about 50 kg for each kilogram landed on the Moon [6]. It was obvious that mass reduction was crucial.

It was realised early on in the design process that the LM would have to mimic the capabilities of a helicopter, in providing the pilot with a good view of the landing site and the ability to hover above it long enough to choose a safe touchdown point. Thus early designs featured four curved windows, giving the LM a 'bug-like' appearance and engendering its 'lunar bug' nickname. Among the earliest mass-saving measures was the reduction of the size and number of the windows, resulting in the two tri-angular windows of the final design. To regain some of the lost visibility, the panes were canted inwards to be closer to the crew's eyes [3].

The design team also reduced the original five legs of the so-called 'spider' to four, removed a second docking hatch on the front of the vehicle (replacing it with the egress hatch above the ladder). The astronauts provided a degree of design input for the LM, as with the CSM, one example of which was changing the round hatch

*Figure 9.4   LM under construction at Grumman plant [Northrop Grumman Corp.,
formerly Grumman Aircraft Engineering Corp.]*

into a square one, through which it was easier to manoeuvre their square-shaped back
packs. The team also eliminated the seats for the crew. Although 'bar stools' were
considered, it was decided that seats would be unnecessary in microgravity conditions
and that restraining straps would suffice [3]. And since there would be little time for
sitting down when the LM was on the Moon, their removal was a fairly obvious
measure.

Less obvious were the lengths to which the structure of the spacecraft could be
minimised. Chemical milling of aluminium alloy shear webs, or stiffeners, was used
extensively to reduce the thickness of the cabin walls and other panels to 0.012 in.
(0.3 mm), and in some places to 0.010 in. (0.25 mm), so thin that one could pierce
them with a screwdriver.[2] According to Joseph Gavin [5], the team also examined

---

[2] Though the relative strengths are quite different, this is a little over the thickness of two sheets of
paper.

(a)

Quadrant I
Erectable antenna
EPS batteries
PLSS spare
battery
stowage

Engine
mount

Aft
interstage
fitting

Fuel tank
Quadrant II
Landing radar
electronics
Scientific
equipment
package
Water tank

SLA adapter
attachment
point (4 ea.)

Oxidiser
tank
Ambient
helium tank

Descent
engine

Structural
skin

Insulation

Thermal and
micrometeoroid
shield

FWD (+z)

Forward
interstage
fitting

Oxidiser tank
Quadrant IV
Cable cutter
electrical power
and explosive devices
batteries

Fuel tank

Oxygen tank

Quadrant III
Descent engine
electronics

Supercritical
helium tank

Note:
Landing gear shown
in retracted position

Descent
engine
skirt

(b)

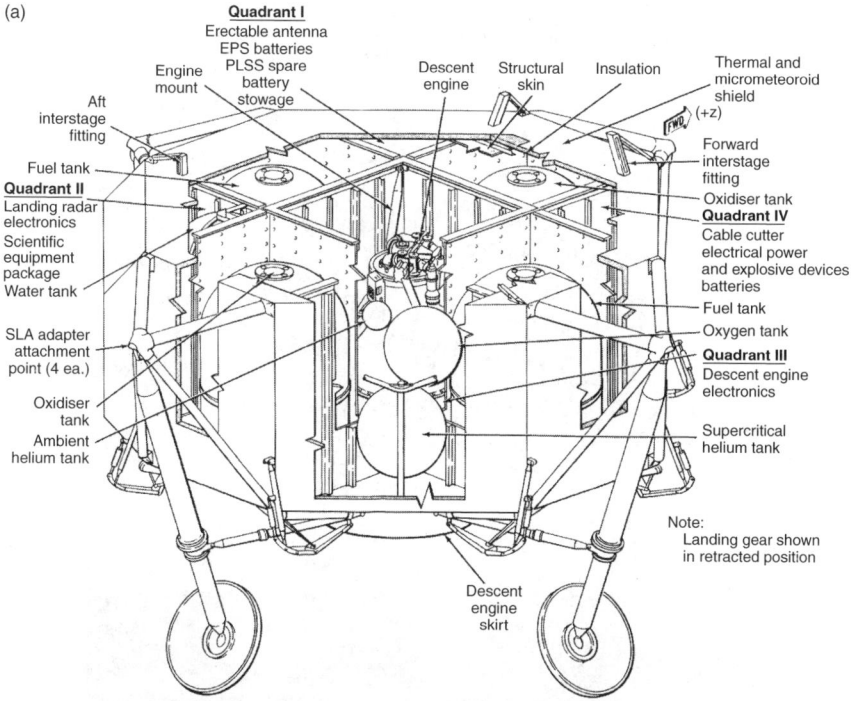

Figure 9.5   LM descent stage [NASA and Northrop Grumman Corp., formerly Grumman Aircraft Engineering Corp.]

(a)

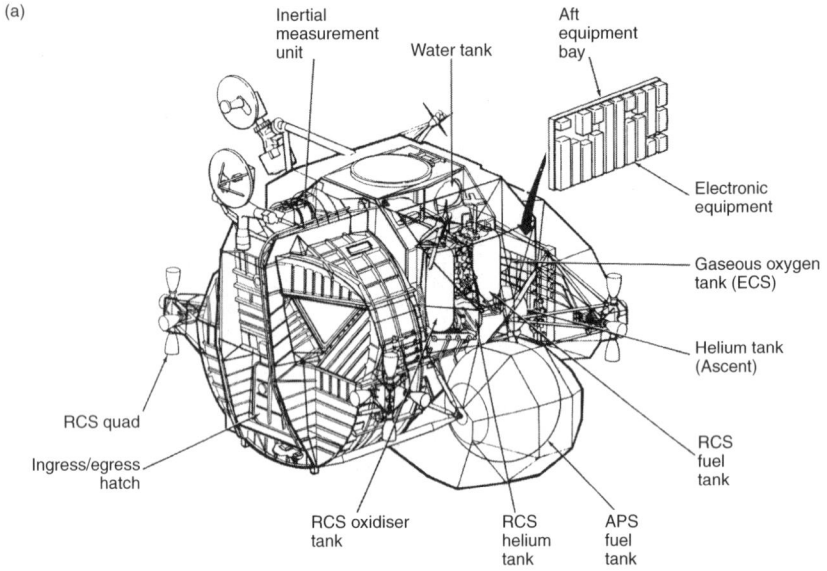

(b)

Figure 9.6   LM ascent stage [NASA and Northrop Grumman Corp., formerly Grumman Aircraft Engineering Corp.]

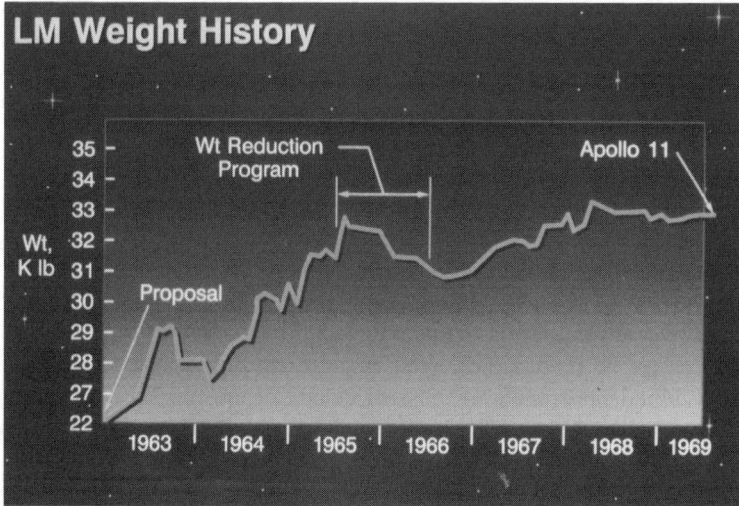

Figure 9.7   Variation of lunar module's weight throughout its development [Northrop Grumman Corp., formerly Grumman Aircraft Engineering Corp.]

the possibility of reducing the thickness to 0.008 in., but ruled against it after counting the number of grains in a cross-section of the material. As it was, the thinness of the material led to overstressing and stress-corrosion, such that fittings on one LM had to be replaced at the launch site.

Propellant lines were also chemically milled to reduce their thickness and the smallest possible gauge of electrical wiring (26-gauge) was used where practicable. There was a penalty to pay in re-work resulting from handling damage, but this was considered worthwhile in view of the mass saving [5]. Then, late in the design process, it became necessary to minimise the mass per unit area of the lunar module's outermost coatings, the thermal and micrometeoroid shields. The large area of the shielding and the desire to retain the simplicity of a mainly passive thermal subsystem made this a difficult task, requiring much analysis and vacuum chamber testing [7]. An example of designing specifically for the 1/6-gravity lunar environment to save mass is provided by the LM's ladder, the structure of which was reduced below what would normally be considered 'fit for purpose' in the 1 g environment of Earth.

Despite these measures, the mass of the lunar module rose inexorably until mid-1965 when a major mass-reduction programme was initiated. This checked the rate of increase, but the mass was to rise again above the 1965 peak, to some 14.6 tonnes, by the time the Apollo 11 LM-5 made the first lunar landing, partly because of the addition of considerable amounts of fire-proofing material in the aftermath of the Apollo 1 command module fire in 1967. It is important to note, however, that only about 4.3 tonnes of the 14.6 tonnes was due to the LM itself [8]; the majority of the mass budget was allocated to propellant, while the remainder was for the crew and their equipment. In other words, the LM's dry weight was only about four times that of a family car.

Weight reduction became a way of life at Grumman, eventually becoming a part of the day-to-day language at the plant, to the extent that the introduction of the 'Super Weight Improvement Program' produced the verb 'to swip', where 'swipping' meant scrutinising every rib, panel, valve and switch in the lunar module to determine where extra weight savings could be made. Grumman was reportedly willing to spend tens of thousands of dollars to save one pound of weight! [9]

This policy had potentially serious repercussions for an arguably crucial part of the lunar module's payload – the television camera. Perhaps more than any other item of space hardware, the TV camera on Apollo 11 sold the space programme to the American public (Figure 9.8). What is most amazing to us now, in this age of palm-sized digital camcorders, is the poor quality of those first, historic images of a man from Earth descending the ladder to another world ... and the fact that they were shot in black-and-white! What is even more amazing, however, is that twice during weight reduction programmes the camera, which weighed just 3.4 kg, was deleted from the payload. Somehow, it managed to find its way back in again each time [9].

Despite the quality of the pictures, the camera was an engineering achievement in its own right. Due to limitations on its size and weight, and the low bandwidth of the available data channel, it was constrained to produce only 325 scan-lines, which reduced its resolution to half of the present-day norm. Moreover, it could operate at only 10 frames per second, compared with the US NTSC standard of 30 (or 25 for PAL in Europe), which reduced its ability to capture movement. Storage circuits on the Earth repeated frames to give continuous pictures but, when the astronauts moved, their images blurred into ghostly apparitions – manifestations of contemporary technology.

Some of the lunar module's structural design solutions seem crude in their simplicity. The shock of touchdown, for example, was absorbed by simple tubes of crushable aluminium-alloy honeycomb in the landing gear (Figure 9.9). In hindsight, this was significantly over-designed in providing a 0.81 m stroke to withstand a maximum vertical sinking speed of about 3 m/s, since operational experience recorded a maximum deflection of 7.6 cm corresponding to a touchdown speed of only about 0.6 m/s. As a consequence, once the LM was on the Moon, the bottom of the ladder was much further from the surface than expected, obliging the astronauts to jump from its lower rung to the LM's footpad – a rather larger 'small step' than Neil Armstrong had anticipated.

It was, however, wise to err on the side of caution, considering the unknown and alien nature of the landing surface. Indeed, a range of surface friction from zero to infinity was assumed in the modelling process, from a theoretical, frictionless surface to one where a foot-pad became jammed in a crater. This and a range of horizontal and vertical landing speeds were combined with possible vehicle attitudes of $\pm 6°$, attitude rates of $\pm 2°/s$ and a $\pm 6°$ surface slope. Some 500 computer runs later – infinitely more impressive then than it sounds today – the design team was reasonably confident that the lunar module would not fall over. A quarter-scale

(a)

(b)

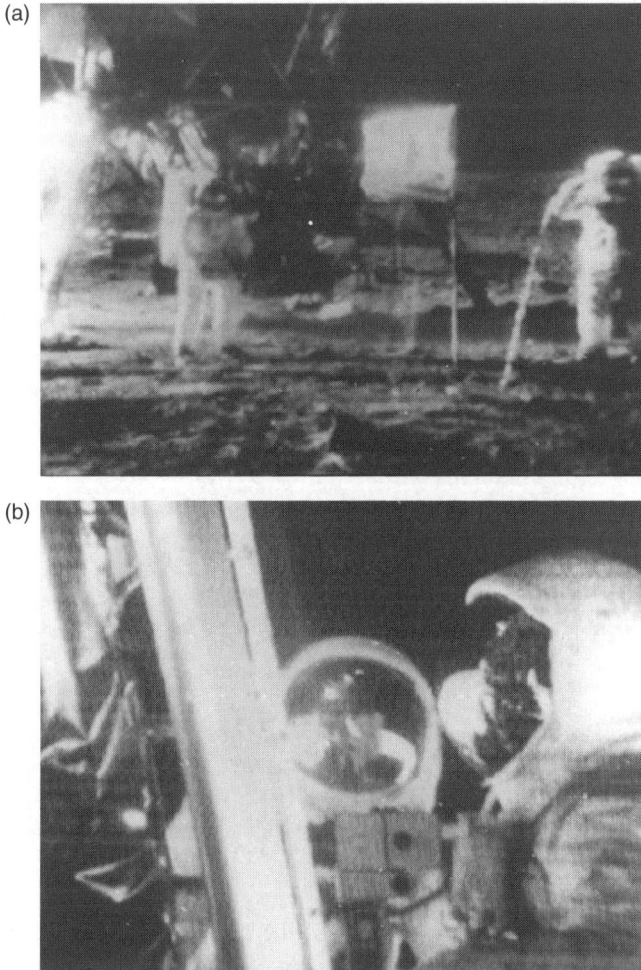

Figure 9.8   Televised pictures from Apollo 11 [NASA]

dynamic model was built to validate the computer model and the final vehicle was designed to be stable up to a 45° tilt. A maximum inclination of 30° was allowed to ensure a successful launch of the ascent stage, but the maximum experienced was only 8.6° [2].

Two drop-tests of the flight models were performed to check the basic structure and the integrity of the electrical system (Figure 9.10). These were conducted to give extra assurance that the limp and flexible lunar module, designed as it was for 1/6-g, would survive the shocks of a landing [10]. The first LM to reach space was a 'boiler-plate' instrumented test article, launched on the Apollo 4 mission of 9 November 1967 on the first Saturn V flight; it carried instruments to record and analyse vibrations,

Figure 9.9   LM-2 landing gear [Mark Williamson]

and demonstrate the structural integrity of the LM. Indeed, the continual weight reduction programmes had made the LM an exceedingly flimsy vehicle, with the ride quality of a vintage bone-shaker. According to astronaut Tom Stafford, commander of Apollo 10, whenever one of the reaction control thrusters fired 'it sounded as if somebody had put a washtub over your head and was banging on it with drumsticks'.

Propulsion subsystem

Although the LM may not have been the most comfortable of spacecraft, the technology incorporated within its spindly frame was nothing less than state-of-the-art. For example, the LM's descent engine, developed by Rocketdyne and Space Technology Laboratories, was the first gimballed, throttleable rocket engine to be used on a spacecraft [11]. It was throttled by injecting helium into the propellant flow to alter the thrust level without altering propellant flow rates. Although this complicated

Figure 9.10   LM drop test [Northrop Grumman Corp., formerly Grumman Aircraft Engineering Corp.]

the design of the engine, it was necessary to provide the manoeuvrability required for a safe landing, in particular the ability to hover.

A concern regarding the descent engine was the ultimate proximity of the nozzle to the lunar surface, but when early missions showed the compression of the landing gear to be minimal the problem resolved itself. In fact the nozzles for the last three missions were lengthened to provide greater engine performance after ground tests showed that the nozzle extension, made of columbium, would fold back on itself as it approached the surface.

Since the ascent engine, produced by Bell Aerosystems, would provide the astronauts' only lifeline to the orbiting command and service module, it was made as simple and foolproof as possible and was neither steerable nor throttleable. Although both the descent engine and the ascent engine were potential single point failures, redundant valves and pipe runs were included to reduce the probability of a system failure. In addition, all engines were critically examined for combustion stability as part of a 'man-rating' programme. This almost doubled the time required to develop the engines, but was considered vital due to the restrictions on redundancy and the human cost of a system failure. Extreme cleanliness of the propulsion systems was crucial, since contaminant particles in the propellant lines could mean the end of a mission. The only way found to clean the lines sufficiently to avoid blocking of the solenoid valves was to vibrate them during flushing [12].

All three lunar module propulsion systems – descent engine, ascent engine and reaction control thrusters – used pressure-fed, storable propellants (Aerozine-50 and nitrogen tetroxide). Since these propellants ignite on contact, there was no need for an ignition system, and since the engines were pressure-fed by helium there was no need for pumps – factors which further reduced mass and complexity, enhancing the reliability of the spacecraft.

It was originally intended to use four propellant tanks in the ascent stage – two for the fuel and two for the oxidiser – which would have allowed designers to balance the stage by placing one on either side, taking into account the fact that the fuel was lighter than the oxidiser. Mass balancing was especially important since the engine was not steerable. However, Grumman became worried about the reliability of the design and its extra pipework, and changed it to a two tank system. This meant that, in order to align the stage's centre of mass with the thrust axis of the engine, the fuel tank had to be set further from the centre-line than the oxidiser tank, giving the ascent stage its asymmetric appearance (see Figure 9.1). Although the redesign saved about 45 kg, it cost $2 million [13].

The propellant tanks for the descent stage were designed, from fracture mechanics considerations, for a lifetime not exceeding 100 pressure cycles and were fabricated in titanium alloy to reduce their mass to a minimum. However, some of the weight-trimming design solutions led to leaks, which were discovered inconveniently late in the programme. They were particularly worrying since the oxidiser alone was extremely corrosive and the fuel and oxidiser together were hypergolic. It turned out that the flanges holding the propellant lines to the tanks were insufficiently rigid to hold the O-ring seals, which would otherwise have prevented the leaks; the solutions were to double the seals and reinforce the flanges [12].

Despite the problems, the lightweight tank design provided an unexpected bonus, which surprised even the designers. The tank walls were so thin that, at first filling, it stretched enough to hold an extra 20 seconds-worth of hovering fuel. This was a particularly useful spin-off, since the hovering time had formerly been limited to between 140 and 150 s [12]. Neil Armstrong and Buzz Aldrin would be glad of the extra margin when making the first manned close approach to the lunar surface: because of the difficulty in finding a smooth touchdown point, Armstrong was forced to manoeuvre for longer than expected and it has since been estimated that only about 20 s worth of fuel remained when the Eagle finally landed.

The first flight of a fully operational LM (LM-1) occurred on 22 January 1968 on the Apollo 5 mission; it was launched by a Saturn IB, since the test did not include a CSM. The flight was used to prove the propulsion system by test-firing the lunar module's engines in space for the first time: it comprised three firings of the descent engine and two of the ascent engine, including a so-called 'fire in the hole' which simulated the emergency separation of the two modules in orbit. In fact the flight of LM-1, the first flight model, was so successful that a follow-up was cancelled and LM-2 was donated to the Smithsonian Air and Space Museum in Washington, where it is displayed today (see Figure 9.2) [14].

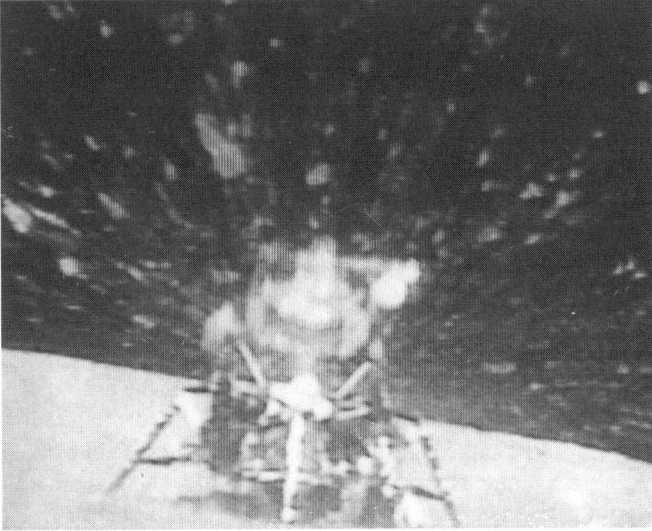

Figure 9.11   Ascent stage lift-off from TV camera mounted on Apollo 16 Lunar Roving Vehicle (LRV) [NASA]

One aspect of the lunar module propulsion system that could not be fully tested on Earth was the operation of the ascent engine *in vacuo* and under 1/6-g, as it separated from the descent stage (although the engine was tested in Earth orbit on the Apollo 5 mission). Brief ascent engine bursts were made in proximity to a descent stage, at the NASA test facility at White Sands, New Mexico, but the true operation of the system could not be witnessed until the TV camera on the lunar rover recorded the lift-off of the Apollo 15 ascent stage. It was, of course, 'picture perfect' (Figure 9.11).

Power subsystem

It was originally suggested that fuel cells should be used as the power source for the lunar module, as with the Apollo CSM, but this was changed at an early stage to batteries, which were less expensive. They were also considered the more conservative option, since batteries with considerable space heritage could be chosen: those used for the LM were 28 V (5.5 Ah/kg) silver–zinc cells with potassium hydroxide electrolyte [11].

Although experience with the Gemini spacecraft had shown that fuel cells would be more mass-efficient than batteries for a relatively long duration mission – a concept that was directly transferable to the CSM – this did not apply to the LM because its mission was short. However, in the continual drive to minimise the mass of the LM, the individual plastic battery cells were enclosed in magnesium casings, designed for just one deep discharge and flight qualified using a sampling procedure. Unfortunately, it was later discovered that the case was abrading the cells as the size of the electrodes altered during discharge. This problem was solved by substituting a lubricant for the potting compound in which the cells were originally seated [11].

The batteries and other heat-producing items were mounted on cold rails in the LM's aft equipment bay which distributed the thermal energy for subsequent radiation into space (see Figure 9.6). A problem in the glycol cooling system, which served both the cold rails and the crew cabin's environmental system, came to light just 90 days before the launch of Apollo 11 when a routine sampling of the fluid revealed the growth of needle-shaped crystals about 40 μm long. Extensive filtering and sampling outside the LM failed to produce a clear liquid; in fact, more filtering seemed to produce more crystals. The system was given a coolant transfusion, but the new coolant developed its own crystals [15].

When none of the bench-analyses produced solutions, the team decided to run a mission simulation using the worst crystal mix available: according to Joseph Gavin, instead of being clear, 'it looked like orange juice'. But the system worked to specification and was flown, complete with crystals, on the first lunar landing. Later on, the crystal mystery was solved: it turned out that an inexpensive anti-corrosion additive had been replaced by a much purer batch of the same chemical, which was ten times as expensive. So, for Apollo 12 and subsequent flights, Gavin's team reverted to the earlier mix. It was a cruel reward for the exacting standards to which the design team adhered, since they had spent precious time on a problem that didn't need fixing.

This example indicates the difficult balance that often had to be struck between *conservative* design using tried-and-tested components and procedures, and *innovative* design intended to improve efficiency and performance. The same quandary remains in engineering design to this day.

The solution to providing power for the Apollo Lunar Surface Experiments Package (ALSEP), which contained seismometers, a magnetometer, solar wind and lunar heat flow experiments, was quite different from that for the LM itself. It used a radioisotope thermoelectric generator (RTG) package, originally developed by the US Atomic Energy Commission under its Systems for Nuclear Auxiliary Power (SNAP) programme.

An RTG derives energy from the thermoelectric effect, whereby an array of thermocouples heated by the decay of a radioactive isotope produces electricity. The radioactive source is typically plutonium 238 in its ceramic form of plutonium dioxide, a non-weapons-grade isotope (Pu-239 is weapons-grade). Thus, although spacecraft using RTGs are sometimes referred to as 'nuclear-powered', RTGs are not nuclear reactors and use neither fission nor fusion processes to produce energy. They have proved crucial to deep space exploration, since solar arrays cease to be effective much beyond the orbit of Mars.

One of the first RTGs used in space was the prototype SNAP-3, which produced a power of just 3 W for the Transit navigation satellites of the early 1960s. The ALSEP SNAP-27 unit, developed for Apollo by General Electric, was specified at 63 W, sufficient for a full year of continuous operation. The ALSEP itself, carried on all Apollo landing missions except Apollo 11, was stored in the descent stage of the lunar module.

Figure 9.12   Radar and communications antennas on LM (top right); note also thruster quads, propellant tank and cold rails [Northrop Grumman Corp., formerly Grumman Aircraft Engineering Corp.]

Guidance, navigation and control

The guidance, navigation and control of the LM relied on the same basic technology and systems as that of the CSM, described in the previous chapter. In addition to the primary guidance, navigation and control system computer (PGNCS), the LM carried a second computer as part of the abort guidance system (AGS) – to be used, by definition, only in case of emergency.

Key sensors in the guidance system were the rendezvous and landing radars, supplied by RCA. As its name suggests, the rendezvous radar, which was mounted on the top of the LM ascent stage, was used to detect and track the command and service module in lunar orbit to assure that a rendezvous and docking could be made. The voice and data communications link between the LM and the CSM was provided, respectively, by a steerable S-band antenna adjacent to the LM's rendezvous radar and the CSM's own deployable high-gain antenna (Figure 9.12).

Figure 9.13   Lunar Landing Research Vehicle (LLRV) at Edwards Air Force Base, California in 1964 [NASA]

The LM's landing radar, mounted beneath the descent stage, was used to gauge the LM's height above the lunar surface, providing a constant read-out on the pilot's instrument panel. Another, relatively simple altitude sensor was provided by a set of surface probes fixed to the base of three of the LM's legs (Figures 9.3 and 9.15). When one or more of them touched the Moon, a 'contact light' in the LM came on to indicate the proximity of the surface.

In addition to the 'fixed' simulators, on which the astronauts trained for every phase of the mission, NASA and Bell Aerosystems developed a Lunar Landing Research Vehicle (LLRV), which allowed a pilot to actually fly a vehicle which simulated the behaviour of the LM (Figure 9.13). Nicknamed the 'flying bedstead', the LLRV had a vertically mounted jet engine, which allowed the vehicle to hover, and a number of thrusters for attitude control. The open cabin at the front incorporated an ejection seat, which was lucky for Neil Armstrong, as it turned out, since he was obliged to use it on one occasion when control of the vehicle was lost. Despite the evident risks involved in flying a 'real vehicle', Donald Slayton, then NASA's chief astronaut, thought that there was no other way to simulate a moon landing except by flying the LLRV [16].

Of course, the ultimate test of the lunar module, and its guidance and navigation system in particular, came with the first lunar landing, which proved to be nerve-wracking for mission controllers and astronauts alike. For a start, when the CSM resumed contact with Earth, having released the LM behind the Moon, communications and telemetry performance was poor. Mission control had about five minutes to decide whether to abort the landing attempt, but after careful checking decided to continue. Then, with Eagle four minutes from landing, it was discovered that the LM's altimeter and velocity gauges were in error, and had to be corrected. Next, the

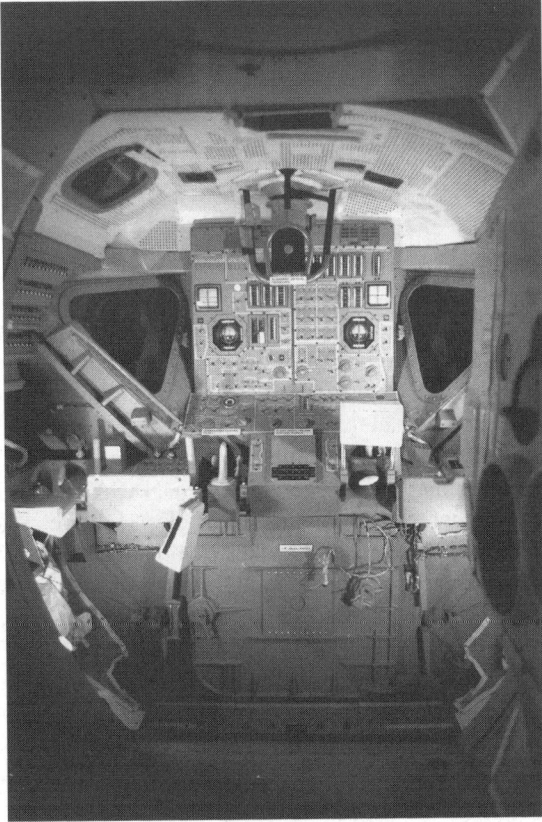

Figure 9.14   LM cabin mock-up at Kennedy Space Center [Mark Williamson]

guidance computer signalled the first of a series of alarms reporting overload conditions, calling for quick judgement on the part of the crew and mission controllers.
And finally, Armstrong had to override the planned landing programme to avoid
an area of rocky and dangerous terrain [17]. The first manned lunar landing was
certainly far from simple, and probably equalled anything the crew had experienced
during their many hours in the simulators. It proved, however, that the time in the
simulators had been well spent, since although it could not prepare the astronauts
for every eventuality, it helped them to remain composed when things did go wrong
(Figure 9.14).

Later analysis of the computer overload problem revealed that the LM's rendezvous radar had been mistakenly left in the active state, causing it to send
unnecessary data to the guidance system, consuming seven per cent of the computer's
capacity. The media and other commentators criticised the computer for being slow,
out-of-date and poorly designed, generally suggesting that it could have caused the
landing to be aborted, or worse. In fact, it was operating as its designers intended,

reporting that it was overworked while also eliminating the least important tasks (such as updating the displays every second) so that it could concentrate on landing the vehicle safely. In today's parlance, the computer would be described as 'fault tolerant' or 'capable of behaviour modification' [18]. The result, during Apollo 11's final approach, was that guidance and control continued uninterrupted and a safe landing was made.

Ultimate reliability

In terms of the engineering development of the lunar module, the Apollo programme's most fundamental problem was the inability to conduct a trial landing without risking the lives of a crew. Since there was no facility for remote-controlled operation of the spacecraft, the lunar module had to work first time.

This stringent requirement for reliability was the responsibility of the lunar module prime contractor, Grumman, whose contract with NASA called for two mission simulators, 15 flight models and 10 ground test articles. However, so concerned were its designers with the ultimate reliability of the LM that they eventually constructed no less than 29 major test articles, including full structural and thermal models of the spacecraft. All flight components were subjected to a range of acceptance testing – including some 6889 vibration tests and 3294 thermal tests – and pre-installation testing. Each assembled flight model was tested at Grumman's plant before shipment to Kennedy Space Center and again upon arrival [2]. So great was the ultimate cost of failure that, even though the spacecraft were designed to survive the high vibration and sound levels of a launch, they were treated as 'fragile' and 'delicate' here on Earth (as indeed are spacecraft today).

An important subsystem, with self-evident reliability requirements, was the environmental control system, which had to support two astronauts for the duration of their mission away from the CSM. This included the astronauts' spacesuits, supplied by International Latex Corp, and their ECS backpacks, or Portable Life Support System (PLSS), designed and developed by United Aircraft's Hamilton Standard Division (Figure 9.15). One of the less obvious challenges in spacesuit design is the outgassing of certain materials in the reduced-pressure environment chosen for American-designed life support systems. Hamilton Standard engineers soon learned that some plastics and paints gave off pungent and unbearable odours, and formed an 'odour panel' of people who met regularly to rate the smell of possible materials [19].

The ECS itself supplied a pure oxygen atmosphere to the cabin and the backpacks at the same pressure as the command module (34.5 kPa or 5 psi). The low pressure was one of the factors which allowed the LM cabin walls to be so thin, thus contributing to the weight-saving design of the vehicle. The ECS was designed to maintain a cabin pressure of 24 kPa for at least two minutes should a 13 cm hole be punched in the wall [20]. This would give the astronauts time to fit their helmets and connect their suits to the ECS, though luckily this was never necessary.

Manned spacecraft are renowned for their demands of high reliability and excellence of design, which typically lead to high costs and extended development

Figure 9.15   *Buzz Aldrin on the Moon, photographed by Neil Armstrong who appears reflected in Aldrin's visor; note surface probe attached to LM footpad in foreground [NASA]*

programmes. The LM design team realised early in the programme that the high priority attached to the reliability of the spacecraft provided little freedom to trade performance for either schedule or cost. What this meant in practice was that the schedule could only be met by hard work and long hours, and that the cost would increase as a result [2]. One NASA cost estimate for the LM placed early versions at $40 m apiece, whereas later versions – for the Apollo 15, 16 and 17 'J-Missions' – rose to $50 m each [21].

It is difficult to recall the state of the art in electrical and electro-mechanical devices of the early-to-mid-1960s, when the lunar module was under development. Suffice it to say that its development was coincident with that of the 'silicon chip' and pre-dated the advent of the commercial pocket calculator by about a decade.

According to Joseph Gavin, state-of-the-art predictions of equipment failure-frequency in the early sixties were 'entirely unacceptable' and led to a highly conservative approach to component and system design. At variance with common engineering practice, the team refused to recognise the concept of the 'random failure', preferring to believe that 'all failures have a cause that can be found and fixed'. Incredibly, in ten years of testing, they tabulated 14,247 test failures or anomalies, of which only 22 defied satisfactory understanding [2].

Figure 9.16   LM on the Moon – front (a) and rear (b) views (Al Bean descending ladder) [NASA]

Figure 9.17   Lunar module in 'magnificent desolation' [NASA]

This was the kind of attention to detail that allowed twelve men to land on the surface of the Moon and return safely to the Earth – undoubtedly one of the key historical events of the 20th century. Although the lunar module was first and foremost an engineering creation, which followed design and development rules accepted throughout the aerospace industry, it was also a source of the type of creative engineering that only a project of this magnitude could have engendered.

Conclusions

The sheer novelty of the Apollo concept is often overlooked in the broad perspective of space technology. Its primary objective – to land a crew on the surface of another planetary body – seems, even now, like science fiction. The great majority of the thousands of spacecraft that have left the Earth since 1957 have been launched to an orbit or interplanetary trajectory. Only a handful, manned or unmanned, have been designed to perform a controlled landing, so the field could still be considered to be in its infancy. Indeed the manned lunar landing itself happened only six times, between July 1969 and December 1972 (Figures 9.16 and 9.17).

The engineering development of the Apollo lunar module was one of the most notable achievements of NASA's manned spaceflight programme. Ten lunar modules were launched into space, three for in-orbit tests and seven for lunar landings. Of the seven intended landers, six were entirely successful and the seventh made history as the first 'space lifeboat' on the Apollo 13 mission.

Today, only one Apollo module remains in space intact: Apollo 10's LM ascent stage is in solar orbit, its descent stage having been allowed to crash onto the Moon. The entire Apollo 13 LM was destroyed on re-entering Earth's atmosphere and the other ascent stages were crashed onto the Moon. Six descent stages remain – to use Buzz Aldrin's description of the Moon, in 'magnificent desolation' – just as the astronauts left them, waiting for the museum curators … and mankind's long-awaited return to our nearest celestial neighbour.

References

1 Dethloff, Henry C.: 'Suddenly Tomorrow Came…A History of the Johnson Space Center' (NASA SP-4307, Lyndon B Johnson Space Center, Houston, USA, 1993) p. 159

2 Gavin, Joseph G.: 'Engineering Development of the Apollo Lunar Module' (IAA-90-633), 41st IAF Congress, Dresden, Germany, 6–12 October 1990, p. 2

3 Collins, Michael: 'Liftoff – the Story of America's Adventure in Space' (Aurum Press, London, 1989) p. 143

4 ibid., p. 142

5 Gavin, Joseph G.: 'Engineering Development of the Apollo Lunar Module' (IAA-90-633), 41st IAF Congress, Dresden, Germany, 6–12 October 1990, p. 4

6 Dethloff, Henry C.: 'Suddenly Tomorrow Came … A History of the Johnson Space Center' (NASA SP-4307, Lyndon B Johnson Space Center, Houston, USA, 1993) p. 173

7 Gavin, Joseph G.: 'Engineering Development of the Apollo Lunar Module' (IAA-90-633), 41st IAF Congress, Dresden, Germany, 6–12 October 1990, p. 5

8 Fleisig,   Ross: 'The   First   Manned   Lunar   Landing   Spacecraft' (IAA-94-IAA.2.1.611), 45th IAF Congress, Jerusalem, Israel, 9–14 October 1994, p. 2

9 Discussion with Joseph Gavin following paper presentation, 11 October 1990

10 Gavin, Joseph G.: 'Engineering Development of the Apollo Lunar Module' (IAA-90-633), 41st IAF Congress, Dresden, Germany, 6–12 October 1990, p. 3

11 ibid., p. 6

12 ibid., p. 8

13 Allday, Jonathan: 'Apollo in Perspective' (IOP Publishing Ltd, Bristol, 2000) p. 189

14 Collins, Michael: 'Liftoff – the Story of America's Adventure in Space' (Aurum Press, London, 1989) p. 145

15 Gavin, Joseph G.: 'Engineering Development of the Apollo Lunar Module' (IAA-90-633), 41st IAF Congress, Dresden, Germany, 6–12 October 1990, p. 10

16 Thompson, Elvia: 'NASA Celebrates 90 Years of Aeronautics Excellence', NASA press release 05-067, NASA Headquarters, Washington, March 3, 2005

17 Dethloff, Henry C.: 'Suddenly Tomorrow Came ... A History of the Johnson Space Center' (NASA SP-4307, Lyndon B Johnson Space Center, Houston, USA, 1993) pp. 176–7

18 Hall, Eldon C.: 'Journey to the Moon: The History of the Apollo Guidance Computer' (AIAA, Reston, 1996) p. 4

19 Smith, Harvey: 'The Apollo Moon Program: Exciting Times at Hamilton Standard', *Quest (The History of Spaceflight Quarterly)*, **12** (1), 2005, p. 36

20 Allday, Jonathan: 'Apollo in Perspective' (IOP Publishing Ltd, Bristol, 2000) p. 194

21 'Manned Spaceflight', NASA press document, May 1972, pp. 6–7

Chapter 10

Electric moon car – the development of the Apollo Lunar Roving Vehicle

It was the finest machine I ever had the pleasure to drive
Astronaut Gene Cernan following his use of LRV-3 on the Apollo 17 mission

Having succeeded in meeting the challenge of landing a man on the Moon, NASA's tightly constrained mission plan allowed the Apollo 11 crew to remain on the lunar surface for just 21 hours and 36 minutes before returning to Earth. This gave Neil Armstrong 2 h and 14 min outside the spacecraft, while Buzz Aldrin's EVA lasted only 1 h and 33 min. Obviously, if truly worthwhile lunar exploration was to occur, the timelines of future missions would have to be extended.

Arguably the most important prerequisite for exploration, on the Earth or on any other planetary body, is transportation. The first three manned lunar landing missions – Apollo 11, 12 and 14 – each delivered two astronauts who were obliged to explore the landing area on foot. Despite the Moon's 1/6-gravity environment, carrying equipment, setting up experiments and even moving around proved extremely tiring for the astronauts. Although their spacesuits had been designed specifically for lunar surface activities, the astronauts found that they constrained movements, made some tasks difficult and caused them to use more oxygen than expected. As far as locomotion was concerned, they soon developed the 'lunar lope' or 'bunny hop', which was easier and more efficient than walking.

In recognition of the need to transport tools, cameras and other equipment to survey sites some distance from the lunar module, the Apollo 14 mission of January 1971 was equipped with a wheeled trolley. In typical NASA practice it was named the Modularised Equipment Transporter (MET), but the astronauts referred to it variously as the 'rickshaw', 'wheel-barrow' or 'caddy cart' [1]. Although the two-wheeled aluminium transporter helped to ease the burden of carrying equipment and lunar samples, the area that could be safely explored was limited by the astronauts' stamina and consumables, most notably their oxygen supplies.

Figure 10.1   The Apollo 17 crew with 1g Lunar Roving Vehicle trainer and Saturn V [NASA]

In fact, the MET was the second wheeled-vehicle on the Moon. In November 1970, following the Apollo 11 and 12 manned lunar landings, the USSR's Luna 17 lander delivered the first of two teleoperated rovers, Lunokhod 1, to the Moon; it was followed by Luna 21's deployment of Lunokhod 2 in January 1973. Significantly, in terms of the history of space exploration, the Lunokhods remained the only teleoperated rovers to be sent to another planetary body until NASA's Sojourner rover was deployed by Mars Pathfinder in 1997.

The Apollo astronauts' ability to explore was improved substantially for the final three missions of the series – Apollo 15, 16 and 17 – by the provision of an electrically powered 'Moon car', known officially as the Lunar Roving Vehicle (LRV) (Figure 10.1). The lunar rover was the first and, so far, only vehicle driven by astronauts on the Moon – or come to that, anywhere else other than the Earth.

As with any other space hardware, the LRV had to be designed to survive the harsh vacuum, thermal and radiation environment, while facing the usual mass and power constraints of a space mission. In addition, it had to transport two astronauts, their equipment and geological samples – safely and reliably – across the relatively unknown surface of an alien world. It was evident from the outset that this was no ordinary vehicle.

Design overview

Although science fiction writers and futurists had long extolled the possibilities of lunar rovers, the first serious consideration appears to date from 1959, when the US Army Ordnance Missile Command, in Huntsville, Alabama, completed its 'Project Horizon' study for a manned Moon base. The design study included a two-man rover with a pressurised cabin, metal wheels and rechargeable batteries. Although it came to nothing, some of the study's participants were transferred, in 1960, to the newly established Marshall Space Flight Center [2], which became the *de facto* NASA centre for rover studies.

In June 1964, Marshall sponsored a design competition – between the Boeing Company and Bendix Corporation – for the so-called Apollo Logistics Support System. Intended to produce a design for a rover capable of supporting two men for 14 days, the study contracts led to the specification of a vehicle known as MOLAB (Mobile Laboratory), an enclosed travelling laboratory complete with atmosphere, provisions and scientific equipment. The Boeing solution, for example, was 11.5 m long, weighed 3.6 tonnes and had six wire wheels individually driven by electric motors [2], a design solution that would later reappear in the LRV.

It soon became clear, however, that MOLAB was too grandiose a concept for the mass-limited Apollo missions. The contracts were therefore extended into 1966–7 and redirected to study a smaller rover called the Local Scientific Survey Module (LSSM). The LSSM was an open vehicle comprising little more than a frame, a single seat and six wheels with woven wire tyres – the progenitor of the LRV [3].

For the next two years, concept studies apart, most of NASA's efforts were channelled into making sure that the agency would meet President Kennedy's directive to land a man on the Moon before the end of the decade. On 23 May 1969, with Apollo 11's July mission in sight, NASA finally decided to develop a lunar rover for the last four flights [4] (the fourth vehicle was eventually delivered in the form of spare parts because later Apollo missions were cancelled).

Since NASA–Marshall had been studying rovers since the early 1960s, it was the obvious NASA centre to manage the task and in June 1969 the LRV Project Office was established there with Saverio F. Morea, a rocket engine specialist, as manager [4]. Within weeks, Morea's office issued a request for proposals to 29 companies, and for the next two months Morea's team evaluated proposals from the four companies that responded – Bendix, Boeing, Chrysler Space Division and Grumman Aerospace. By 30 September, only Bendix and Boeing were in the running and finally, on 28 October 1969, NASA announced Boeing as the LRV's prime contractor and a cost plus incentive fee contract was signed [5].

The original contract was valued at $19m, but by the end of the project, costs had risen to $38m. In a scenario that would be repeated many times in the American space programme, the contract was underbid by Boeing, which did not fully appreciate the complexity of the LRV and was forced to spend money on overtime and extra shifts to maintain the schedule. However, given that nothing like the LRV had ever been built before, this was understandable, especially since the contract called for the parallel development of certain hardware options and the completion of all development testing before manufacturing the 1/6-g qualification models [6].

As with other hardware in the Apollo programme, the development schedule for the LRV was punishing: the preliminary design review was due within 10 weeks of contract award and the critical design review within 22 weeks, at which point the

Figure 10.2   LRV on the Moon [NASA]

design would be frozen for manufacturing. Delivery of the first rover was due in April 1971, just 18 months after contract award. This was a record for the space industry, especially when compared with the 26-month development schedule of the Lunar Orbiter series, which in 1963 was considered a 'near miracle' [7].

The LRV itself was deceptively simple in appearance (Figure 10.2). In fact, in Boeing's eyes, it was a specialised 'spacecraft on wheels', designed to function in the vacuum and wide temperature extremes of space on the rugged and difficult terrain of the lunar surface [8]. Since it was a part of the manned Apollo programme, it was built to the exacting specifications of all Apollo hardware and underwent a similar development, qualification and acceptance test procedure to qualify it as a manned spacecraft. Undoubtedly, it was this that made it very reliable and extremely expensive.

The LRV was composed of a number of subsystems:

- structure (including chassis and crew station)
- thermal protection
- electrical power (from non-rechargeable batteries)
- mobility (including wheels, suspension and steering)
- control and display
- communications
- navigation
- vehicle deployment.

Although some of the LRV's design and development engineering was under-taken in Huntsville, the majority of the work, including manufacturing and testing,

Figure 10.3   LRV at Boeing [Boeing]

was conducted at the Boeing Space Center near Seattle, Washington (Figure 10.3). Boeing's major subcontractor was General Motors' Delco Electronics Division in Santa Barbara, California, which supplied the vehicle's mobility system, comprising wheels, suspension and traction drive elements.

To ensure that any development problems were ironed out at an early stage, eight engineering and test models of the LRV were built before the first flight model, six by Boeing and two by Delco:

- a full-scale static mock-up for design and development related to human factors
- a vibration model for structural testing
- a qualification model tested for vibration, temperature extremes and vacuum to prove it could withstand the lunar environment
- an LM–LRV test model used to determine whether the LRV would cause stresses or strains on the lunar module structure
- two one-sixth weight models for development of the deployment system
- a mobility-test engineering model to prove the power/propulsion system (Delco)
- a 1g astronaut trainer model, strengthened to operate under Earth's gravity (Delco) [9,10].

In view of the short timescale of the programme, design, development testing and qualification testing were conducted essentially in parallel. Special test facilities included vacuum chambers to simulate space conditions, a simulated lunar surface made from crushed basalt and NASA's KC-135 aircraft, which was employed to fly a series of trajectories simulating 1/6-g to test the LRV's wheels and suspension.

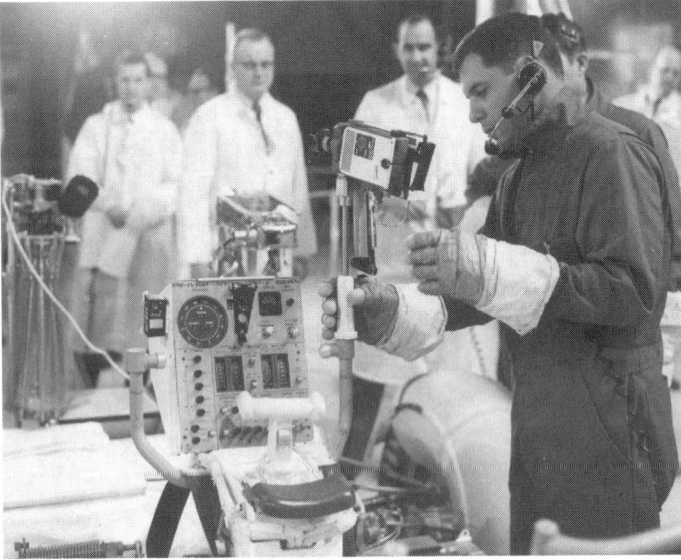

Figure 10.4   Qualification model LRV being used to prove astronauts could operate equipment with gloved hands [NASA]

Delco's 1g trainer was test-driven by the astronauts over a simulated lunar landscape at Houston and at training areas at Kennedy Space Center, Florida, and near Meteor Crater, Arizona [11]. Astronauts were also involved in the preliminary design review and made several helpful suggestions, particularly with regard to the 'human factors' aspects of the design (most notably the fact that they would have to operate the vehicle in a bulky spacesuit) (Figure 10.4).

Structure subsystem

The LRV was 3.1 m long and 1.8 m wide, with a 2.3 m wheelbase and a ground clearance of 0.35 m. Its structure was designed to be as simple as possible, comprising a chassis of square-section frame members joined together with elbows and tees, although it had to be divided into three hinged sections to allow the vehicle to be folded for storage in the lunar module (LM). The 0.8 m forward and rear chassis supported the wheels and equipment platforms, while the 1.5 m centre chassis housed the so-called Crew Station (Figures 10.5 and 10.6).

The forward chassis housed the Lunar Communications Relay Unit and high gain antenna (for communications with the orbiting command and service module), a TV camera, drive control electronics, signal processing unit, directional gyro unit and two batteries. The centre chassis contained two seats with under-seat stowage, a display console, 'control stick', low gain antenna, 16 mm camera and bag dispenser. The

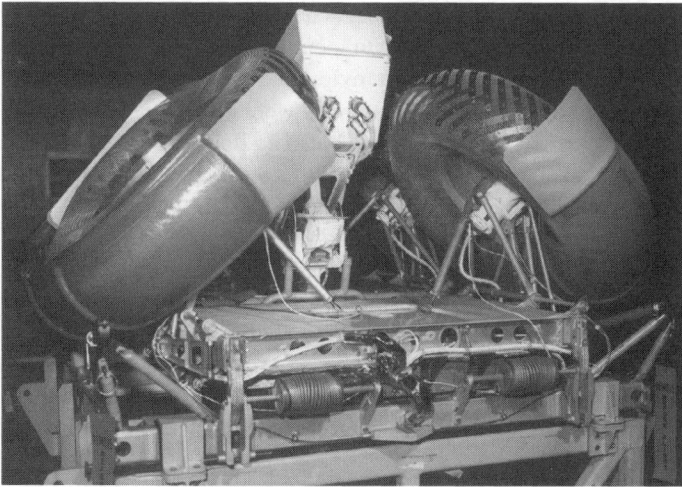

Figure 10.5   LRV wheels folded over chassis for stowage in LM; note console in centre [NASA]

Figure 10.6   LRV deployment simulation (showing chassis construction) [NASA]

aft chassis provided accommodation for science and crew equipment (including a lunar drill and tool carrier), and sundries such as sample collection bags. Much of the equipment was stored inside the LM until landing and fixed to the LRV following deployment.

The simplicity of the Crew Station is evidenced by the official Boeing equipment list [12], which includes seats, footrests, inboard and outboard handholds, a fibreglass armrest, beaded aluminium floor panels, seat belts made from nylon webbing, fenders and toeholds; there was very little else to mention. The seats were constructed from tubular aluminium frames spanned by nylon and were folded flat onto the centre chassis for stowage in the LM. The footrests were also folded down and even the fibreglass fender extensions were deployable. The floor panels were designed to be removable and could support the full weight of a standing astronaut. The fenders, or 'mud-guards', were included as part of the Crew Station since they helped to protect the crew, and their spacesuits, from dust thrown up by the wheels.

The side-by-side seating arrangement was designed to make both front wheels visible to both astronauts, while the central position of the 'joy-stick' hand controller, used to drive and steer the LRV, meant that either astronaut could take control; in practice, however, the mission commander, in the left-hand seat, tended to be the driver. The hand controller was originally a pistol grip, similar to that in the Apollo command module, but the astronauts suggested a T-shaped handle for better grip in pressurised gloves [13].

The functional design of the controller made driving the LRV simple: it was pushed forward to move the vehicle forward, pulled backward to slow or stop it, and moved from side to side for steering (the movement being proportional to the result in all cases). A backward pull beyond the neutral position and a flick of a switch allowed the vehicle to reverse and, if the handle was pulled back about 8 cm, a spring-loaded catch would engage to lock it in 'park'; a 'turn left' command released the parking brake [14]. As a testament to its design, in 1982 the Johnson Space Center demonstrated a version of the hand controller for disabled drivers, as part of NASA's technology 'spin-off' effort [15].

During 1/6-g tests in the KC-135, conducted in January and March 1970, the astronauts also discovered that toe-holds and hand-holds would be necessary for entering and leaving the vehicle, but deleted the option of a swivelling seat and a rollbar, thus saving a considerable amount of mass. As an example of mass efficiency, the toeholds were made from parts of the tripod assemblies used to attach the LRV to the LM. If necessary, a toehold could also be used as a tool to decouple the wheels in case of a wheel-drive failure [16].

As with most other space programmes, the main preoccupation, apart from cost, was the mass budget. The specification demanded that the LRV weigh less than 200 kg while being able to carry more than twice that in payload; by comparison, the average family car can carry only a third to half its own weight. The LRV Statement of Work (see Box), specified a maximum weight, including tie-down and deployment systems, of 181.6 kg. Despite the best efforts of engineers and managers, the first flight model of the LRV was overweight by some 28 kg at 209.6 kg [8]. Its typical 517 kg payload included two astronauts and their life support equipment (at 181.4 kg each) plus 154.2 kg of experiments, tools and samples (Figure 10.7).

Although not strictly comparable with the unmanned Lunokhod rover, because of their widely differing payloads, even fully loaded the LRV was lighter than Lunokhod 2, which weighed about 840 kg [17]. Of course, the LRV's operational weight on the Moon was only one-sixth of its Earth weight, making it about 86 kg fully loaded. A key advantage in designing lunar vehicles is the inherent mass saving which results from

---

Selective specifications from lunar roving vehicle statement of work [39]

Configuration: four wheels individually powered by electric motors fed by storage batteries.

Stowage: in one bay of the Extended LM.

Deployment: minimum activity by one astronaut.

Operation: by one astronaut.

Sterilisation: not required, but contractor to indicate approach to reduce level of biological contamination consistent with LM requirements.

Crew safety: ensure safety from all identified hazards (e.g. solar glare from reflective surfaces, lunar surface roughness, vehicle instability, etc.).

Emergency aids: e.g. hand holds to help free the vehicle.

Dust: critical surfaces or components designed to minimise degradation by dust; located such that dust coverage is difficult.

Single point failures: no SPF shall abort the mission and no second failure shall endanger the crew.

Weight: 182 kg max. including tie down and deployment systems.

Payload capacity: 45 kg of science experiments and two astronauts (168 kg each), plus 32 kg samples.

Range: four 30 km traverses in a 78 h period (total 120 km).

Lifetime: 78 h (min).

Speed: 16 km/h fully loaded on a smooth surface; continuously variable from 0 to 16 km/h.

Acceleration: should be specified.

Obstacle negotiation: 30 cm steps and 70 cm crevasses with both wheels in contact at zero velocity.

Slope negotiation: climbing and descending 25° slopes fully loaded.

Lateral and longitudinal static stability: minimum pitch and roll angles of 45° with full load; 15 cm jounce, 10 cm rebound.

Reversing: driver to have visibility.

Ground clearance: 35 cm on flat surface.

Turn radius: approximately one vehicle length.

Power system: margin of 150 W over requirements while driving.

Additional design requirements dealt with mobility and chassis, power supply, controls and displays, scientific equipment, stowage, thermal control, caution and warning, crew station, and tie down and deployment.

---

the vehicle not having to withstand the loads of the 1 g environment. A consequent advantage was that the LRV's unladen weight of 35 kg on the Moon – about the same as a petrol-engined garden mower on Earth – made it extremely easy to move. It also goes some way to explain the bouncing characteristics of the ride as shown by Apollo mission footage.

Figure 10.7   Astronaut Dave Scott drives fully loaded LRV [NASA]

Thermal subsystem

The LRV made use of both passive and semi-passive thermal control to ensure that its electronic equipment and moving parts did not reach temperatures that could cause damage. Although the rover was only operational on the lunar surface, it also had to survive the launch and three-day flight to the Moon packed inside the lunar module. The overall temperature constraint at lift-off was $21 \pm 10°$C, a reflection that the nominal temperature of the equipment was the 'room temperature' environment in which it was designed and built. More specifically, batteries had to be maintained between 4.5°C and 52°C, while other equipment could survive a range from $-34°$C to 85°C [18].

Heat loss during the trans-lunar flight and landing was kept to a minimum using multi-layer insulation (MLI), the primary passive control solution for spacecraft. It was composed of alternating layers of aluminised Mylar and nylon netting, with an outer layer of 'beta cloth', a polished-glass material. The insulating layers and reflective coating also helped to keep the LRV cool on the surface.

In addition to the MLI, the thermal subsystem used radiative surfaces, special surface finishes, thermal mirrors and straps, and fusible-mass heat sinks (boxes containing a wax-like material which melted as it absorbed heat). After LRV deployment, the latter, semi-passive control system came into operation to dissipate heat from equipment in the forward chassis section and maintain the control and display console within its operating limits. Aluminium thermal straps connected to electronic equipment transferred heat to the batteries and heat sinks, where it was stored [18]. When the ambient temperature dropped, heat would be given out by the sinks as the wax cooled and solidified.

At the conclusion of a sortie, the LRV was parked with a heading that would limit thermal exposure and powered down. The astronauts opened fibre-glass dust covers to expose the fused-silica thermal radiators (otherwise known as second surface mirrors) which were mounted on the batteries, electronics boxes and heat-sinks. When the batteries reached their lower operating limit, the covers would close automatically to prevent further cooling and to keep out lunar dust.

The all-important instrument console was isolated from the rest of the vehicle by fibre-glass mounts and its external surface was painted with heat-resistant paint, the faceplate itself being black anodised for temperature control, as well as to reduce reflection.

Power and mobility

The LRV's power was provided by a pair of non-rechargeable 36V/121Ah silver–zinc batteries, either of which alone could power the rover if necessary. The 25-cell batteries, developed by Eagle Picher Industries, were of Plexiglas monobloc construction (with silver–zinc plates in potassium hydroxide electrolyte) and housed in a magnesium case for lightness. An instrument called an ampere-hour integrator was used to measure battery discharge, accumulating total current drawn from the batteries and relaying the information to the display console between the seats [19].

The required range of the vehicle, quoted in the statement of work, was four 30 km traverses in a 78 hour period (a total of 120 km). A Boeing document quotes the range as only 91.7 km [20], but this was quite sufficient for the actual distances covered in the three missions (a maximum of 36.1 km on Apollo 17), and would have been sufficient even if one of the two batteries had failed. Although theoretically, the operational design lifetime of 78 hours could be divided into any number of sorties, the physical limitations of the astronauts and their life support systems restricted use of the vehicle to three sorties during the three-day stay on the Moon.

Vehicle mobility was provided by a traction drive assembly on each of the four wheels, comprising a harmonic drive reduction unit mounted on the outer hub, a drive motor and a brake assembly. The wheels were driven by individual 0.25 hp series-wound, brush-type DC motors (rotating at up to 10,000 rpm) through gears with an 80:1 step-down ratio. The harmonic drive arrangement allowed continuous application of power to the wheels at all speeds without gear shifting [21]. A version of the drive unit had previously been used on Marshall Space Flight Center's Pegasus satellite and was later used on the Skylab space station and the Hubble Space Telescope [22].

The motors operated from a nominal input voltage of 36V DC, as provided by the batteries, and speed control was derived by pulse-width modulation from the drive controller electronics. The LRV specifications called for a top speed of 16 kph, fully loaded, on a smooth, hard surface. Speed was continuously variable in either forward or reverse and a fully loaded LRV could climb and descend slopes as steep as 25°.

Each traction drive unit had a mechanical parking brake, which could hold it on slopes of up to 35°. It was actuated by a cable connected to the hand controller and applied by moving the controller backwards, an action which both de-energised the motors and applied a drum brake on each wheel. The units also contained an odometer pickup which transmitted nine pulses to the navigation system for each wheel revolution.

The drive units were hermetically sealed to allow the motor to run in non-vacuum conditions, thus avoiding dust contamination and possible damage to the motor brushes. They were pressurised with nitrogen gas at 7.5 psi to help transfer heat to the motor's outer casing. Each motor contained a thermistor to measure temperature (which was displayed on the console) and a thermal switch to de-activate the motor should its upper operating temperature be reached; this in turn was displayed on a caution and warning panel at the top of the control console. Each wheel could be uncoupled from the traction drive system and allowed to 'free-wheel' as necessary [21].

The wheels themselves were developed by General Motors under a major sub-contract to Boeing. Dr M.G. Bekker, a pioneer in vehicle mobility and locomotion, headed the team which, surprising though it may seem, applied a principle dating back to the nineteenth century. In 1858, a metallic wheel design originally applied to 'locomotive engines and other carriages to facilitate their transit', had been patented by Thomas Rickett, a mechanic at the Castle Foundry in Buckingham, England [23]. In this case, the LRV programme had effectively re-invented the wheel!

The LRV's 'tyres' comprised an outer surface made from hand-woven strands of high-tensile steel (zinc-coated piano wire) with a titanium 'inner tube' or 'bump stop' connected to a spun aluminium hub (Figure 10.8) [24,25]. Each wheel contained 800 strands – 81.3 cm long and 0.83 mm in diameter – which were subjected to X-ray inspection prior to use [23]. Titanium tread cleats, riveted together in a chevron pattern covering 50 per cent of the tread area, were added following reports from Apollo 11 and 12 which suggested that the added traction on harder surfaces and 'flotation in deep dust' would be useful [25].

Although it seems unnecessary to go to these lengths when rubber tyres could have been used, in common with many other engineering solutions this was a mass-limitation exercise. The rover's 0.85 m-diameter, 0.22 m-wide wheels weighed 5.4 kg each (0.9 kg on the Moon), and the design saved over 45 kg compared with conventional rubber-tyred wheels [25]. Everything else being equal, this alone would mean that the lunar module would be able to carry an extra 45 kg of useful payload.

The rover's wheels were each connected to the chassis, via the drive units, by a pair of triangular suspension arms, torsion bars and dampers in a conventional arrangement. Loads were transmitted to the chassis through the suspension arms to the torsion bars, while the dampers, mounted between the chassis and upper suspension arms, limited the extent and rate of the wheels' vertical travel. Deflection of the

Figure 10.8   LRV wheel showing internal 'bump-stop' and chevron cleats for tread [NASA]

suspension system and wheels allowed 35 cm of chassis ground clearance fully loaded and 43 cm unloaded.

Moreover, the vehicle was designed to ride over obstacles up to 30.5 cm in height and to do so from a standing start with both front wheels in contact. It could cope with pitch and roll angles of up to 45° with a full load and could also cross 71 cm wide crevasses, from a standing start, even if both front wheels were resting across the crevasse. The 85 cm wheels and low gravity environment were, of course, crucial in meeting these rigorous specifications.

Power-assisted four-wheel steering was supplied by two 0.1 hp, series-wound, 500 rpm motors mounted centrally between the front and rear wheels. The identical front and rear systems were independent, providing steering redundancy, and could be disconnected manually. The systems were a modification of the standard Ackerman geometry, used in terrestrial cars, which allows the inner wheel to pivot slightly more during turns than the outer wheel to prevent scuffing. The LRV had a 6.2 m turning circle and took 5.5 seconds to go from lock to lock but, according to its users, with both sets of wheels steering the vehicle had 'excellent responsiveness' [14].

Control and display

The LRV's control and display console, mounted in the centre of the vehicle between the astronauts, was divided into two main parts: navigation on the upper section and

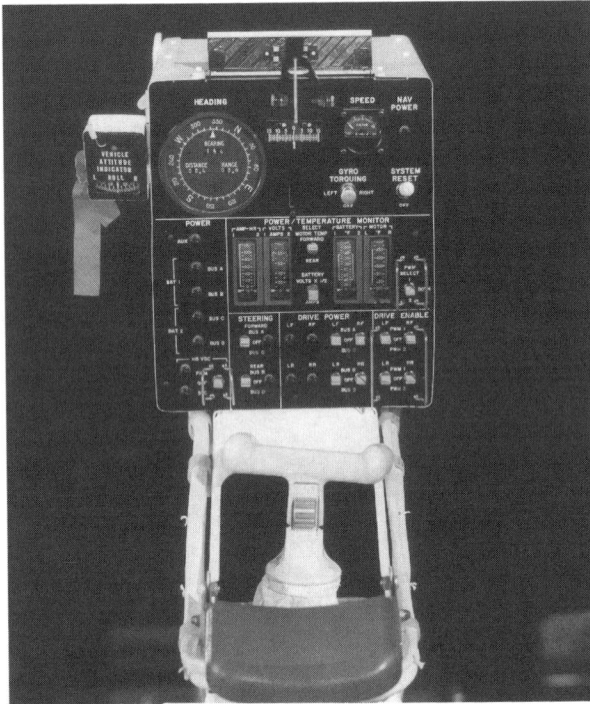

*Figure 10.9   LRV control and display console. Note attitude indicator at left, cau-
tion and warning indictor on top, and hand-controller below console
[NASA]*

electrical switching and monitoring below [26]. The panel labelling, which appeared
green against a black anodised background, contained radioluminescent Promethium
147 allowing it to be read in darkness (Figure 10.9).

The console also featured an unusual and somewhat Heath–Robinson indicator
known as the Caution and Warning System, which provided a visual warning that
one of the batteries or motors was overheating. It consisted of a spring-loaded flap,
on top of the control console, held in place by an electromagnet. If any of the units
overheated, the circuit would break and the panel would be released, thereby 'flagging
up' the problem.

Navigation was of crucial importance to an LRV crew. They would be travelling
in an alien environment with few recognisable landmarks and their limited supply of
consumables meant that they had to know exactly where they were in relation to the
lunar module in case of emergencies.

The Moon has no magnetic field, so compasses could not be used for direction-
finding. Boeing originally proposed an inertial navigation system based on the
strapped-down gyroscopes and integrating accelerometers used in long-range guided
missiles, but NASA and the astronauts objected that it was too expensive and that

the potential errors would be too great. Since such a system had never been built and tested on the rough lunar surface, they feared that the gyros might not remain accurate [13].

So a simple 'dead reckoning' navigation system, with which astronauts as pilots would be familiar, was developed: with knowledge of direction and distance to the LM at any point during the sortie, a speedy return could be made should the need arise. The system included a directional gyroscope, odometers on each traction drive unit and a signal processing unit (SPU), which was essentially a small solid-state computer for navigational trigonometry calculations. The SPU calculated E–W and N–S distances based on inputs from the odometer on the third fastest wheel (thus compensating for wheel slippage) and heading data from the gyro. These distances were added to those already in the registers, and range and bearing to the LM was calculated and displayed [27].

The console itself indicated the LRV's heading (relative to lunar north), the range and bearing of the LM, and the distance travelled in 0.1 km increments. System specs dictated that the bearing should be displayed with an accuracy of ±6° and range to ±600 m at a radius of 5 km from the LM, and that the total distance travelled should be displayed to an accuracy of ±2 per cent [28].

However, a fair amount of manual intervention was required to ensure that the LRV's position was known accurately. Regular navigation updates involved lifting a so-called Sun Shadow Device into position above a graduated scale. The shadow cast by the device revealed the vehicle's heading with respect to the sun, which was compared with that derived from the gyro as a check against gyro drift. But this would only be accurate if the vehicle was level. To take this into account, the crew also had to read an attitude (pitch and roll) indicator, pivoted off the left hand side of the main panel, and transmit its reading to Mission Control along with the sun shadow reading [29]. The basic nature of the navigation system is an indication of the low processing power available on the lunar missions of the early 1970s.

The navigation console also featured a gyro torquing switch, to adjust the gyroscope when necessary, a system reset button to 'zero' the bearing, distance and range displays before embarking on another sortie, and a speed indicator (registering 0 to 20 km/h) driven by odometer pulses from the right rear wheel.

Deployment

The lunar rover was stowed in a quadrant of the lunar module's descent stage to the right hand side as viewed from the ladder (see Figure 9.3). To fit the vehicle into the space available, the forward and aft chassis sections were folded over the centre section and the wheels were folded over their respective sections. Springs ensured that the vehicle unfolded and that its wheels deployed and locked into their proper positions (Figure 10.10).

Deployment was possible with the LM tilted at any angle up to 14.5° and with the bottom of the descent stage between 0.35 m and 1.57 m from the surface. The worst-case deployment time was intended to be 15 minutes, but with the LM almost

Figure 10.10   Observing Apollo 16 LRV deployment test [NASA]

level this could be reduced to about 5 minutes [30]. The operation itself was semi-automated so that a single astronaut could deploy, activate, check and operate the vehicle quickly and easily. Originally, the system was to be fully automatic with a manual redundancy feature, but it soon became evident that there was no way to test a process that would take several minutes under 1/6-g.

Operation and performance

The stated objectives of the Lunar Roving Vehicle were to expand the astronaut's capability to explore the lunar surface and to increase the scientific return from the Apollo missions. This it certainly did.

In fact, the LRV proved to be an important and integral part of the Apollo programme. It more than doubled the time astronauts could spend on the lunar surface, since they expended less energy riding than walking and used less oxygen and water. It also allowed them to venture further from the LM: up to 7.6 km in the case of Apollo 17. Moreover, in terms of total distance travelled, while the astronauts on the Apollo 11, 12 and 14 missions covered 0.6 km, 2.3 km and 4.2 km, respectively, those on Apollo 15, 16 and 17 covered 27.9 km, 27.1 km and 36.1 km.

Not surprisingly, the availability of a vehicle also increased the weight of lunar material that could be collected. The sample load returned by Apollo 15 with the aid of the LRV was 77.1 kg – almost double the previous record of 42.6 kg for Apollo 14,

Table 10.1   LRV mission summary [24,34]

	Apollo 15	Apollo 16	Apollo 17
Launch date	26 July 1971	16 April 1972	7 December 1972
Landing date	30 July 1971	21 April 1972	11 December 1972
Landing site	Hadley Rille	Descartes Crater	Taurus-Littrow
Astronauts	Scott and Irwin	Young and Duke	Cernan and Schmitt
No. of sorties	3	3	3
Sortie dates	31 July 1971	21, 22, 23	11, 12, 13
	1, 2 August 1971	April 1972	December 1972
Total distance (km)	27.9	27.1	36.1
Average speed (kph)	9.2	7.9	8.1
Max. speed (kph)	14	17	18
Duration (h:m)	18:33	20:14	21:30
Driving time (h:m)	3:08	3:17	4:27
Longest traverse (km)	12.5	11.5	20.4
Max. range fr. LM (km)	5.0	4.5	7.6
Mass of samples (kg)	77.1	97.5	112.9

aided by the hand cart – and the hauls for Apollo 16 and 17 were even greater (see Table 10.1). In addition, using the LRV's navigation system, the astronauts were able to chart the exact location of geological features and sample sites on each sortie (Figure 10.11).

In fact, the navigation system proved to be reliable and accurate to within 100 m, when measured over the three sorties of Apollo 15 (Figure 10.12). Since their landing site could be located with precision, astronaut Jim Irwin reported that he and Dave Scott had no hesitation in deviating from the planned route, since they were confident that the system would guide them back to the LM [31]. Here on Earth, except under extreme circumstances, this would be seen as a convenience; on the Moon, where an accidental tear in a spacesuit would mean the difference between life and death, it could have been a life saver. Scott and Irwin termed the LRV a 'remarkable machine' [32] and were 'very pleased with the vehicle's performance, particularly the speed (of up to 14 kph) and hill-climbing capability' [33].

Performance on Apollo 16 was similar, although astronauts John Young and Charles Duke reported a lunar land-speed record of 17 kph while driving downhill on their return to the LM. The astronauts were also allotted an eight-minute speed trial, during which Duke filmed Young driving 'flat out' around a triangular course and handling characteristics were reported as 'excellent' [34]. More importantly, Young estimated that, since the landing area was far more rugged than expected, they would have been unable to accomplish more than five per cent of their assignment without the LRV.

Apollo 17's LRV-3 broke all previous records (see Table 10.1), being driven further, faster and longer than LRVs 1 and 2. Its second sortie covered

Figure 10.11   Astronaut Duke investigates crater on Apollo 16 mission [NASA]

a significant 20.4 km, an Apollo record, providing an indication of how important vehicles would be to any future missions, on either the Moon or Mars. During EVA 3, commander Gene Cernan reported climbing a hill with an inclination greater than 25° and later travelling at 18 kph to set a new, and so far unchallenged, lunar land-speed record. Although Cernan could be forgiven for any exaggeration following an almost flawless final Apollo mission, his comments after parking LRV-3 for the last time must have pleased the designers at Boeing: 'it was the finest machine I ever had the pleasure to drive', he said [35].

Photography was an important part of all the Apollo missions, although severe weight restrictions suffered by the early flights meant that only a monochrome TV camera could be carried; colour TV cameras of the day were simply too heavy. The LRV's colour TV camera not only enhanced the science return of the missions but also added to the spectacle enjoyed by TV viewers at home, including live views of the lunar module at the moment of lift-off (see Figure 9.11) .

More importantly, since the camera was controlled remotely from Earth, ground controllers became part of the exploration team and were able to advise the astronauts in real time. Voice communications and TV transmissions were made via the S-band high gain parabolic antenna mounted on the front of the LRV, while the voice link between the LRV and the astronauts was at VHF. Since the high gain antenna would have needed constant repointing while the LRV was in motion, TV relay was restricted to times when the vehicle was stationary. When driving, voice communication was maintained via the low gain antenna.

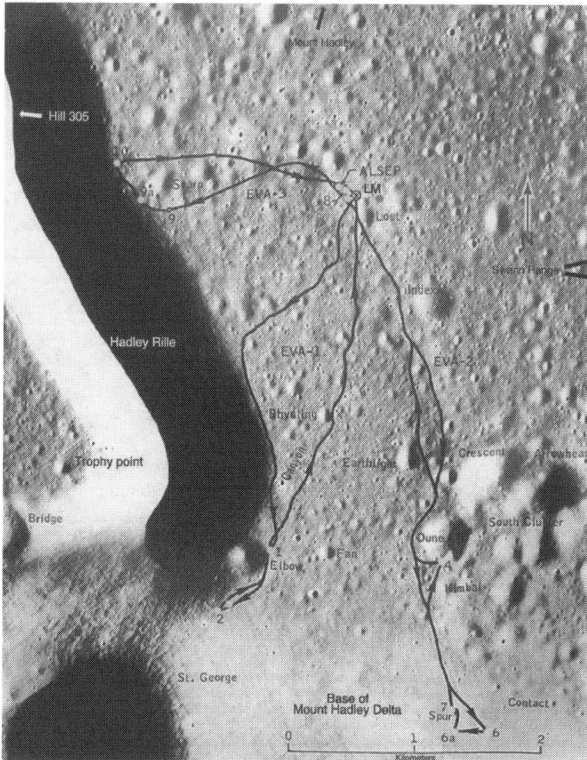

Figure 10.12   Apollo 15 Hadley Rille sortie map. The LRV allowed astronauts to conduct three separate sorties on three consecutive days [NASA]

One of a long list of Apollo subcontractors was Motorola, perhaps better known at the time for its car radios. In addition to supplying the command system hardware for the Mercury, Gemini and Apollo spacecraft, it supplied the S-band receiver for the LRV, in a sense the first car radio on the Moon.

Considering the difficulties of the lunar environment, very few operational problems were experienced by the three crews, as the minor nature of the reported anomalies shows. For example, the utility of the redundant steering system was proven during the first sortie of LRV-1 on the Apollo 15 mission, when the front-wheel steering failed temporarily. The astronauts completed the sortie using rear-wheel steering only, but the problem cleared itself the following day and operations were normal for the second and third sorties. They also found the LRV's seat belts difficult to fasten, so the design was improved for subsequent missions.

LRV-2 lost a rear fender extension, broke its yaw indicator during a rough ride [36], suffered a temporary loss of rear-wheel steering and experienced higher-than-expected battery temperatures, but none of these threatened the mission. Experience with LRV-3 was similarly uneventful: although its chassis locking pins failed to

Figure 10.13   The creative use of a lunar map [NASA]

engage fully after deployment, they were easily seated using a deployment tool. Learning from earlier missions, fender extension stops were added to LRV-3, but this did not stop the commander knocking one of the extensions off before the first traverse! The creative use of a lunar map repaired the damage (Figure 10.13) [33].

Conclusions

Despite the problems, the first LRV flight model (FM-1) was delivered to NASA on 15 March 1971, two weeks ahead of schedule and less than 17 months from contract signature. According to Boeing, this was the shortest development and manufacturing programme for any major item of Apollo equipment. For comparison, the astronauts' spacesuits took 60 months to develop and their portable life support systems 72 months, while the more complex command and service module and lunar module took 52 and 66 months respectively from contract award to FM-1 delivery [37].

In the perspective of a programme to land men on the Moon, all of these schedules are impressive, but this should not detract from Boeing's achievement. The prime contract for the LRV had been awarded just one year and nine months before the resulting hardware was flown for the first time on Apollo 15 – a development schedule which compares favourably with that of conventional terrestrial vehicles.

In fact, a letter from a terrestrial vehicle manufacturer gave Boeing pause for thought in June 1971, when the Rover Company of Solihull, UK, pointed out that the

Figure 10.14   Apollo 17 'pre-owned' LRV still on the Moon [NASA]

words 'Rover' and 'Land-Rover' were copyrighted by the firm. It stated that although Rover had 'no foreseeable intention ... of producing a vehicle for use in any space programme, and therefore [was] not really in competition with [Boeing] ...', the company felt that there was 'some risk of confusion'. Further on, however, the letter stated that the company could not 'possibly object to such words as Lunarover ...' [4] and it appears that no legal action was taken (probably because of its evident futility).

To say that the Apollo lunar rover was a unique vehicle would be an understatement. Given that its design is now several decades old, it is unlikely ever to be used again. However, it is worth pointing out that three 'second-hand models' remain to this day on the lunar surface, in need of new batteries but more than likely in serviceable condition (Figure 10.14). In fact, following the Apollo 15 mission, NASA received several offers, mainly from American used car companies, to buy LRV-1: the offers ranged from $100 to $1000 (the low bids probably reflecting the difficulty in returning the vehicle to Earth). According to one source, 'NASA artfully dodged the issue in a typical bureaucratic maneuver: the offers were shuffled from center to center and NASA Headquarters until the would-be purchasers gave up' [38].

Under the largely inert conditions of the lunar environment, the LRVs and other Apollo hardware could theoretically remain untouched and unchanged for centuries. But if mankind is to bestow a fitting tribute on the first vehicle driven 'off-world', it could do no better than to incorporate the Apollo Lunar Roving Vehicle in the first lunar museum, a development that would preserve the memory and the technology of the first lunar explorers.

References

1 Froehlich, Walter: 'Apollo 14: Science at Fra Mauro' (NASA EP-91, Washington, 1971) p. 3

2 Burkhalter, Bettye B., and Sharpe, Mitchell R.: 'Lunar Roving Vehicle: Historical Origins, Development, and Deployment' (IAA.2.2-93-673), 44th IAF Congress, Graz, Austria, 16–22 October 1993, p. 3

3 Boeing Background Information: 'Lunar Roving Vehicle (LRV)', A-1018 (Updated), (Boeing Aerospace Company, Seattle, WA, April 1981) pp. 2–3

4 Burkhalter, Bettye B., and Sharpe, Mitchell R.: 'Lunar Roving Vehicle: Historical Origins, Development, and Deployment' (IAA.2.2-93-673), 44th IAF Congress, Graz, Austria, 16–22 October 1993, p. 5

5 ibid., p. 7

6 ibid., p. 12

7 ibid., p. 6

8 Boeing Background Information: 'Lunar Roving Vehicle (LRV)', A-1018 (Updated) (Boeing Aerospace Company, Seattle, WA, April 1981) p. 3

9 Burkhalter, Bettye B., and Sharpe, Mitchell R.: 'Lunar Roving Vehicle: Historical Origins, Development, and Deployment' (IAA.2.2-93-673), 44th IAF Congress, Graz, Austria, 16–22 October 1993, pp. 7–8

10 Boeing Background Information: 'Lunar Roving Vehicle (LRV)', A-1018 (Updated) (Boeing Aerospace Company, Seattle, WA, April 1981) pp. 7–8

11 ibid., p. 9

12 Boeing Press Information brochure: 'LRV' (Boeing Aerospace Company, Seattle, 1972) pp. 5–7

13 Burkhalter, Bettye B., and Sharpe, Mitchell R.: 'Lunar Roving Vehicle: Historical Origins, Development, and Deployment' (IAA.2.2-93-673), 44th IAF Congress, Graz, Austria, 16–22 October 1993, p. 10

14 Boeing Press Information brochure: 'LRV' (Boeing Aerospace Company, Seattle, WA, 1972) p. 10

15 Burkhalter, Bettye B., and Sharpe, Mitchell R.: 'Lunar Roving Vehicle: Historical Origins, Development, and Deployment' (IAA.2.2-93-673), 44th IAF Congress, Graz, Austria, 16–22 October 1993, p. 14

16 Boeing Press Information brochure: 'LRV' (Boeing Aerospace Company, Seattle, WA, 1972) p. 5

17 Thompson, Tina D. (Ed.): 'TRW Space Log 1996' (TRW Space and Electronics Group, Redondo Beach, Volume 32, 1997) p. 151

18 Boeing Background Information: 'Lunar Roving Vehicle (LRV)', A-1018 (Updated) (Boeing Aerospace Company, Seattle, WA, April 1981) p. 6

19 ibid., p. 4

20 ibid., p. 5

21 Boeing Press Information brochure: 'LRV' (Boeing Aerospace Company, Seattle, WA, 1972) pp. 8–9

22 Burkhalter, Bettye B., and Sharpe, Mitchell R.: 'Lunar Roving Vehicle: Historical Origins, Development, and Deployment' (IAA.2.2-93-673), 44th IAF Congress, Graz, Austria, 16–22 October 1993, p. 19

23 ibid., p. 9, 19

24 Bonsack, Richard F.: 'A Brief Look at Lunar Rover' (Boeing Aerospace Company, Seattle, WA, undated presentation by Manager, Robotic Systems Technology, Boeing Aerospace Company)

25 Boeing Background Information: 'Lunar Roving Vehicle (LRV)', A-1018 (Updated) (Boeing Aerospace Company, Seattle, WA, April 1981) p. 8

26 Boeing Press Information brochure: 'LRV' (Boeing Aerospace Company, Seattle, WA, 1972) pp. 11–15

27 ibid., pp. 14–15

28 Burkhalter, Bettye B., and Sharpe, Mitchell R.: 'Lunar Roving Vehicle: Historical Origins, Development, and Deployment' (IAA.2.2-93-673), 44th IAF Congress, Graz, Austria, 16–22 October 1993, p. 11

29 Boeing Press Information brochure: 'LRV' (Boeing Aerospace Company, Seattle, WA, 1972) p. 11

30 ibid., pp. 17–20

31 Boeing Background Information: 'Lunar Roving Vehicle (LRV)', A-1018 (Updated) (Boeing Aerospace Company, Seattle, WA, April 1981) p. 10

32 Boeing Press Information brochure: 'LRV' (Boeing Aerospace Company, Seattle, WA, 1972) p. 22

33 Burkhalter, Bettye B., and Sharpe, Mitchell R.: 'Lunar Roving Vehicle: Historical Origins, Development, and Deployment' (IAA.2.2-93-673), 44th IAF Congress, Graz, Austria, 16–22 October 1993, p. 13

34 Boeing Background Information: 'Lunar Roving Vehicle (LRV)', A-1018 (Updated) (Boeing Aerospace Company, Seattle, WA, April 1981) pp. 10–11

35 ibid., p. 11

36 Furniss, Tim: 'One Small Step' (Haynes Publishing Group, Yeovil, 1989) p. 63

37 Boeing Background Information: 'Lunar Roving Vehicle (LRV)', A-1018 (Updated) (Boeing Aerospace Company, Seattle, WA, April 1981) p. 7

38 Burkhalter, Bettye B., and Sharpe, Mitchell R.: 'Lunar Roving Vehicle: Historical Origins, Development, and Deployment' (IAA.2.2-93-673), 44th IAF Congress, Graz, Austria, 16–22 October 1993, p. 15

39 'Lunar Roving Vehicle Statement of Work, July 3, 1969, RFP #IL-LRV-1' (NASA Marshall Space Flight Center, Huntsville, 1969) pp. 15–16

Chapter 11

An epilogue to the space race

The impossible of today will become the possible of tomorrow

Konstantin Tsiolkovsky

The early years of the Space Age, from the late 1950s to the early 1970s, are seen by many as the 'Golden Age' of space exploration, chiefly because the period saw such incredible advances in space technology and culminated in the landing of men on the Moon. By the end of the 1960s, all of the technology necessary to send unmanned spacecraft to the planets and astronauts to our nearest astronomical neighbour had been designed, developed and successfully flown in space . . . and it had been done in little more than a decade.

While the race for the Moon provided a boost to many aspects of space technology, there were other social, political and technical reasons for space-related development, as previous chapters have attempted to show. This chapter draws those strands of technological heritage together and looks at the broader effects of space technology on our culture.

Space technology development

The development of space launch vehicles from sounding rockets and converted ballistic missiles opened the gateway to space. And once it was open, there was no turning back. Political competition between the Cold War superpowers saw to that.

Continued developments in rocket technology allowed larger and heavier payloads to be launched into higher and longer-lived orbits, resulting in a host of communications, remote sensing and science applications. Although many of the launch attempts were failures, it seems remarkable today that so much effort was made in those early years: between October 1957 and the end of 1969, more than a thousand payloads were launched into space [1].

Since then, the performance and reliability of launch vehicles has improved and launching payloads into space has become an industry in its own right. Manufacturing companies build rockets for launch vehicle operators, which then sell launch services to commercial satellite owners and government organisations. The relentless increase in size and complexity of spacecraft in the last two decades of the twentieth century led to the development of more powerful rocket variants, in a process driven largely by commercial communications satellite operators, but also by demands from the space science community for larger astronomical payloads.

Although America's development of the partially reusable Space Shuttle, first launched on 12 April 1981, appeared to offer a new era of less expensive space transportation, it soon became clear that it could not live up to the promise. President Richard Nixon had formally instructed NASA to develop the Shuttle in March 1972, but the programme met with many difficulties and its maiden launch slipped from 1978 to 1981. The system was too complex and too labour intensive to launch more than eight or nine missions a year, and ultimately too sensitive to hardware failures to be regarded as an operational, as opposed to experimental, vehicle. The two most important failures, which led to the destruction of the Challenger and Columbia orbiters and the loss of 14 lives, were far from complex. The Challenger accident of 1986 was caused by the failure of a pressure seal in a solid rocket booster, while the Columbia incident of 2003 was the result of damage to the orbiter's thermal protection system caused by a piece of insulating foam that detached from the External Tank during launch.

Figure 11.1   Space Shuttle launch [NASA]

Although the Space Shuttle is a remarkable spacecraft, and has flown many successful missions, its legacy is open to question. It seems that more conventional launch vehicles will continue to provide both manned and unmanned access to space in the early decades of the twenty-first century.

Meanwhile, the technological development of successive generations of commercial satellites and science spacecraft has allowed a fast-paced evolution of their capabilities. In a continual battle to reduce launch mass, the development of lighter but stronger spacecraft structures has allowed the specification of larger and more complex payloads. Where once honeycomb aluminium panels were the height of weight-saving technology, man-made plastic composites, such as carbon fibre and Kevlar, are now among the basic materials for satellite structures.

Of course, stabilisation and attitude control will always be fundamental requirements for spacecraft and this subsystem has undergone continual improvement. While spin stabilisation is used for some small spacecraft and for larger ones during transfer orbit, three axis stabilisation now dominates the field. It not only allows better attitude control and finer pointing, but also – and more importantly – larger and more powerful satellites. Improvements in attitude control have, for example, allowed space telescopes to make very long exposures (hours at a time over several days), producing fine-detailed images of distant galaxies and other astronomical objects, and Earth imaging spacecraft to produce high resolution pictures of the planet, showing features as small as 50 cm under certain conditions.

Meanwhile, the development of satellite communications payloads has seen higher power satellites carrying greater numbers of transponders; increasingly flexible payloads that can be reconfigured in orbit, by switching between transponders and steering antenna beams for example; and satellites that can communicate with each other via intersatellite radio links. The antennas themselves have evolved from small fixed dishes, via larger deployable reflectors (Figure 11.2), to 9 m-diameter unfurlable antennas capable of focusing their coverage beams on cities rather than whole countries.

To a large extent, these advances have been due to the continual development of space qualified electronics systems and computer software, often performed in parallel with terrestrial development. The general improvement in electronics-based subsystems has tended to make space hardware more reliable and flexible in its operation; unfortunately software development has yet to meet the same standards and is often blamed for systems failures and anomalies.

All of these improvements have required parallel developments in spacecraft power subsystems which, in addition to increasing raw power levels from a few hundred watts to around 20 kW, have led to longer and more productive operating lifetimes. Although most unmanned spacecraft still use the photovoltaic cell, the sheer size of solar arrays deployed from satellites has increased from the few square metres of the early satellites to arrays that could easily span a football pitch. As far as the solar cells themselves are concerned, the silicon cell has been superseded – in terms of conversion efficiency – by the gallium arsenide cell and multilayer cells which squeeze the most out of the different wavelengths of light that illuminates them. Likewise, battery technology has progressed from the standard nickel–cadmium technology to

Figure 11.2   Communications satellite for New Skies Satellite company [EADS-Astrium]

nickel–hydrogen and lithium-ion units. Perhaps the only area of spacecraft power technology that has seen limited progress is the radioisotope power source, largely because of its image as 'nuclear' technology, while true in-space nuclear reactors have been 'off-limits' for years. However, space nuclear power and propulsion systems may be due for a renaissance with the renewed focus on deep space exploration of the early 2000s.

Finally, spacecraft propulsion itself has moved on from cold-gas and hydrazine thrusters to more efficient bipropellant systems, using separate fuel and oxidiser, and a variety of electric propulsion systems that provide low thrusts with relatively small amounts of propellant. This has allowed spacecraft designers to offer a flexibility of operation that the early designers could only have dreamed of, and has helped to increase the design lifetimes of commercial geostationary satellites from a few years to as many as twenty.

Industrial development

In addition to the technological advances, the foundation of dedicated space agencies and national space industries to support them made space technology development a field of human endeavour in its own right. 'Space' became a government tool, a source of employment and a driving force for innovation all rolled into one.

In the United States in particular, it was realised by the early 1960s that the space technology requirements of the nation could not be met by government laboratories

Figure 11.3   Ariane 5 Vulcain engine at Paris Air Show 1995 [Mark Williamson]

alone and a nascent space industry was formed as an outgrowth from the existing aerospace industry. For defence contractors, which were obliged to keep much of their work secret, a civilian space agency, NASA, offered a 'shop window' for their capabilities. Likewise, companies that were known mainly for earthbound products, such as cars and refrigerators, used their involvement in space to indicate their 'advanced', 'high-tech' status. For example, it is not widely known outside the space industry that Ford, renowned for its motor vehicles, was one of America's top three satellite manufacturers until 1990, when its space division was bought by the Loral Corporation. And Bendix, which among other things manufactures white goods, was a bidder for the Apollo Lunar Roving Vehicle. In the 1960s and 1970s, corporations of all different shapes and sizes wanted to be a part of the space industry.

Although America was the undisputed leader in most aspects of space technology in the 1960s, it was challenged later by other nations as they too developed a space industry. For example, although Russia operated under an entirely different political and industrial system, it took the lead in manned spaceflight in the 1970s and 1980s with its deployment of the Salyut and Mir space stations. Meanwhile, Europe's introduction of the Ariane launch vehicle in 1979 led to its domination of the commercial satellite launch industry in the late 1980s and 1990s, while the development of its satellite industry made it a worthy competitor to established US manufacturers (Figure 11.3).

In common with the United States, other nations developed a working relationship between a civilian space agency and a domestic space industry. A good example is the French space agency CNES (Centre National d'Etudes Spatiales), a government organisation founded in March 1962 and responsible for the implementation of French space policy. The fact that CNES was established early in the Space Age – in fact three months prior to the formation of the European Space Research Organisation (ESRO) – afforded France an advantage and helped to ensure its continuing lead in European space affairs. It is no coincidence that the headquarters of the European Space Agency (ESA), which was formed in May 1975, is in Paris. Likewise, while the space sector's industrial capability is distributed around Europe, its key players are based in France.

Several other nations have also developed their own space industries, predominantly based on government programmes. Japan, for example, has an industry founded on science applications which has benefited from US technology transfer programmes for both satellites and launchers. Many of its satellites have been built jointly with US contractors and its early launch vehicles were effectively American rockets built under license by domestic contractors. By the end of the twentieth century, this had placed Japan in a position to embark on the commercialisation of its satellite industry and its H-IIA launch vehicle.

India, too, is a nation that has long recognised the advantages of space systems, particularly communications satellites which interconnect villages that will probably never be linked by terrestrial telecommunications cables. At the same time, its satellites broadcast television signals to the entire nation, an advantage used by India to provide education services and health advice, as well as entertainment, to millions of people who would otherwise remain deprived. By the same token, remote sensing satellites which provide wide geographical coverage on a repeatable basis are used for weather forecasting, flood monitoring, crop-health monitoring and many other applications in support of a growing population. Despite its status as a technologically less developed country with a large under-privileged and under-educated population, India made a political decision to develop both satellites and launch vehicles manufactured by a domestic space industry.

A similar strategy has been followed by China, in that it has developed its own satellite and launch vehicle industry. Indeed, China even supplied its launch services in the world commercial market for a time, until US Government sanctions made it impossible for any American-built technology (which appeared in some form in all commercial satellites) to be launched from Chinese soil. Despite the setbacks, China has developed its own communications, navigation, remote sensing and science satellites and undertaken joint programmes with both ESA and European industry. Moreover, with the launch of Shenzhou 5 carrying astronaut Yang Liwei on 15 October 2003, China became only the third nation to place one of its own citizens in orbit using indigenous technology.

Space technology and its applications

Although the Space Age began as a scientific and political endeavour, the 'space race' and the 'race for the Moon' were over by the end of the 1960s. Since then, the key

Figure 11.4   Replicas of satellites used to televise the 19th Olympic Games from Mexico City to audiences in Europe and Japan (left to right: ATS-3, Intelsat II, Early Bird/Intelsat I) [NASA]

developments have been in the application of space technology to the needs of an ever more demanding global society.

Since the first privately owned communications satellite, Telstar 1, was launched in 1962, this type of satellite has had the most significant impact on society, not least in transforming the concept of the 'Global Village' into reality. As a result of their broad view of the planet, communications satellites have breached political boundaries, enabled disparate cultures to learn more about each other and, generally, helped to engineer the phenomenon of 'globalisation'. Indeed, one of the earliest applications of the communications satellite was to transmit broadcasts of the Olympic Games (Figure 11.4).

As a result of the development of the communications satellite, it is possible to remain in touch with the rest of the world from the middle of a desert, an ocean or a polar cap, and from the summit of a mountain or the depths of a rain forest. Global business communications, satellite television and live news reporting via satellite phone are now taken for granted, while broadcast satellite radio is making inroads in the US market (Figure 11.5). Meanwhile, communications signals relayed via satellite from a variety of distress beacons have allowed search and rescue organisations to save lives that would otherwise have been lost.

The development of the navigation satellite, most notably the Global Positioning System (GPS), has had a significant impact, first among the military forces and more

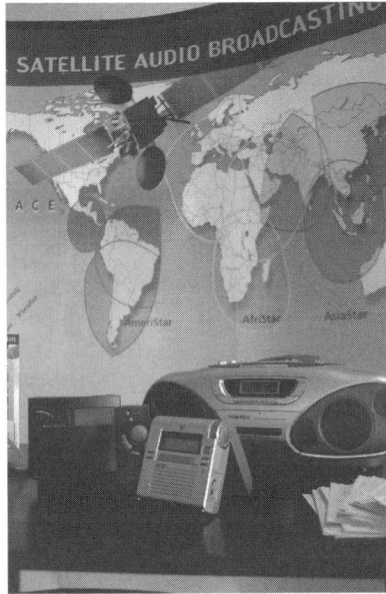

Figure 11.5   Broadcast satellite radio is a fast-developing application (Worldspace stand at IAF congress exhibition in 2001) [Mark Williamson]

recently among the general public. Now that the receiver technology has become commercialised, it is only a matter of time before satellite navigation becomes an integral part of modern-day society. Indeed, GPS receivers are already fitted to many cars as standard and serious ramblers are as likely to carry a GPS device in their rucksacks as they are a map and compass.

Whereas most people are familiar with the satellite images that accompany TV weather forecasts, few will realise the fundamental effect this imagery has had on professional weather forecasting since its introduction. The Tiros satellites, launched in the early 1960s, laid the groundwork for the revolution in meteorology that was to follow and, since 1966, the entire Earth has been photographed at least once a day. Most importantly, by providing early warning of hurricanes and other storm systems, satellites have helped to safeguard property and save lives.

The sibling to the meteorological satellite, the Earth observation satellite, has permitted an accuracy in map-making, urban planning and agricultural management that would be impossible without it. Satellites which monitor pollution, deforestation and the growth of deserts help to preserve the planet's ecosystems for future generations, thus having direct and indirect effects on the global population.

Meanwhile, the majority of satellite technologies also have military applications, since communications, meteorology, surveillance and reconnaissance are crucial to the operation of armed services deployed around the globe. In addition, dedicated military satellites are used for aspects as varied as ballistic missile defence, global arms treaty verification and crisis management.

Figure 11.6 NASA's Skylab space station [NASA]

In terms of automated space exploration, the early concentration on our nearest astronomical neighbour, the Moon, was later matched by missions to far more distant bodies of the solar system. Following the largely abortive attempts of the 1960s, the 1970s turned out to be the golden years for planetary probes, with spacecraft despatched to all of the planets except Pluto. Obvious highlights among the many programmes were the American Pioneer, Viking and Voyager missions. For example, Pioneer 10, launched on 2 March 1972, became the first spacecraft to escape the solar system by passing the orbit of Pluto (on 13 June 1983), and Viking 1, launched on 20 August 1975, was the first spacecraft to soft land on Mars (on 20 July 1976). Voyager 1, launched on 5 September 1977, flew by Jupiter and Saturn and went on to become the most distant of mankind's creations when it overtook Pioneer 10 on 17 February 1998; it was 70 astronomical units, some 10.4 billion km, from Earth.

Finally, of course, there is the application of technology to manned spaceflight. Following Apollo, America used its remaining hardware for a series of three missions to the Skylab space station in 1973 (Figure 11.6) and a one-off orbital docking mission with a Russian spacecraft in the Apollo–Soyuz Test Project (ASTP) of 1975. The nation then concentrated its efforts on the Space Shuttle and, in the 1990s, began the development of the International Space Station. The first space station, however, was the Soviet Salyut 1, launched on 19 April 1971. It was the first of several Soviet stations and led, as mentioned above, to the Mir station which dominated manned orbital

Figure 11.7   Russia's Mir space station [NASA]

activity from its launch on 19 February 1986 until it was de-orbited in March 2001. With the exception of a brief period in 1989, it was permanently occupied from the arrival of the first long-term crew in early 1987 until its decommissioning in 2000 (Figure 11.7).

In the context of the history of technology, it is worth repeating the observation that by mid-1966 both Russia and America had succeeded in landing spacecraft on the Moon. Less than nine years after the Space Age began, technology from Earth was operating, not only in an extraterrestrial environment, but on the surface of another planetary body. This was a hugely impressive cultural achievement, brought about largely by the industrialisation of space technology.

Space technology and culture

The practical applications of space technology have had an undeniable impact on society, and will continue to do so, but 'space' has had an even deeper impact on our culture. Indeed, its ability to influence our culture is equal to that of the 'great works' of art or literature.

In the 1950s and 1960s, when the Space Age was in its formative phase, society was undergoing a period of tremendous change, and science and technology were in the ascendant. The Western nations had largely overcome the effects of World War II and were entering a period of relative prosperity. Towns and cities, particularly those ravaged by the War in Europe, were undergoing modernisation and much of their architectural heritage was being replaced by clean, new buildings, structured in the 'modern' style. There could be no greater manifestation of the 'modern age' than

Figure 11.8   Image of Space Age engineering: Apollo 12 Saturn V leaves VAB for launch pad [NASA]

mankind's ventures into space and, later, to the Moon. Thus 'space' became a cultural icon of the 1960s.

Interestingly, the progression of the space race was paralleled by the technological and commercial development of television, which became a unifying and all-pervading window on world events. Images of Space Age engineering were transmitted into people's homes on a scale never seen before: the Saturn V rocket, the Vehicle Assembly Building at Kennedy Space Center, the launch pads and the infrastructure that supported them became familiar to the TV-viewing public (Figure 11.8). The image of the curiously attired astronaut was lifted from the realms of science fiction, while the form of the even less believable Apollo lunar module became, through photos, films and TV animations, an icon of the age. Even the word 'module' began to creep into common usage, as consumers began to buy modular furniture, kitchens and even homes. In fact, at least one company incorporated the word 'modules' into its title because its founders wanted to convey a modern and forward-looking image [2].

In retrospect, there was every reason to expect that a programme as well publicised as Apollo would have a significant effect on society and on the people who were

Figure 11.9   Space is part of our culture 1: children's space handkerchiefs [Mark Williamson]

shaping western culture in the early years of the Space Age. Thus the trappings of space technology became an inspiration for fashion and product designers, advertisers and toy manufacturers alike (Figures 11.9 and 11.10).

Among those influenced by the Space Age was André Courrèges, a French fashion designer who, in 1964, presented his 'Space-Age Collection'. The clothes were clean-cut, starkly plain and chiefly in white and silver. Other designers followed suit and simple, space-age clothes in modern materials, including PVC, became the rage [3]. Perhaps the ultimate in space-age designs was the famous 1967 mini-dress by Paco Rabanne, constructed as a chain mail of silver plastic disks [4], the high street version of the spacesuit.

Like fashion designers, product designers gain much of their creative inspiration from the world around them and, in the 1960s, it was difficult to avoid the influence of the gleaming perfection of the rockets, modules and other space paraphernalia. In the early 1960s, for example, Cadillac cars still sported magnificent tail fins, while the Ford Thunderbird exhibited rear light housings reminiscent of rocket engines. Although their extravagant configurations were influenced more by the rockets of 1950s science fiction films and TV series than the space capsules of the time, marketing departments were not slow to draw comparisons.

Indeed, advertisers used space technology as a reliability comparison, to imply by association, if not to state directly, that theirs was a product of the Space Age. Although no sensible advertiser would associate itself with the space technology of the late 1950s, when rocket explosions were all too common, by the time Sony's 1962 magazine advert for its sleek and modern 'all-transistorized, truly portable TV' was released, the reliability of launch vehicles had improved significantly. The advert, which featured the Saturn I launch vehicle, boasted of a television 'built like a

Figure 11.10   Space is part of our culture 2: rocket soap set from 1996 [Mark Williamson]

SPACE AGE rocket' (their capitals) and 'engineered [with] space age reliability and compactness' [5]. It was far from the only consumer product to associate itself with space technology and the high-tech image it engendered.

Not surprisingly, the entertainment industry was also influenced by the events of the early Space Age, particularly where astronauts were involved. The space-related films and TV series of the 1960s were thus an avenue through which an impression of space technology entered our culture. While much of the output was, at best, romantic rather than realistic, the film that broke the mould was Stanley Kubrick's *2001: A Space Odyssey*, released in 1968. By and large, the classical aerodynamic rocket shapes were gone and spacecraft were more functional, almost industrial in appearance. The Earth-orbiting, wheel-shaped space station, a visual SF classic in its own right, was not only realistic in form, it was unfinished (Figure 11.11) [6]. Despite, or possibly because of the lack of romance, it is argued that 2001 influenced more people to read and learn about space, and even seek employment in the space industry, than any of its forerunners.

Meanwhile, on the 'small screen', influenced by the introduction of real space technology, writers were looking even further into the future. Most notable was

Figure 11.11   The film that broke the mould: 2001: A Space Odyssey [NASA/GRIN DataBase Number: GPN-2003-00093]

Gene Roddenberry's *Star Trek*, which was set in the twenty-third century; it first aired in America in 1966 and came to British TV, coincidentally, at the time of the Apollo 11 mission. Among the best-remembered 'technology' from the series is the Communicator, a portable 'satellite telephone' the shape and size of an electric shaver. Although it was impossible to communicate with an orbiting spacecraft by means of a hand-held device in the 1960s, the Iridium satellite system made this fictional device possible before the close of the twentieth century, thus beating science fiction by at least a couple of centuries. If nothing else, this shows how difficult it is to predict the development of technology.

In the final analysis, however, one of space technology's greatest contributions to contemporary global culture is its ability to show the Earth as a planet. The images returned from space – particularly those of the Earth – have had a significant and long-lasting effect. Although the weather satellites of the early 1960s produced revealing

Figure 11.12   Earthrise from Apollo 8. The lunar horizon is about 780 km from the spacecraft and the width of the image at the horizon is about 175 km [NASA]

and extremely useful images of the Earth, it was not until Apollo 8 ventured far from Earth to lunar orbit, in 1968, that the general public became aware of the Earth as a planet in space. The now-famous 'Earthrise' photographs showed the blue-green Earth, surrounded by a sea of space, rising above the barren and cratered surface of the Moon (Figure 11.12).

The image was a poignant legacy of the Space Age: it showed the Earth as beautiful, but insignificant and fragile, ably illustrating the finite nature of the planet. As if by way of revelation, it instilled a palpable sense of 'global awareness' and is often cited as an inspiration for the Green movement [7]. Since then, of course, successive images of the Earth have shown us large-scale deforestation, pollution and the ozone hole – not to mention the thinness of the atmospheric layer that surrounds our world.

Beyond our own planet, spacecraft designed for space astronomy have also made significant contributions to the human equation by influencing the way we view ourselves and our relationship with the Universe. In addition to their practical aspects, such as helping us to understand how the Sun affects the Earth's climate, these spacecraft help to answer fundamental questions about the Universe relating to its size, age, composition and possibly even its inhabitants. Meanwhile, the spacecraft that venture out beyond Earth orbit address the human need for discovery and exploration by producing images of distant astronomical objects. Together they help to place our planet in a wider context and, by illustrating our insignificance on the Universal scale, offer an opportunity to grow beyond our current parochial, geocentric culture (Figure 11.13).

Figure 11.13   First portrait of Earth and Moon together (imaged by Voyager 1 on 18 September 1977 when it was 11.7 million km from Earth) [NASA]

Learning from history

However important and useful the many varieties of unmanned spacecraft are to mankind, it is difficult to ignore the impact of manned spaceflight – both fictional and factual – on our culture (Figure 11.14) . The involvement of individual people, whether vicariously through the eyes of astronauts or personally as space tourists, is seen by many as the key to the future development of space technology. The first crack in the dam of pent up ambition appeared on 28 April 2001, with the flight of the first space tourist, Dennis Tito, to the International Space Station. The completion of the X-Prize competition in late 2004 was a sign that the floodgates of space tourism may be about to open.

A wealth of predictive science fiction charts the possible avenues of this development, some leading to the planets of our own solar system, but many leading well beyond. It is inevitable that the first such journey of reality – embodied within

(a)

(b)

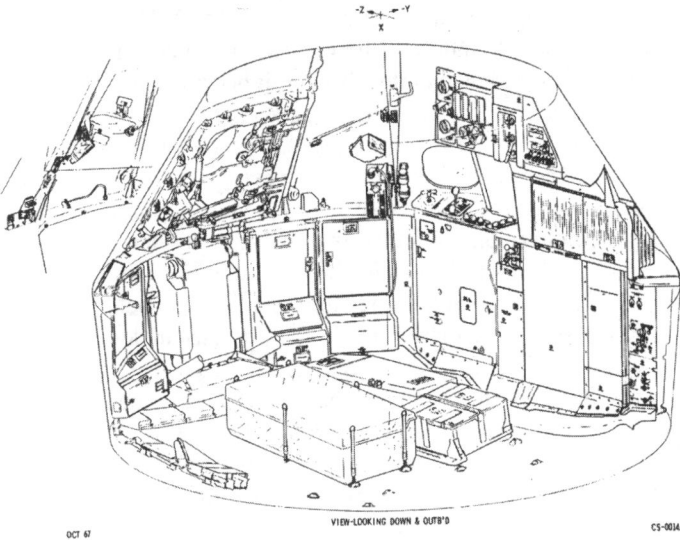

Figure 11.14   *Science fiction versus engineering fact (a) cabin of Jules Verne's spaceship [US Space & Rocket Center] compared with (b) the Apollo command module [NASA]*

the Apollo programme – should provide the model, if only by comparison, for what follows. This is part of the reason for devoting a proportion of this book to manned lunar exploration.

The Apollo 11 mission of July 1969 represented a watershed in the development of space technology. It was the culmination of a race whose final heat involved a manned lunar landing (Figure 11.15). However, from the moment that goal had been won, the Apollo programme was little more than a TV re-run, at least to non-technical eyes. As has been well documented, US politicians and public alike lost interest in sending men to the Moon, funding for the programme dwindled and the last three missions, Apollo 18, 19 and 20, were cancelled.

In retrospect, it is clear that Apollo was a 'blip' in the development of space technology – a highly significant blip, but a blip nonetheless. So much effort was put into developing the systems that would deliver men to the Moon that US manned space exploration effectively 'burnt itself out'. When the programme ended with Apollo 17 in December 1972, there was no clear course for future space exploration. Starved of the necessary funds to pursue either a continuing presence on the Moon, or a trip to Mars, NASA could do little more than redevelop some of the remaining Apollo hardware for the Skylab space station missions and place the rest in museums. It seemed that manned space exploration on the Apollo model was unsustainable, so the hard-won technologies which enabled the lunar missions were abandoned.

In the realm of technology development, this was an unusual occurrence. The great majority of technology developments – such as those that have enabled global air travel, telecommunications and power generation – have been improved and refined throughout their history, not discarded shortly after their first use. The abandonment of a technology did not happen again on this scale until 2004, when British Airways finally abandoned Concorde and the technology of the supersonic passenger aircraft. Once a technology is abandoned in this fashion, it becomes increasingly difficult to resurrect as time passes. As far as manned space exploration is concerned, there is now no hope, and no point, in resurrecting the technology of Apollo; when it comes to the time, it will be necessary to reinvent the wheel.

In this case, the lessons of history appear to include not basing a programme of space exploration entirely on a political imperative, ensuring that a programme is sustainable, and not abandoning a technology as soon as it has been proved. This is, of course, much easier to suggest than to implement. It could be argued that the technology required to place a man on the Moon was a technology ahead of its time, since the mission was based on a political rather than practical premise. An alternative view is that the political push accelerated the normal development process, allowing designers to leap-frog their way to a conclusion.

Whatever the case, the results of abandoning the technology of Apollo, rather than evolving it to suit future requirements, are clear for all to see. Between the end of the third Skylab mission in February 1974 and the first flight of the Space Shuttle in April 1981, Americans astronauts ventured into space only once – on the nine-day Apollo–Soyuz Test Project of July 1975. NASA's concentration on the Shuttle, and later the International Space Station, effectively capped manned space exploration by confining astronauts to low Earth orbit. Those policy decisions are illustrated by

(a)

(b)

Figure 11.15   (a) Buzz Aldrin's boot print, the ultimate iconic image of the twentieth century, and (b) many others at the Apollo 11 landing site – surprisingly, this is the only reasonably clear image of Neil Armstrong on the Moon [NASA]

the fact that no human being has ventured more than a few hundred kilometres from Earth since 1972.

Flawed and unsustainable as it was, the Apollo programme remains an achievement which, most observers believe, could not be reproduced in the same timescale today. This view is supported by the schedule for the return to the Moon, announced by US President George Bush on 14 January 2004, which even under the most optimistic planning would not see astronauts on the lunar surface before 2015. Without the inevitable delays, this is already two to three years longer than it took the first time.

It is easy to see why many of those who experienced the era of space exploration and development of the 1960s consider it a Golden Age, but that does not mean they see its events through rose-tinted spectacles. As this book has attempted to show, the methods were often imperfect, the decisions were sometimes misguided and the advances were made only with great sacrifice. But the technological developments stand as testament to the vision and dedication of those involved, and as a source of inspiration to those who celebrate the human skills of inventiveness and perseverance.

The Apollo 17 mission ended this historic first phase of manned planetary exploration in December 1972, but the legacy of the Apollo programme remains – in science, technology and culture – even decades later. As has often been said, when the authoritative history of the second millennium is written, a few hundred years hence, the twentieth century will be remembered not for its wars, its politicians and its poets, but as the century when mankind broke the gravitational bonds of the home planet and made its first tentative steps into the Universe beyond. The individual steps may well have been small, but together they undoubtedly represented a giant leap for mankind.

References

1 Heyman, Jos: 'Spacecraft Tables 1957–1990' (Univelt, San Diego, CA, 1991) p. 171

2 Douglass, Stuart (SB Modules): personal communication, 1984

3 Yarwood, D.: 'The Encyclopaedia of World Costume' (B.T. Batsford, London, 1988), p. 175

4 Nunn, J.: 'Fashion in Costume 1200–1980' (The Herbert Press, London, 1984) p. 235

5 National Geographic, 1962, **121** (6), June 1962 (rear advertisement section)

6 Williamson, Mark: 'Space Technology in Films and Television – Portrayal And Perception' (IAA.8.2-93-767), 44th IAF Congress, Graz, Austria, 1993

7 Williamson, Mark: 'Space Age Images – Part of our Culture' (IAA-94-IAA.8.2.713), 45th IAF Congress, Jerusalem, Israel, 1994

Abbreviations and acronyms

ABMA	Army Ballistic Missile Agency
AC	Alternating current
AGC	Apollo Guidance Computer
AGS	Abort Guidance System
AIAA	American Institute of Aeronautics and Astronautics
AIS	American Interplanetary Society
ALSEP	Apollo Lunar Surface Experiments Package
APT	Automatic Picture Transmission
ARPA	Advanced Research Projects Agency
ARS	American Rocket Society
ASTP	Apollo-Soyuz Test Project
AT&T	American Telephone and Telegraph Company
ATS	Applications Technology Satellite
AVCS	Advanced Vidicon Camera System
BIS	British Interplanetary Society
BNCSR	British National Committee for Space Research
BTL	Bell Telephone Laboratories
Caltech	California Institute of Technology
CCD	Charge-Coupled Device
CCIR	International Radio Consultative Committee (French abbreviation)
CM	Command Module
CNES	Centre National d'Etudes Spatiales
Comsat	Communications Satellite Corporation
comsat	Communications satellite
COSPAR	Committee on Space Research
CSM	Command and Service Module
DC	Direct current
DCTL	Direct-coupled transistor logic
DMSP	Defense Meteorological Satellite Program

DoD	Department of Defense
DSIF	Deep Space Instrumentation Facility
DSKY	Display and keyboard interface ('disky')
DSN	Deep Space Network
ECM	Electronic counter-measures
ECS	Environmental Control System
EHF	Extra high frequency
ELDO	European Launcher Development Organisation
EPC	Electronic power conditioner
ERTS	Earth Resources Technology Satellite
ESA	European Space Agency
ESMR	Electrically Scanning Microwave Radiometer
ESRO	European Space Research Organisation
ESSA	Environmental Science Services Administration
EVA	Extra-vehicular activity
FAI	Fédération Aéronautique Internationale
FCC	Federal Communications Commission
FM	Flight model
GaAsFET	Gallium arsenide field effect transistor
GALCIT	Guggenheim Aeronautical Laboratory, California Institute of Technology
GARP	Global Atmospheric Research Programme
GDL	Gas Dynamics Laboratory
GEO	Geostationary orbit
GfW	Gesellschaft für Weltraumforschung
GNC	Guidance, navigation and control
GOES	Geostationary Operational Environmental Satellite
GPS	Global Positioning System
HPA	High power amplifier
IAA	International Academy of Astronautics
IAC	International Astronautical Congress
IAF	International Astronautical Federation
IAU	International Astronomical Union
IC	Integrated circuit
ICBM	Intercontinental ballistic missile
ICSU	International Council of Scientific Unions
IDSCS	Initial Defense Satellite Communications System
IEE	Institution of Electrical Engineers
IEEE	Institute of Electrical and Electronics Engineers
IGCY	International Geophysical Cooperation Year
IGY	International Geophysical Year
Intelsat	International Telecommunications Satellite organisation
IQSY	International Quiet Sun Year
IR	Infrared
IRBM	Intermediate range ballistic missile

IRIS	Infrared Interferometer Spectrometer
ITU	International Telecommunication Union
JATO	Jet-assisted take-off
JPL	Jet Propulsion Laboratory
KPA	Klystron power amplifier
KS	Korabl Sputnik
LEM	Lunar Excursion Module
LEO	Low Earth orbit
LES	Launch Escape System
LH_2	Liquid hydrogen
LLRV	Lunar Landing Research Vehicle
LM	Lunar Module
LNA	Low noise amplifier
LOR	Lunar orbit rendezvous
LOX	Liquid oxygen
LRV	Lunar Roving Vehicle
LSS	Life Support System
LSSM	Local Scientific Survey Module
Mascon	Mass concentration
MET	Modularised Equipment Transporter
MIDAS	Missile Defense Alarm System
MIPS	Million instructions per second
MIT	Massachusetts Institute of Technology
MLI	Multi-layer insulation
MMH	Monomethyl hydrazine
mm-wave	Millimetre-wave
MOLAB	Mobile Laboratory
MRBM	Medium range ballistic missile
MSS	Multi-Spectral Scanner
N_2O_4	Nitrogen tetroxide
NACA	National Advisory Committee for Aeronautics
NASA	National Aeronautics and Space Administration
NiCd	Nickel–cadmium
NNSS	Navy Navigation Satellite System
NOAA	National Oceanic and Atmospheric Administration
NTSC	National Television System Committee
OAO	Orbiting Astronomical Observatory
OSO	Orbiting Solar Observatory
PAL	Phase Alternating Line
PAS	Perigee-Apogee System
PGNCS	Primary Guidance, Navigation and Control System
PLSS	Portable Life Support System
psi	Pounds per square inch
RAE	Royal Aircraft Establishment
RAM	Random access memory

RBV	Return Beam Vidicon
RCA	Radio Corporation of America
RCS	Reaction control system
RF	Radio frequency
RFNA	Red fuming nitric acid
RFP	Request for proposals
RNV	Rakyeta Nosityel Vostok
ROM	Read-only memory
RTG	Radioisotope thermoelectric generator
s/c	Spacecraft
SCORE	Signal Communication by Orbiting Relay Experiment
SHF	Super high frequency
SIRS	Satellite Infra-Red Spectrometer
SIRTF	Space Infrared Telescope Facility
SMS	Synchronous Meteorological Satellite
SNAP	Systems for Nuclear Auxiliary Power
SPS	Service Propulsion System
SPU	Signal processing unit
SR	Scanning Radiometer
SSCC	Spin–Scan Cloud-cover Camera
SSPA	Solid state power amplifier
STC	Standard Telephones and Cables
TAT	Thrust Augmented Thor
TC&R	Telemetry, command and ranging
Tiros/TIROS	Television Infrared Observation Satellite
TOMS	Total Ozone Mapping Spectrometer
TOS	Tiros Operational System
TT&C	Telemetry, tracking and command
TWT	Travelling wave tube
TWTA	Travelling wave tube amplifier
UDMH	Unsymmetrical dimethylhydrazine
UHF	Ultra high frequency
USAF	US Air Force
USGS	US Geological Survey
VfR	Verein fur Raumschiffahrt
VHRR	Very high resolution radiometer
VISSR	Visible and Infrared Spin Scan Radiometer
VSAT	Very small aperture terminal
WMO	World Meteorological Organisation

Selected bibliography

The following books cover specific programmes and events in space history in more detail than is possible in this volume. Although they are arranged under the chapter headings to which they have most relevance, many of them contain information relevant to many different types of spacecraft and missions. Books which cover material in the three Apollo chapters are grouped together. In addition to newer sources, the list contains a selection of older books, some of them classics in their own right.

Chapter 1: A prologue to the Space Age

Allward, Maurice, (Ed.): 'The Encyclopedia of Space' (The Hamlyn Publishing Group, London, 1968–69)

Bilstein, Roger, E.: 'Orders of Magnitude: A History of the NACA and NASA, 1915–1990' (NASA, Washington, 1989)

Ciancone, Michael, L.: 'The Literary Legacy of the Space Age: An Annotated Bibliography of pre-1958 Books on Rocketry and Space Travel' (Amorea Press, Houston, 1998)

Gatland, Kenneth: 'The Illustrated Encyclopedia of Space Technology' (Salamander Books, London, 1981)

Hardy, Phil: 'The Encyclopedia of Science Fiction Movies' (Octopus Books, London, 1986)

Miller, Ron: 'The Dream Machines: A Pictorial History of the Spaceship in Art, Science and Literature' (Krieger, Malabar, 1993)

Ordway, Frederick, I., and Liebermann, Randy (Eds): 'Blueprint for Space' (Smithsonian Institution, Washington, 1992)

Ordway, Frederick, I.: 'Visions of Spaceflight' (Four Walls Eight Windows, New York, 2001)

Winter, Frank, H.: 'Rockets into Space' (Harvard University Press, Cambridge, 1990)

Winter, Frank, H.: 'The First Golden Age of Rocketry' (Smithsonian Institution, Washington, 1990)

Chapter 2: Highway to space – the development of the space launch vehicle

Baker, David: 'The Rocket' (Crown Publishers Inc, New York, 1978)

Bilstein, Roger, E.: 'Stages to Saturn' (University Press of Florida, Gainesville, 1996)

Burgess, Eric: 'Long Range Ballistic Missiles' (Chapman and Hall, London, 1961)

Chapman, John, L.: 'Atlas: The Story of a Missile' (Harper and Brothers, New York, 1960)

Dougherty, Kerrie and James, Matthew, L.: 'Space Australia: The Story of Australia's Involvement in Space' (Powerhouse Publishing, Haymarket, 1993)

Editors of Fortune: 'The Space Industry: America's Newest Giant' (Prentice-Hall, Englewood Cliffs, 1962)

Godwin, Robert: 'Rocket and Space Corporation Energia' (Apogee Books, Burlington, 2001)

Hill, C. N.: 'A Vertical Empire: the History of the UK Rocket and Space Programme', 1950–1971' (Imperial College Press, London, 2001)

Koppes, C. R.: 'JPL and the American Space Program' (Yale University Press, New Haven, CT, 1982)

Ley, Willy: 'Rockets and Space Travel' (The Viking Press, New York, 1948)

Martin, Charles, H.: 'De Havilland Blue Streak' (British Interplanetary Society, London, 2002)

McLaughlin Green, Constance, and Lomask, Milton: 'Vanguard: A History' (Smithsonian Institution Press, Washington, 1971)

Millard, Douglas: 'The Black Arrow Rocket' (NMSI Trading Ltd, London, 2001)

Ordway, Frederick, I. and Sharpe, Mitchell, R.: 'The Rocket Team' (William Heinemann, London, 1979)

Pardoe, G. K. C.: 'The Challenge of Space' (Chatto and Windus, London, 1964)

Piszkiewicz, Dennis: 'Wernher von Braun: The Man Who Sold the Moon' (Praeger, Westport and London, 1998)

Stoiko, Michael: 'Pioneers of Rocketry' (Hawthorn Books, New York, 1974)

Stoiko, Michael: 'Soviet Rocketry: The First Decade of Achievement' (David and Charles, Newton Abbot, 1970)

Stuhlinger, Ernst and Ordway, Frederick, I.: 'Wernher von Braun: Crusader for Space' (Krieger Publishing Co, Malabar, 1994)

Ulanoff, Stanley: 'Illustrated Guide to US Missiles and Rockets' (Doubleday and Co Inc, New York, 1959)

Williamson, Mark: 'The Cambridge Dictionary of Space Technology' (Cambridge University Press, Cambridge, 2001)

Chapter 3: Looking at space – the development of the space science satellite

Bester, Alfred: 'The Life and Death of a Satellite' (The Scientific Book Club, London, 1967)

Caprara, Giovanni: 'The Complete Encyclopedia of Space Satellites' (Portland House, New York, 1986)

Corliss, William, R.: 'Scientific Satellites' (NASA SP-133, Washington, 1967)

Fraser, Ronald: 'Once Around The Sun: the Story of the IGY 1957-58' (Hodder and Stoughton, London, 1956)

Gatland, Kenneth, W.: 'Astronautics in the Sixties' (Iliffe Books Ltd, London, 1962)

Heppenheimer, T. A.: 'Countdown: A History of Space Flight' (John Wiley and Sons, New York, 1997)

Leverington, David: 'New Cosmic Horizons: Space Astronomy from the V2 to the Hubble Space Telescope' (Cambridge University Press, Cambridge, 2000)

Lewis, Richard, S.: 'The Illustrated Encyclopedia of Space Exploration' (Salamander Books, London, 1983)

Naugle, John, E.: 'Unmanned Space Flight' (Holt, Rinehart and Winston, New York, 1965)

Sharpe, Mitchell, R.: 'Satellites and Probes' (Aldus Books, London, 1970)

Shelton, William: 'Soviet Space Exploration: The First Decade' (Arthur Barker Ltd, London, 1969)

Shternfeld Ari: 'Soviet Space Science' (Basic Books Inc, New York, 1959)

Smolders, Peter, L.: 'Soviets in Space' (Lutterworth Press, Guildford, 1973)

Chapter 4: Looking at Earth – the development of the Earth observation satellite

Baker, D. James: 'Planet Earth – The View from Space' (Harvard, Cambridge, 1990)

Day, Dwayne, A., Logsdon, John, M., and Latell, Brian: 'Eye in the Sky: The Story of the Corona Spy Satellites (Smithsonian Institution Press, Washington, 1998)

Fishlock, David (Ed.): 'A Guide to Earth Satellites' (Macdonald, London, 1971)

Peebles, Curtis: 'Guardians: Strategic Reconnaissance Satellites' (Ian Allan Ltd, London, 1987)

Porter, Richard, W.: 'The Versatile Satellite' (Oxford University Press, Oxford, 1977)

Temple, L. Parker: 'Shades of Gray: National Security and the Evolution of Space Reconnaissance' (AIAA, Reston, 2005)

Van Allen, James, A. (Ed.): 'Scientific Uses of Earth Satellites' (Chapman and Hall, London, 1956)

Chapter 5: Keeping in touch – the development of the communications satellite

Clarke, Arthur, C.: 'Voices from the Sky' (Harper and Row, New York, 1965)

Evans, Barry, G.: 'Satellite Communication Systems' (The Institution of Electrical Engineers, London, 1999)

Gittins, J. F.: 'Power Travelling Wave Tubes' (The English Universities Press Ltd, London, 1965)

Hansen, James, W.: 'Hughes TWT/TWTA Handbook' (Hughes Aircraft Company Electron Dynamics Division, Torrance, 1993)

Pelton, Joseph, N., Oslund, Jack, and Marshall, Peter: 'Communications Satellites: Global Change Agents' (Lawrence Erlbaum Associates, New York, 2004)

Pierce, John: 'Traveling Wave Tubes' (Van Nostrand, 1950)

Solomon, Louis: 'Telstar: Communication Break-through by Satellite' (Constable Young Books Ltd, London, 1963)

Williamson, Mark: 'The Communications Satellite' (Adam Hilger/IOPP, Bristol and New York, 1990)

Chapter 6: Probing the Moon – the development of the lunar science spacecraft

Burgess, Eric: 'Assault on the Moon' (Hodder and Stoughton, London, 1966)

Caidin, Martin: 'Race for the Moon' (William Kimber, London, 1960)

Kopal, Zdenek: 'Exploration of the Moon by Spacecraft' (Oliver and Boyd, Edinburgh and London, 1968)

Kopal, Zdenek: 'The Realm of the Terrestrial Planets' (Institute of Physics, Bristol, 1979)

Reeves, Robert: 'The Superpower Space Race' (Plenum Press, New York, 1994)

Sharpe, Mitchell, R.: 'Satellites and Probes' (Aldus Books, London, 1970)

Smith, R. A. and Clarke, Arthur, C.: 'Exploration of the Moon' (Frederick Muller Ltd, London, 1954)

Troebst, Cord-Christian: 'Reaching for the Moon' (Hodder and Stoughton, London, 1961)

Wilson, Andrew: 'Solar System Log' (Jane's, London, 1987)

Chapter 7: Man in space – the development of the manned capsule

Abramov, Isaak, P., and Skoog, A. Ingemar: 'Russian Spacesuits' (Springer-Praxis, Chichester, 2003)

Arnold, H. J. P. (Ed.): 'Man in Space: An Illustrated History of Spaceflight' (Smithmark, New York, 1993)

Burrows, William, E.: 'This New Ocean: The Story of the First Space Age' (Random House, New York, 1998)

Catchpole, John: 'Project Mercury: NASA's First Manned Space Programme' (Springer-Praxis, Chichester, 2001)

Clark, Philip: 'The Soviet Manned Space Programme' (Salamander Books Ltd, London, 1988)

Dethloff, Henry, C.: 'Suddenly Tomorrow Came ... A History of the Johnson Space Center' (NASA SP-4307, Lyndon B Johnson Space Center, Houston, TX, 1993)

Furniss, Tim: 'Manned Spaceflight Log' (Jane's publishing Company, London, 1983)

Grimwood, James, M.: 'Project Mercury: A Chronology' (NASA SP-4001, Washington DC, USA, 1963)

Harvey, Brian: 'The New Russian Space Programme: From Competition to Collaboration' (Wiley-Praxis, Chichester and New York, 1996)

Newkirk, Dennis: 'Almanac of Soviet Manned Space Flight' (Gulf Publishing Co, Houston, TX, 1990)

Shayler, David, J.: 'Gemini: Steps To The Moon' (Springer-Praxis, Chichester, 2001)

Siddiqi, Asif, A.: 'Challenge to Apollo: The Soviet Union and the Space Race, 1945–1974' (NASA SP-2000-4408, Washington, 2000)

Swenson, Loyd, S., Grimwood, James, M., and Alexander, Charles, C.: 'This New Ocean: A History of Project Mercury' (NASA SP-4201, Washington DC, USA, 1966)

Turnill, Reginald (Ed.): 'Jane's Spaceflight Directory, 1987' (Jane's Publishing Co Ltd, London, 1987)

Wolfe, Tom: 'The Right Stuff' (Farrar, Straus and Giroux Inc, New York, 1979)

The Apollo Chapters

Chapter 8: First and only moonship – the development of the Apollo Command and Service Module

Chapter 9: Lunar lander – the development of the Apollo Lunar Module

Chapter 10: Electric Moon car – the development of the Apollo Lunar Roving Vehicle

Allday, Jonathon: 'Apollo in Perspective: Spaceflight Then and Now' (IOP Publishing, Bristol and Philadelphia, 2000)

Benson, Charles, D., and Faherty, William, B.: 'Gateway to the Moon' (University Press of Florida, Gainesville, 2001)

Benson, Charles, D. and Faherty, William, B.: 'Moon Launch!' (University Press of Florida, Gainesville, 2001)

Collins, Michael: 'Liftoff – the Story of America's Adventure in Space' (Aurum Press, London, 1989)

Cooper, Henry, S. F.: 'Moonwreck – 13: The Flight That Failed' (Granada/Panther, St Albans, 1975)

Gainor, Chris: 'Arrows to the Moon: Avro's Engineers and the Space Race' (Apogee Books, Burlington, 2001)

Godwin, Robert (Ed.): 'Apollo 13: The NASA Mission Reports' (Apogee Books, Burlington, 2000)

Gray, Mike: 'Angle of Attack: Harrison Storms and the Race to the Moon' (W W Norton and Co, New York, 1992)

Hall, Eldon, C.: 'Journey to the Moon: The History of the Apollo Guidance Computer' (AIAA, Reston, 1996)

Kluger, Jeffrey: 'The Apollo Adventure: The Making of the Apollo Space Program and the Movie Apollo 13' (Boxtree Ltd, London, 1995)

Lindsay, Hamish: 'Tracking Apollo to the Moon' (Springer-Verlag, London, 2001)

Lovell, Jim, and Kluger, Jeffrey: 'Lost Moon: The Perilous Voyage of Apollo 13' (Houghton Mifflin Co, Boston, MA, 1994)

Shepard, Alan, and Slayton, Deke: 'Moon Shot: The Inside Story of America's Race to the Moon' (Virgin Publishing Ltd, London, 1995)

Chapter 11: An epilogue to the space race

Cooper, Henry S. F.: 'A House in Space' (Angus and Robertson, London, 1977)

Freeman, Marsha: 'Challenges of Human Space Exploration' (Springer-Praxis, Berlin and Chichester, 2000)

Harland, David M.: 'The Space Shuttle: Roles, Missions and Accomplishments' (Wiley-Praxis, Chichester, 1998)

Harvey, Brian: 'The Chinese Space Programme: From Conception to Future Capabilities' (Wiley-Praxis, Chichester and New York, 1998)

Harvey, Brian: 'The Japanese and Indian Space Programme: Two Roads into Space' (Springer-Praxis, Chichester and Berlin, 2000)

Holder, William G., and Siuru, William D.: 'Skylab: Pioneer Space Station' (Rand McNally, Chicago, IL, 1974)

Krige, John, and Russo, Arturo: 'Europe in Space 1960–1973' (European Space Agency, Noordwijk, 1994)

Index

Note: page numbers in **bold** refer to illustrations; entries in *italics* are titles of publications.